The Management of Water Quality and the Environment

OTHER INTERNATIONAL ECONOMIC ASSOCIATION PUBLICATIONS

MONOPOLY AND COMPETITION AND THEIR REGULATION
THE BUSINESS CYCLE IN THE POST-WAR WORLD
THE THEORY OF WAGE DETERMINATION
THE ECONOMICS OF INTERNATIONAL MIGRATION
STABILITY AND PROGRESS IN THE WORLD ECONOMY
THE ECONOMIC CONSEQUENCES OF THE SIZE OF NATIONS
ECONOMIC DEVELOPMENT OF LATIN AMERICA
THE THEORY OF CAPITAL
INFLATION
THE ECONOMICS OF TAKE-OFF INTO SUSTAINED GROWTH
INTERNATIONAL TRADE THEORY IN A DEVELOPING WORLD
ECONOMIC DEVELOPMENT WITH SPECIAL REFERENCE TO EAST ASIA
ECONOMIC DEVELOPMENT FOR AFRICA SOUTH OF THE SAHARA
THE THEORY OF INTEREST RATES
PROBLEMS IN ECONOMIC DEVELOPMENT
THE ECONOMICS OF EDUCATION
THE ECONOMIC PROBLEMS OF HOUSING
PRICE FORMATION IN VARIOUS ECONOMIES
THE DISTRIBUTION OF NATIONAL INCOME
ECONOMIC DEVELOPMENT FOR EASTERN EUROPE
RISK AND UNCERTAINTY
ECONOMIC PROBLEMS OF AGRICULTURE IN INDUSTRIAL SOCIETIES
INTERNATIONAL ECONOMIC RELATIONS
BACKWARD AREAS IN ADVANCED COUNTRIES
PUBLIC ECONOMICS
ECONOMIC DEVELOPMENT IN SOUTH ASIA
NORTH AMERICAN AND WESTERN EUROPEAN POLICIES
PLANNING AND MARKET RELATIONS
THE GAP BETWEEN RICH AND POOR NATIONS
LATIN AMERICA IN THE INTERNATIONAL ECONOMY
MODELS OF ECONOMIC GROWTH
SCIENCE AND TECHNOLOGY IN ECONOMIC GROWTH
THE ECONOMICS OF HEALTH AND MEDICAL CARE
ALLOCATION UNDER UNCERTAINTY
TRANSPORT AND THE URBAN ENVIRONMENT
AGRICULTURAL POLICY IN DEVELOPING COUNTRIES
THE ECONOMIC DEVELOPMENT OF BANGLADESH
ECONOMIC FACTORS IN POPULATION GROWTH

CLASSICS IN THE THEORY OF PUBLIC FINANCE

The Management of Water Quality and the Environment

Proceedings of a Conference held by
the International Economic Association
at Lyngby, Denmark

EDITED BY
J. ROTHENBERG AND IAN G. HEGGIE

First published 1974 by
THE MACMILLAN PRESS LTD
London and Basingstoke
Associated companies in New York
Dublin Melbourne Johannesburg and Madras

Published in the U.S.A. and Canada
by Halsted Press, a Division of
John Wiley & Sons, Inc., New York

SBN 333 15692 7

Printed in Great Britain by
R. AND R. CLARK LTD *Edinburgh*

Contents

List of Participants   vii
Copyright Permissions   viii
Introduction   J. Rothenberg   ix

PART I: AIR AND WATER POLLUTION

1 The Optimum Management of Social Overhead Capital   3
     H. Uzawa
   Discussion   18
2 Industrial Structure, Growth and Residual Flows   21
     Finn R. Førsund and Steinar Strøm
   Discussion   70
3 The Application of Economic Analysis to the Management   73
     of Water Quality: Some Case Studies   Allen V. Kneese
4 A Linear Decision Model for the Management of Water   104
     Quality in the Ruhr   Rainer Thoss and Kjell Wiik
   Discussion   134

PART II: PROBLEMS OF AND APPROACHES TO PUBLIC POLICY

5 Qualitative Returns to Scale and the Optimum Financing   151
     of Environmental Policies   Serge-Christophe Kolm
   Discussion   172
6 Effluent Charges versus Effluent Standards   189
     Karl-Göran Mäler
   Discussion   213
7 The Management of the Quality of the Environment   224
     Clifford S. Russell, Walter O. Spofford, Jr., and
     Edwin T. Haefele
   Discussion   273

PART III: EVALUATION AND CONSOLIDATION PANEL

8 Urbanisation and Environment: Retrospective and Pro-   281
     spective Views   Ian G. Heggie, Henry Tulkens, Rainer
     Thoss, Karl-Göran Mäler and Edwin S. Mills

Index   301

List of Participants

Dr G. Albers, Institute of Town and Regional Planning, Technical University, Munich, F.G.R.

Mrs Bodil Nyboe Andersen, Institute of Economics, University of Copenhagen, Denmark

Professor Tibor Bakacs, National Institute of Public Health, Budapest, Hungary

Mr Jean-Philippe Barde, Environment Directorate, O.E.C.D., Paris, France

Professor W. Beckerman, Department of Economics, University College, London, U.K.

Professor Niels G. Bolwig, Institute of Economics, University of Aarhus, Denmark

*Professor E. von Böventer, Department of Economics, University of Munich, F.G.R.

Professor C. Cameron, Department of Social and Economic Research, University of Glasgow, U.K.

Mr Ulf Christiansen, Building Research Institute, Copenhagen, Denmark

Mr Alan W. Evans, Centre for Environmental Studies, London, U.K.

Professor Luc Fauvel, Secretary-General, I.E.A., Paris, France

Professor Finn R. Førsund, Institute of Economics, University of Oslo, Norway

Mr C. D. Foster, London School of Economics, London, U.K.

*Mr Edwin T. Haefele, Resources for the Future, Washington, D.C., U.S.A.

Professor Niles M. Hansen, Centre for Economic Development, University of Texas, Austin, U.S.A.

Mr Ian G. Heggie, Nuffield College, Oxford, U.K.

Professor Sir John Hicks, All Souls College, Oxford, U.K.

Lady Hicks, Linacre College, Oxford, U.K.

Mr Chr. Hjorth-Andersen, Institute of Economics, University of Copenhagen, Denmark

Dr Irving Hoch, Resources for the Future, Washington, D.C., U.S.A.

Dr Erik Hoffmeyer, Governor, Danish National Bank, Copenhagen, Denmark

Mr N. J. Kavanagh, Department of Industrial Economics and Business Studies, University of Birmingham, U.K.

Mr A. V. Kneese, Resources for the Future, Washington, D.C., U.S.A.

Professor S.-Ch. Kolm, CEPREMAP, Paris, France

Professor Lester B. Lave, G.S.I.A., Carnegie-Mellon University, Pittsburgh, U.S.A.

Mrs Judith R. Lave, G.S.I.A., Carnegie-Mellon University, Pittsburgh, U.S.A.

*Professor H. Lévy-Lambert, Société Générale, Paris, France

Professor Fritz Machlup, Department of Economics, Princeton and New York University, U.S.A.

Mr Karl-Göran Mäler, Department of Economics, University of Stockholm, Sweden

Professor Niels Meyer, Technical University of Denmark, Lyngby, Denmark

Professor Edwin S. Mills, Department of Economics, Princeton University, U.S.A.

*Professor Robert Mossé, University of Grenoble, France

Mr Anders Müller, General Planning Directorate, Copenhagen, Denmark

   * Presented a paper but did not attend the Conference.

Professor Frank E. Münnich, University of Dortmund, G.F.R.
Professor Knud Østergard, Technical University of Denmark, Lyngby, Denmark
Professor Remy Prud'homme, BETURE, Puteaux, France
Professor P. Nørregaard Rasmussen, Institute of Economics, University of Copenhagen, Denmark
Professor Jerome Rothenberg, Department of Economics, Massachusetts Institute of Technology, U.S.A.
Dr Clifford S. Russell, Resources for the Future, Washington, D.C., U.S.A.
Professor Eugene P. Seskin, G.S.I.A., Carnegie-Mellon University, Pittsburgh, U.S.A.
Mr Alessandro Silj, The Ford Foundation, New York, U.S.A.
Mr Irving Silver, Ministry of Urban Affairs, Ottawa, Canada
*Mr Walter O. Spofford, Resources for the Future, Washington, D.C., U.S.A.
Professor Steinar Strøm, Institute of Economics, University of Oslo, Norway
Professor Dr Rainer Thoss, Department of Economics, University of Münster, G.F.R.
Professor Henry Tulkens, Centre for Operations Research and Econometrics, Catholic University of Louvain, Heverlee, Belgium
Professor H. Uzawa, Faculty of Economics, University of Tokyo, Japan
Mr Kjell Wiik, Department of Economics, University of Münster, G.F.R.

Secretariat and Editorial Staff

Miss Mary Crook
Mr Ian G. Heggie
Mrs Elizabeth Majid
Mr Karsten Peterson

Programme Committee

Jerome Rothenberg, U.S.A. (*Chairman*)
Peter Bohm, Sweden
E. von Böventer, G.F.R.
Sir John Hicks, U.K.
Allen V. Kneese, U.S.A.
Shigeto Tsuru, Japan

Copyright Permissions

A number of publishers have kindly allowed us to make use of material for which they hold the copyright. We are most grateful for their co-operation and would specifically like to thank the Johns Hopkins University Press for permission to reproduce material.

* Presented a paper but did not attend the Conference.

Introduction

J. Rothenberg

1 CONFERENCE OVERVIEW

This is a second volume drawn from the International Economic Association-sponsored Conference on Urbanisation and the Environment, held near Copenhagen on 20–24 June 1972. The first volume, *Transport and the Urban Environment*, included the materials relating primarily to the process of urbanisation, its causes and environmental consequences, with special attention to urban transport. The present volume contains the materials more closely focused on the environmental quality issue itself. These are of course interconnected with urban environment questions; but they stand independently as well, as a provocative set of closely related examinations in an area becoming important to both academic inquiry and public policy concern. Seven of the fifteen conference papers are included here, along with the general discussion they generated, and the whole is concluded with the proceedings of the summary and evaluation session for the conference as a whole.

The papers were for the most part given by economists, but a variety of other disciplines were represented at the conference – biology, sociology, planning and political science. This is entirely appropriate, because the essential substance of the field involves a variety of specialised subject-matter from physical science, engineering and the several social sciences. Some of the conference's proceedings underscore how strongly progress in the field depends on the ability to mobilise and co-ordinate contributions from various disciplines. The materials in this book show chiefly how economists apply their distinctive analytic and statistical viewpoints to an area whose subject-matter has stemmed largely from practical experiences and research outside of economics.

2 ENVIRONMENTAL QUALITY AND ECONOMIC ANALYSIS

The growing seriousness of economic research on environmental quality has led to a fundamental question being asked: what *is* the environment, or environmental quality, from the point of view of the kinds of variable and behaviour relations used in economic analysis? This is not trivial, since we have become sensitised to the many kinds of role played by the environment in the economy: as final consumption, as a set of productive inputs, or as an input into other productive inputs (as, for example, air quality affecting the health and therefore productivity of workers).

A major direction in formulating how the environment enters into economic activities is to subsume the relationships under the theory of public goods. The concept of common property resources has been elaborated: in his paper, Uzawa develops a related concept of social overhead capital; Kolm, in his paper, develops further aspects in discussing qualitative returns to scale. The environment presents economic agents with a variety of aspects: free goods, resources owned in common, joint production, scale economies, congestion, positive and negative externalities, among others. Conventional paradigms are suggestive for some of these, but none so far is consistent with all. Thus, new efforts towards a complete formulation of this are welcome, and both Uzawa and Kolm have made contributions in this volume.

Some of the elements that matter are as follows. The environment can be considered as a set of resources that render both consumption and production services to a variety of users. Most environmental resources render these services in a non-exclusive – but not in principle non-excludable (think of admission fees to beaches or national parks or barred entry to a sewage system) – way. Many can 'simultaneously' share services without anyone thereby obtaining less or of lower quality than desired. In addition, many such resources are not subject to private ownership, and so can be shared at zero fee. But there is a finite assimilative capacity for each kind of resource, short of which untrammelled sharing of usage is possible, but beyond which higher usage, or usage by different types of agents, deteriorates the quality at which various of the users can obtain the services. For many portions of the environment one can delineate a 'natural' assimilative capacity which specifies the boundaries 'given by Nature' (in level, type, time and space of use) within which shared use can avoid quality deterioration, and the schedule of deteriorations that occur at various stages beyond these boundaries.

This natural assimilative capacity is not immutable: human intervention can change it. There are various forms of investment in assimilative capacity. They are based on the different respects in which use of each environmental resource 'pollutes' its ability to maintain quality and quantity of services for its clients. Probably the most important – but not the only – mode of 'pollution' is that economic agents, in carrying on their primary consumption and production activities, generate various wastes (residuals) as by-products and use the environmental resources as inexpensive channels in which to discharge these wastes for removal. This use interferes with other types of use of the same resources. Investment in assimilative capacity for this mode can take broad forms in which the environmental resource is itself restructured or augmented (e.g. lengthening a beach, increasing the rate of water flow in a river), narrower forms in which a given level of waste discharge is made less noxious through public (large-scale) treatment of the wastes, to forms in which individual waste dischargers themselves either denature their own wastes before discharge or recycle them back into primary production (or consumption).

All these investments have in common that a given level and composition of primary consumption and production are made to have a smaller adverse effect on quality than hitherto. Yet the last of these is not so much an operation on environmental resources as on the activities that impinge on them. Research often makes no intrinsic distinction within the category of treatment: public or private forms are simply distinguished in terms of scale economies. Indeed, treatment activities are often bracketed with substitution of inputs and even technological change in a very broad 'treatment' category. Underlying such usage is that the environment and its capacity are in fact deeply, possibly inextricably, intertwined with the pattern and level of human activities. The concept of environment as durable, useful resources that have a quantity capable of being modified, and thus as a form of capital, can be developed, but their extreme (non-linear) sensitivity to the pattern and level of the human activities must be heeded in the formulation, both in defining units of measurement of capital and in the relations describing its productive properties.

Given the assimilative capacity of environmental resources with respect to various activities (residuals) at a given time, the actual quality level of environmental services rendered at that time depends on the relation between the actual pattern and level of activities and this assimilative capacity. The basic economic problem is that the quality of environmental services can be enhanced by increasing the assimilative capacity of the environment, or changing the composition of primary activities to one with less obtrusive impacts, or decreasing its overall level. All these involve a social cost: assimilative capacity investment uses up resources for treatment or innovation, compositional changes use less efficient input combinations or involve less preferred compositions of output, lower activity levels directly decrease the rate of final consumption.

This raises two questions: (1) how much improvement in environmental quality is it worth while making, considering that every additional degree invariably entails a social cost; (2) how best shall any quality improvement be brought about? As is well known from general resource allocation considerations, the two questions are interrelated. Different adjustments involve different schedules of marginal cost of improvement, and the most desirable degree of improvement is that for which the marginal benefit from improvement equals the marginal cost of improvement of the most efficient mode of adjustment at the relevant level.

These two questions are closely related to a third: how should the authorities attempt to bring about the desired degree of improvement by the desired pattern of adjustment? Some research treatments have acted as though it were possible to solve the first two questions independently of the third. Growing sophistication about the empirical and behavioural complexities in the field – in particular, the extremely demanding requirements of knowledgeable administration of public policies – makes clearer that the constraints, the imperfections, the sheer substantial resource costs involved in implementing public policy, should be an intrinsic component of the calculation of what should be the goals and instrumentalities of that policy.

In the present collection, Uzawa and Kolm have examined various aspects of these fundamental considerations. Uzawa develops the concept of social overhead capital: resources which contribute to production and depreciate as a function of the ratio of level of usage to capacity (size of stock) – a kind of 'crowding'. As we have argued, this is an important feature of environmental resources. Uzawa considers appropriate resource allocation criteria between investment in social overhead capital and all other uses, both in static equilibrium and growth contexts. The paper and the ensuing discussion raise significant issues about what features of environmental resources need be introduced into resource allocation models in order to avoid misleading analytic inferences.

Kolm studies in depth another set of basic aspects of the environment. He is concerned with the formulation of optimal public policy in terms of degree of environmental quality achieved and the financing of improvement projects. These will be discussed further below. At this point, what is relevant is the enlightening way these efficiency results derive from characteristics of environmental resources and their susceptibility to pollution and treatment. His discussion of qualitative returns to scale relates the magnitude of primary activities which give rise to detrimental residual discharges (or other modes of 'pollution') to the net degree of environmental improvement that can be accomplished by treating these activities and their inescapable by-products as a tax base on which to levy pollution charges, and using the resulting tax revenues to finance efficient treatment. Technical relations among primary processes, treatment and environmental degradation are at the heart of the inquiry.

The general discussion evoked by this paper considered further how fundamental characteristics of the environment should be used in analytic models. The discussion is notable especially for an extensive contribution by Førsund which both reviews a number of such models in the literature and critically examines difficult theoretical problems arising from them.

3 MODELS OF POLLUTION AND ECONOMIC ACTIVITY

Environmental research has not waited for definitive answers to fundamental questions about the nature of the environment. One important direction taken by recent research is the attempt to capture the actual quantitative pattern of interaction between primary activity and residual production for different productive sectors. An extension is to link primary activities together in terms of their functional interrelationships in the economy, so that a tableau of primary inter-industry relationships generates as well a tableau of the unwanted by-product residuals of these industrial activities. Such tableaux are invaluable for developing quantitative models of the economy, in which both the pattern and the magnitudes of the dependence of residual discharges on primary activities are revealed. Not only are the relative magnitudes of residual flows traced to specific industries, but also the different types of residuals generated by different industries.

Models of this kind embed the problem of environmental quality into the broad context of the overall allocation of scarce resources, because their positive associations between primary activities and residual generation make explicit necessary tradeoffs between environmental and non-environmental outputs. They are useful, therefore, for understanding the complexity of the environmental quality phenomenon; for predicting the consequences of various economic changes, like growth, innovation or change in the composition of output; and for examining the effects of different intensities and types of public policy.

The present volume contains discussions of four models of this sort: (1) Førsund and Strøm; (2) Thoss and Wiik; (3) Kneese; and (4) Russell, Spofford and Haefele. The first three refer to real-world data, the fourth to a hypothetical system. Kneese's paper presents the model used by the Delaware River Project as part of a review of the character and findings of important quantitative area studies. The first two develop large-scale models, of Norway and the Ruhr Basin respectively, in which primary activities are directly linked to variables relating to environmental quality. They are ambitious works, and appear to be capable of a variety of applications.

The Førsund–Strøm and Thoss–Wiik models have important differences, some of which make them complementary, but others of which make them difficult to be directly compared. The former is a model of a national economy, the latter of one region in a much larger economy. Moreover the Norwegian model is a predictive equilibrium model, while the Ruhr model is normative.

The model by Førsund and Strøm is based on an input–output tableau comprising many production sectors, where each sector is linked to other sectors through input purchases and output sales, and to the generation of residuals via productive activity. The emphasis is on the different pattern of residuals for each sector, and the pattern of unequal growth experience among sectors as the economy expands. The consequences of economic growth on residual levels are drawn from the model and these are then compared with the consequences from instituting sectoral effluent charges based on relative magnitudes of effluent generation by the different sectors.

The model by Thoss and Wiik is also based on an assumed fixed linear (input–output) technology. But here the technology connecting each sector to the others and to the production of effluents is employed as a constraint in the context of maximising final consumption levels while maintaining minimum specified standards of environmental quality. Effluents are associated with overall ambient standards, and so each configuration of sectoral resource use gives rise to a set of ambient conditions that can be compared to the minimum standards specified. A linear programming maximising solution is used to make explicit the production–environmental trade-offs implicit in the system.

These models are illuminating, but they raise a number of important issues which were discussed at the conference. One concerns the use of general equilibrium versus partial equilibrium contexts. A regional model gives greater detail and localisation of environmental impacts but omits significant interrelationships with other parts of the national – and even international – economy. A national model enables broader sectoral relationships to be revealed for primary activities but neglects the extreme importance of localisation of impact in determining environmental impairment. Total national amounts of certain residuals convey no dependable meaning for environmental quality without being supplemented by information about their spatial – and temporal – distribution. Even the regional aggregate is too broad to convey such meaning, since substantial inequalities exist within a region and it is the precise distribution that matters.

Another issue is the use of linear, fixed-proportions technology within and between sectors, and an absence of technological adaptation and change. Technological heterogeneity and flexibility exist in the real world and are quite relevant to the picture of tradeoffs between production and environmental quality for the economy in both an undisturbed state and under public policy interventions.

Other issues are the use of linear programming techniques for deriving optimal resource-use patterns, and the absence of explicit links between either effluent flows or ambient conditions and the real economic costs (damages) they are presumed to generate. The appropriateness of linear programming to delineate optimal situations for an economy assumes certain structural and operating features for the economy that are controversial. The absence of links between outcome variables of the model and welfare variables prevents a full embeddedness for environmental issues within larger resource allocation considerations.

Kneese reviews two important water quality studies, that of the Delaware river and that of the Potomac. The latter is a cost–benefit analysis of policy intervention to improve water quality, the former an examination of the relative costs of achieving water quality improvement by a variety of different public policy techniques. The former, in particular, shows how positive (descriptive, predictive) models of water pollution can be used to illuminate matters of public policy. It represents for the conference – and for this volume – the transition between understanding the environmental quality problem as part of the problem of overall efficiency in resource use and brandishing public policy instruments to do something constructive about it.

4 PUBLIC POLICY

As hinted at above, the problem of public policy is not merely an appendix to the specification of exact intervention goals, because the degree of effectiveness and cost of different policy approaches are of the same order of concern as the private pollution process itself, and must be balanced against what is desired to be achieved along with the real resource costs involved in private adjustments. There are many issues here. A major one is whether reliance on specific adjustments should be based on centralised specifications and directive – for example, investment and operation of public treatment facilities, and/or central formulation and enforcement of emission standards for individual economic agents – or whether public action should be confined to tilting private incentives in desired directions by taxes and subsidies.

A second issue concerns the substance of policy goals. Is this to be derived from valuations implicit in purely private impacts and adjustments (for example, from aggregate damage functions built up out of measurable welfare losses to individual economic agents); or, because of intrinsic unmeasurability within such a private category or because there exists a 'public' or 'social' dimension to environmental preferences, should these goals be politically determined and enunciated?

A third issue concerns the heavily – but by no means, as is sometimes thought, completely – redistributional question of whether rights to environmental services should be clarified; whether they should be granted in whole or in part to benign users or to obtrusive users (polluters) or divided between them; whether any initial distribution of rights should be subject to modification only by central determination or by a private exchange process akin to a market.

A fourth issue involves the nature and size of the implementation costs for different types of policy instruments, and the extent to which different approaches can achieve similar goals. Chief among the administrative costs are those for obtaining the kind and amount of information necessary for properly formulating public action, the cost of overseeing the degree of private compliance with public regulations (standards, charges, etc.), and the cost of enforcing compliance at desirable levels.

Kneese's examination is not so much to describe the two studies for their own sake as to throw light on some of these salient issues concerning public policy. He deals perceptively with the relative evaluation of effluent charges and effluent standards, with the merits of subsidies versus charges, with the distribution and transfer of environmental rights, with matters of compensation for pollution damages suffered. A central judgment underscored in the work is that the effluent charge approach, with its attendant decentralisation, flexibility and balanced heterogeneity of adjustment, has considerable advantages over the much more centralised effluent standards approach. The inherent interest in such materials provoked extensive, deeply probing general discussion with a further advance in understanding.

Mäler's paper carries this evaluation further, concentrating on the informational requirements of effluent charges versus effluent standards. He develops this rigorously, by formulating a model of pollution and business firm behaviour and examining the micro- and macro-behaviour of the system under alternative sets of constraints representing the effects of different public policies. He projects the informational requirements needed to achieve optimal resource use by alternative policies within this context. His res lts offer modified additional support from a different perspective for effluent charges as against effluent standards. In addition to this, the same method is illuminatingly applied to the comparison between charges and subsidies, thereby extending an earlier conference discussion.

The special approach used does genuinely add new perceptions about the structure of impact of different types of public policy, by integrating pollution and pollution management problems even more integrally with a traditional analytical framework. An opposite consequence of the attempt at rigorous demonstration, however, is that the richness of the complex of issues must be sacrificed; that the particular way in which some of the issues are actually dealt with seems rarified and over-simple; and that significant complications have to be omitted altogether. Thus, the paper should not be viewed as a sufficient ground for selecting among public policies but as an important contribution to unfolding the deeper characteristics of the impact of public intervention on decentralised resource allocation systems.

The paper by Kolm mentioned earlier adds unusual dimensions to the inquiry about public policy. He deals with both allocative and distributional matters, and integrates them profoundly. As noted before, he examines the polar processes of primary activities that generate environmental deterioration and treatment activities that remedy such deterioration. He links the two through the intermediation of public policy, whereby the former becomes a tax base on which to levy effluent charges, and these charges become a lever in changing private incentives and a fund for financing explicit treatment activities. He derives the conditions for optimal use of the taxing instrument under various technological and structural constraints on the economic system modelled, especially under differing qualitative returns to scale situations. These optimal conditions are characterised in terms of whether the balance of primary activities and treatment leaves the environmental quality higher, lower or unchanged for each unit increase in primary activities.

The normative allocational results are shown to have important distributional consequences, since the environmental quality implied by optimal policy use determines the degree and distribution of damages throughout the system. Kolm engages in significant discussion of distributional concerns, tied closely with his innovative optimising schema. He too deals with environmental rights and with their transfer, with subsidies versus charges, and other of the policy issues already mentioned, and the general discussion elicited at the conference developed them further. The clustering of attention to these issues in the present book reveals differing perspectives and presumptions and is highly instructive.

The last paper in this book, that by Russell, Spofford and Haefele, is quite different from the rest. It is in two parts. The first presents a model of water residual generation, superficially not unlike those presented earlier. The second presents a schema for public policy. But this latter is different in kind from other policy discussions. And this difference makes the first part different as well.

The policy section does not deal with public policy as a normative question but as a positive one. The discussion concerns not what public policy ought to be but what it is likely to be. The difference is especially important because the problem of policy goals is made extremely rich, approximating that of the real world. The pollution model generates consequences that have many dimensions: like the level of G.N.P. or unemployment, various types of pollution and different patterns of primary activity affect the various dimensions non-proportionally. Not only is it that some members of the population are harmed by pollution and others by what has to be sacrificed to decrease pollution, but members of the population differ in how they are affected by the different non-proportional dimensions of the model's outcomes. Conflict becomes multifarious. Group solidarity is much less obvious and dependable. Moreover the character of conflicting interests makes for the possibility of agreements via complex interdimensional exchanges of advantage.

In such a setting, the problem of public policy is the behavioural *political* problem of agglomerating decisive legislative support for packages of public action that compromise the differences among the supporting coalition. Russell, Spofford and Haefele present the framework for a predictive model of legislative support in the context of multi-dimensional political goals relating to residuals and primary resource-using activities. The model indicates how log-rolling processes can help to generate legislative choices. The novelty of the work is not so much in the character of this legislative choice model since, despite possessing some unique features, other researchers have treated such phenomena extensively. It is rather in adjoining the environmental quality issue to the processes of practical politics – linkages which appear in disquieting bold face in the daily newspapers – and in formulating those processes with the rich complexity of non-homogeneous, multi-dimensional outcomes from the interaction of human activities with the environment. Their contribution opens up important – if disturbing – vistas.

5 EVALUATION AND CONSOLIDATION PANEL

Most of the final day of the conference was devoted to an Evaluation and Consolidation Panel. This panel was formed to enable the participants to draw together and structure the accumulating experience of the meeting. The formal papers themselves interrelated, but even more the full and wide-ranging general discussions extended certain issues beyond that of the papers and even raised and developed issues that had been at most implicit in the papers. The format of the panel was for Ian Heggie, who had been in charge of taping and rapporteur activities, to begin with a restrospective summary of the conference. He also enunciated and described the set of ten topics which, because of recurring attention or intrinsic importance, were to be examined further by the other four members of the panel.

The topics discussed were as follows:

1. *Preferences* – the actuality and discoverability of the community's preferences respecting environmental and non-environmental commodities or ends.

2. *Objectives* – the criteria set for characterising desirable social situations, when environmental conditions are considered, and procedures for achieving them.

3. *Distribution* – the non-neutral distributional consequences of environmental degradation and of measures to remedy it.

4. *Micro- and Macro-Models* – advantages and disadvantages of micro- versus macro-models of environmental impact.

5. *Environmental Effects* – the relation between objectively observable ambient standards and economic gains and losses, including well-being impacts.

6. *Growth and Environmental Quality* – the effects of economic growth (including technological change) on environmental quality.

7. *Environmental Policies* – the characteristics and relative usefulness of a broad range of environmental public policies.

8. *The Role of the Economist in Influencing Public Policy* – how the economist can influence the political process to act, and act appropriately, on environmental issues.

9. *Causes and Consequences of Changes in Urban Structure* – to what extent environmental problems are the causes or consequences of changes in the urban structure.

10. *Agenda for Further Empirical Research* – problems, issues and topics in the general area where empirical research is especially necessary to resolve difficulties and make for intellectual and policy progress.

The first three of these were led by Tulkens, items 4–6 by Thoss, items 7–8 by Mäler, and items 9–10 by Mills. The introduction of each group of items was followed by general discussion. As evidenced by these pages, the discussion was vigorous and searching.

It will be noticed that a paraphrase of the general discussion following each paper and the panel presentations is included in this volume. This is not an afterthought. The format of the conference proceedings was designed to devote the bulk of the time to designated critique and general discussion, not to presentation of the papers themselves. Thus, the weight of the meeting lay in the discussions. As will be seen, these are not dutiful acquiescences, but controverting, probing, wide-ranging, informed and concerned deliberations. They sometimes raise important issues not included in the papers; they tie together issues from different papers; they comment, condemn and extend. They are an integral part of the record of the conference.

Finally, we must acknowledge the critical contributions which gave the conference its very existence, arranged for its superb accommodations, and ensured its orderly functioning: Professor Fritz Machlup (President, I.E.A.), Professor Luc Fauvel (Secretary-General, I.E.A.), Professor P. Nørregaard Rasmussen (Vice-President, I.E.A.) and Miss Mary Crook (Secretariat, I.E.A.). The substance of the conference is heavily indebted to the work of the Programme Committee: Professor Peter Bohm, Professor Edwin von Böventer, Sir John Hicks, Mr. Allen V. Kneese and Professor Shigeto Tsuru.

Part 1

Air and Water Pollution

1 The Optimum Management of Social Overhead Capital

H. Uzawa

1.1 *PREFACE*

The purposes of this paper are twofold. Firstly, an attempt is made to formulate a theoretical model in which social overhead capital plays a significant role in the processes of resource allocation and in which the distribution of real income is affected by the manner in which the use of such social overhead capital is regulated. Secondly, the pattern of investment in social overhead capital which results in an optimum path of economic growth is examined within the framework of the Ramsey [1] theory of optimum growth.

1.2 *INTRODUCTION*

The phenomenon of environmental disruption, which is widespread and has recently intensified in the more industrialised countries, has created a number of menacing social problems. This is largely attributable to the mismanagement of social resources during the period of rapid post-war economic growth which has led to a significant increase in their scarcity. The phenomenon of environmental disruption may be regarded as an outcome of the misuse and depletion of social overhead capital, such as natural resources and social infrastructure, which cannot be privately appropriated from either a technological or an institutional point of view. In order to analyse the phenomenon of environmental disruption, and to devise means by which such a phenomenon may be effectively controlled, it is necessary to analyse the role played by social overhead capital in the processes of economic growth and the impact it may have upon the economic welfare of a national economy.

In this paper, I shall present a theoretical framework in which the mechanisms of resource allocation and income distribution are significantly altered by the presence of social overhead capital and by the way in which the use of these social resources is controlled. I shall then analyse the measures for regulating the use of social overhead capital which result in an optimum allocation of social, as well as of private, means of production. Finally, I shall briefly discuss the pattern of accumulation of social overhead capital which results in an optimum allocation of scarce resources from a dynamic point of view and the criteria by which public investment should be allocated to attain an optimum path of economic growth.

1.3 *SOCIAL OVERHEAD CAPITAL*

In standard economic analysis, it is often postulated that the means of production which limit the processes of production and consumption are privately owned and that the ownership of these scarce resources can be decided by means of the market mechanism. However, in most contemporary industrialised societies, a significant portion of these scarce resources is not privately appropriated. The resources which are not privately appropriated may, in a sense, be commonly owned by the society and the distribution of such resources may be decided on social grounds instead of by means of a decentralised market mechanism; such resources may be conveniently termed social overhead capital. The term 'social overhead capital' has been chosen in preference to the more widely used [2] 'common property resources', because the latter term occasionally implies natural resources only and thus excludes any reproducible means of production.

Social overhead capital may not be privately appropriated and may be used by or on behalf of members of a society either free of charge or at a negligible price. It may be collectively produced by the society, as in the case of so-called social capital like roads, etc., or simply inherited from Nature, as in the case of natural capital like water, air, etc.

Social overhead capital can be classified into two categories, depending upon the nature of the services it provides. The first category produces the services which are used in the productive process; the second category directly influences the level of economic welfare. However, it is difficult to separate social overhead capital in this way since identical capital may serve as both overhead capital for production as well as for consumption.

An introductory analysis to social overhead capital, and the role played by it in economic processes, has been presented in Uzawa [3]. I should like to summarise the approach it adopted before proceeding to discuss the problems indicated in the introduction.

At a given point of time, the means of production available to a society consist of private means of production and social overhead capital. Private means of production are appropriated by individual members of the society, and the disposal and management of these means of production are determined by individuals who are concerned with maximising private profits, or utilities, according to the rules that prevail in the society. The production and management of social overhead capital is, on the other hand, determined collectively by the society. In principle, the services from the social overhead capital are provided to any member of society either free of charge or at a negligible price, and the construction and maintenance of social overhead capital is delegated to the government, who meet these expenses out of general revenue.

The services provided by social overhead capital have two distinct features which are not appropriately handled in the standard Samuelsonian [4] analysis of pure public goods. The first aspect is concerned with the possibility of individual choice which enables individuals to decide to what extent such services are used. The second is related to the phenomenon that we often term 'congestion', namely, the effectiveness of the services provided by social overhead capital to the individual is affected not only by his own usage of such services but by the extent to which other members of society also use them. Indeed, most environmental disruption and pollution can be interpreted as congestion in relation to the use of natural and other social overhead capital. In order to analyse the role of social overhead capital, one must thus devise a framework in which these two aspects of social overhead capital are adequately represented. Let me first consider the case in which social overhead capital is exclusively used as productive capital which affects the marginal productivity of private factors of production.

Let β represent productive units. The level of output produced by each producer β depends upon the extent to which he utilises private factors of production and the services of social overhead capital. To simplify, it will be assumed that private factors of production and social overhead capital are both homogeneous; the following analysis can, with slight modification, also be extended to the general case where private factors of production or social overhead capital are heterogeneous. It is also assumed that the services provided by social overhead capital can be measured on an appropriate scale.

The level of output Q_β of producer β is determined by the input of private factors of production K_β and the extent to which the services of social overhead capital are employed. However, the effectiveness of the services provided by social overhead capital is related to the extent to which it is already congested. Thus, the level of output Q_β by producer β can be written in the following functional form:

$$Q_\beta = F^\beta (K_\beta, X_\beta, X, V) \tag{1}$$

where K_β and X_β represent the private factors of production and the services of social overhead capital respectively employed by producer β; X represents the aggregate services of social overhead capital employed by all productive units in the society; while V represents the stock of social overhead capital endowed to society. Hence

$$X = \sum_{\beta} X_\beta \qquad (2)$$

It can be assumed that the production function, F^β, is linear homogeneous with respect to K_β, X_β, X and V, indicating that these factors exhaust all the factors of production which limit the output of producer β. The production function's dependence on X and V indicates that congestion occurs in the use of social overhead capital. It can be assumed that the larger the aggregate level of use of social overhead capital, the lower will be the productivity of either the private factors of production or of social overhead capital, and that an increase in the stock of social overhead capital V results in an increase in the marginal productivity of scarce resources. Furthermore, it may be assumed that the private factors of production and social overhead capital are complementary to each other. It can therefore be assumed that the production function satisfies the following conditions:

$$F_X{}^\beta < 0, \ F_V{}^\beta > 0 \qquad (3)$$

$$F_{K_\beta}{}^\beta X_\beta \geq 0 \qquad (4)$$

The present formulation can thus be regarded as a generalisation of the analytical approach introduced by Rothenberg [5].

If, for the sake of simplicity, we assume that production units all produce identical goods (or that the relative prices of the goods produced by different producers remain constant throughout the course of this analysis), then real national product can be indicated by

$$Q = \sum_{\beta} Q_\beta \qquad (5)$$

At each point in time, the stock of social overhead capital V and the supply of private factors of production K are given. The allocation of both private factors of production and social overhead capital can be regarded as efficient if the corresponding level of real national product Q is maximised (subject to it being a feasible solution). If the private factors of production are allocated through a perfectly competitive market mechanism, and the services of social overhead capital are distributed to individuals free of charge, then the resulting allocation of scarce resources is not efficient. This proposition follows from the assumption that congestion occurs in the use of social overhead capital. In order to achieve an efficient allocation of scarce resources, the rules concerning the use of social overhead capital (taking account of the relative scarcity of social resources) have to be modified so that an optimum level of congestion results for use of social overhead capital. Such a modification of the market mechanism can be readily achieved if the administrative cost of enforcing a pricing scheme for the use of social overhead capital is negligible (i.e. the allocation of private factors of production can be done through a perfectly competitive market mechanism as in a decentralised market economy, while the services of social overhead capital can be charged according to their marginal social costs). The marginal social costs associated with the use of social overhead capital can be defined as the aggregate sum of the marginal gains of all the productive units due to a marginal reduction in the aggregate level of the services provided by social overhead capital. In symbols, the marginal social costs MSC can be defined as follows:

$$MSC = \sum_{\beta}(-F_X{}^{\beta}) \qquad (6)$$

As shown in Uzawa [3], if the services of social overhead capital are priced at marginal social costs, and if private factors of production can be allocated in a perfectly competitive market, then the resulting allocation of scarce resources is efficient and the level of real national product will be maximised.

The magnitude of the marginal social costs depends upon the private factors of production K and the amount of social overhead capital V available to society. It can be shown that, if social overhead capital becomes scarce relative to the private factors of production K, then marginal social costs will increase. Each producer will therefore reduce his use of social overhead capital and the aggregate level, X, of the services of social overhead capital will be decreased, resulting in an optimum level of congestion in the use of social overhead capital.

1.4 *AN OPTIMUM ALLOCATION OF SOCIAL OVERHEAD CAPITAL*

The modified market mechanism for achieving an efficient allocation of social overhead capital can now be extended to the general case where social overhead capital can be used directly in the process of consumption as well as that of production. Let consumers be denoted by the symbol α. The utility level of consumer α depends upon the services of social overhead capital X_{α} which he chooses to use, in addition to the private goods and services he consumes. The effectiveness of the services provided by social overhead capital which consumer α uses is affected by the extent to which social overhead capital is congested. Hence, the utility U_{α} enjoyed by consumer α depends upon the amount C_{α} of private goods and services, the amount X_{α} of the services of social overhead capital used by consumer α, and the degree of congestion which is determined by the aggregate demand X and the existing stock V of social overhead capital. This can be formalised as:

$$U_\alpha = U^\alpha (C_\alpha, X_\alpha, X, V) \tag{7}$$

The aggregate demand, X, for the use of social overhead capital can be defined by

$$X = \sum_\alpha X_\alpha + \sum_\beta X_\beta \tag{8}$$

It is again assumed that

$$U_x^\alpha < 0, \qquad U_v^\alpha > 0 \tag{9}$$

If private goods and services are allocated through a perfectly competitive market, then each consumer α will choose private consumption C_α and the services of social overhead capital X_α in such a way that the marginal rate of substitution between the services of social overhead capital and private consumption is zero, i.e. the condition

$$MRS_\alpha (\equiv U_{x_\alpha}{}^\alpha / U_{c_\alpha}{}^\alpha) = 0 \tag{10}$$

is satisfied subject to the ordinary budgetary constraints also being satisfied.

It is easy to see that the resulting allocation of scarce resources by means of a perfectly competitive market mechanism is not efficient. However, when consumers are involved, it becomes important to address the question of income distribution as well as that of the efficient allocation of resources. Suppose, for the moment, that individual utility levels are interpersonally comparable and that social welfare can be represented by the aggregate sum of these utility levels. An allocation of scarce resources and the corresponding income distribution will be optimum if aggregate social welfare U, defined by

$$U = \sum_\alpha U_\alpha \tag{11}$$

is maximised (subject to it being a feasible solution).

In addition to the special case discussed in the previous section, another modified market mechanism suggests itself as a means of obtaining an optimum allocation of resources. Private goods and services are allocated through a perfectly competitive market, while the use of social overhead capital is charged at a price equal to marginal social costs. In the present situation, the marginal social cost MSC is defined by

$$MSC = \sum_\alpha \left(\frac{-U_x^\alpha}{U_{c_\alpha}{}^\alpha} \right) + \sum_\beta (-F_x^\beta) \tag{12}$$

instead of as in equation (6).

In order to obtain an optimum allocation of resources, it is necessary to introduce a transfer mechanism in such a way that the marginal rate of distribution between any pair of individual consumers becomes one, namely,

$$MRD_{\alpha'\alpha''}\left(\equiv\frac{U_{c_{\alpha'}}{}^{\alpha'}}{U_{c_{\alpha''}}{}^{\alpha''}}\right)=1 \tag{13}$$

It is again possible to show that pricing social overhead capital at marginal social costs, together with a scheme for transferring income, will result in an optimum allocation of scarce resources.

1.5 *OPTIMUM CONDITIONS FROM A DYNAMIC POINT OF VIEW*

I have so far been concerned with the optimum allocation of scarce resources from a static point of view: the stock of private factors of production and social overhead capital have been assumed constant. To discuss the impacts of social overhead capital upon the allocative mechanism, it is necessary to examine the dynamic process of accumulating social overhead capital and to discuss pricing principles from a dynamic point of view.

If we assume that the rate at which future utility levels are discounted is identical for all members of society, say equal to δ, then the level of social utility U can be represented by the following utility integral of the Ramsey type:

$$U=\int_0^\infty U(t)e^{-\delta t}dt \tag{14}$$

where $U(t)$ is the level of social utility at time t, defined as the aggregate of each individual's utility, $U_\alpha(t)$, i.e.

$$U(t)=\sum_\alpha U_\alpha(t) \tag{15}$$

Suppose that the stock of social overhead capital at time zero is V_0. In the simplest case it is assumed that the supply of private factors of production is kept at a certain fixed level, K. The problem is now that of allocating private and social resources over time, and of dividing aggregate output between consumption and investment, so that the resulting path of economic growth is optimal in the sense that the level of social utility U is maximised among the set of feasible paths.

Let $V(t)$ be the planned stock of social overhead capital at time t, while $K_\beta(t)$ and $X_\beta(t)$ are the private factors of production and the amount of social overhead capital used by producer B at time t, respectively. Then, for each producer β, the level of output $Q_\beta(t)$ is given by

$$Q_\beta(t) = F^\beta[K_\beta(t),\ X_\beta(t),\ X(t),\ V(t)] \tag{16}$$

On the other hand, if, for each consumer α, $C_\alpha(t)$ and $X_\alpha(t)$ stand for the level of private consumption and the services of social overhead capital used by consumer α at time t, then the level of α's utility, $U_\alpha(t)$, is given by

$$U_\alpha(t) = U^\alpha [C_\alpha(t),\ X_\alpha(t),\ X(t),\ V(t)] \tag{17}$$

The aggregate level $X(t)$ for the use of social overhead capital at time t is defined by

$$X(t) = \sum_\alpha X_\alpha(t) + \sum_\beta X_\beta(t) \tag{18}$$

Since the supply of private factors of production is assumed to be fixed at K, the following condition is satisfied at each moment of time t:

$$K = \sum_\beta K_\beta(t) \tag{19}$$

The aggregate level $Q(t)$ of output is given by

$$Q(t) = \sum_\beta Q_\beta(t) \tag{20}$$

while the aggregate level $C(t)$ of private consumption is given by

$$C(t) = \sum_\alpha C_\alpha(t) \tag{21}$$

Let $I(t)$ be the real level of public investment, i.e. the amount of goods and services devoted to the construction of social overhead capital. Then

$$Q(t) = C(t) + I(t) \tag{22}$$

The rate of increase $V(t)$ in social overhead capital is related to the volume of net investment:

$$I(t) - \mu V(t) \tag{23}$$

where μ is the average rate of depreciation. The average rate of depreciation μ is related to the extent to which social overhead capital is currently used. It can be assumed that

$$\mu = \mu[X(t)/V(t)] \tag{24}$$

in which the following conditions are satisfied:

$$\mu'(x) > 0, \qquad \mu''(x) > 0 \tag{25}$$

The problem then is to find the path of capital accumulation $V(t)$, the allocation of private goods $C(t)$ and $I(t)$, and the services of social overhead capital $X(t)$ which maximises the level of social utility, U. This problem can be easily solved using the Lagrange method in the calculus of variations. An examination of the conditions for an optimum solution readily indicates that it can be obtained by modifying the perfectly competitive market mechanism in a similar way to that discussed for the static case. At each moment in time t, social overhead capital is priced at marginal social costs, while private goods and services are allocated through a perfectly competitive market. For a dynamic case, however, the concept of marginal social costs is defined as follows:

$$\theta = \sum_\alpha \frac{-U_{X^\alpha}}{U_{c_\alpha}} + \sum_\beta - F_{X^\beta} + \mu'\left(\frac{X}{V}\right) \tag{26}$$

where the marginal cost of depreciation μ is added to the static concept of marginal social costs.

The pricing scheme thus obtained is closely related to what is usually referred to as the principle of 'he who pollutes should pay'. Whoever uses the services of social overhead capital should thus be required to reimburse society for the external diseconomies he has imposed on other members of society together with the value of the depreciated social overhead capital (evaluated at the discounted present value of the marginal losses in net benefits society will suffer due to the marginal depletion of social overhead capital).

1.6 *OPTIMUM INVESTMENT IN PRIVATE AND SOCIAL CAPITAL*

The propositions concerning the optimum path of economic growth can be extended to the situation where we are concerned with the accumulation of private as well as social overhead capital. In this general case, I should also like to include those factors which limit the process of capital accumulation for both private and social capital. In order to bring out the essential characteristics of this phenomenon I shall confine myself to the simple case where both consumers and producers are additive, so that I can characterise the model in terms of only one consumer and one producer. I shall also assume that the services provided by social overhead capital exhibit the feature of Samuelsonian [4] public goods; the general case for the type of social overhead capital discussed in the previous sections can be derived by slightly modifying the conceptual definitions.

Let me consider the situation where both private and social capital are homogeneous; let K and V stand for the stock of private and social overhead capital at a particular moment of time. The level of utility for a (representative) consumer, U, depends on the level of private consumption, C, and the stock of social overhead capital, V:

$$U = U(C, V) \tag{27}$$

while the output produced by the (representative) producer, Q, is a function of the stock of private capital, K, and social overhead capital, V:

$$Q = F(K, V) \tag{28}$$

For the sake of exposition, it will be assumed that the utility function $U(C, V)$ and the production function $F(K, V)$ are both linear homogeneous, so that

$$U = u(c)\, V, \qquad c = C/V \tag{29}$$

$$Q = f(k)\, V, \qquad k = K/V \tag{30}$$

Part of the output Q may be devoted to the accumulation of private and social capital; let I_K and I_V represent the amount (in real terms) of investment in private and social capital. It is assumed that the factors which limit the processes of capital accumulation are proportional to the stock of private and social capital. Hence, the rates of increase of K and V are related to the rates of real investment through the following functional relationships:

$$I_K/K = \phi_K(\dot{K}/K) \tag{31}$$

$$I_V/V = \phi_V(\dot{V}/V) \tag{32}$$

where both functions $\phi_K(.)$ and $\phi_V(.)$ exhibit increasing marginal costs:

$$\phi''_K(.) > 0 \tag{33}$$

$$\phi''_V(.) > 0 \tag{34}$$

Both functions $\phi_K(.)$ and $\phi_V(.)$ can be expressed net of depreciation, so that one can assume that

$$\phi_K(0) = 0, \qquad \phi_K(.) > 0 \tag{35}$$

$$\phi_V(0) = 0, \qquad \phi_V(.) > 0 \tag{36}$$

These functions $\phi_K(.)$ and $\phi_V(.)$ are analogous to the Penrose function used in the analysis of investment behaviour. A detailed discussion of the nature of these functions is set out in Uzawa [6].

The problem of optimum investment in private and social capital can now be formulated as follows. Let the stock of private and social capital at time zero be given by K_0 and V_0. The problem is then to find a path of consumption and capital accumulation $[K(t), V(t)]$ for which the level of social utility (of the Ramsey type):

$$U = \int_0^\infty U[C(t),\, V(t)]e^{-\delta t}\, dt \tag{37}$$

is optimal among the set of all feasible paths of capital accumulation.

A path of consumption $C(t)$ and capital accumulation $[K(t), V(t)]$ is called feasible if it satisfies the following conditions:

$$Q(t) = F[K(t), V(t)] \tag{38}$$

$$Q(t) = C(t) + I_K(t) + I_V(t) \tag{39}$$

$$\frac{I_K(t)}{K(t)} = \phi_K[\alpha_K(t)], \qquad \frac{I_V(t)}{V(t)} = \phi_V[\alpha_V(t)] \tag{40}$$

$$\frac{\dot{K}(t)}{K(t)} = \alpha_K(t), \qquad \frac{\dot{V}(t)}{V(t)} = \alpha_V(t) \tag{41}$$

$$K(0) = K_0, \qquad V(0) = V_0 \tag{42}$$

It is extremely difficult to find the optimum path for such a maximisation problem, but it is possible to show that it can be approximated by utilising a simpler structure to establish the path of capital accumulation.

Let the ratio of private capital to social capital be k:

$$k = K/V \tag{43}$$

The marginal product r_K of private capital is defined as

$$r_K = f'(k) \tag{44}$$

while the marginal product r_V of social capital is given by

$$r_V = f(k) - kf'(k) \tag{45}$$

Furthermore, for the level of the consumption to social capital ratio $c = C/V$, the marginal utility p of private consumption is defined by

$$p = u'(c) \tag{46}$$

and the marginal rate of substitution, s_V, between private consumption and social capital is given by

$$s_V = \frac{u'(c)}{u(c)} - c \tag{47}$$

The optimum rates of increase in private capital and social capital, α_K and α_V, can now be determined so that the following conditions are satisfied:

$$\frac{r_K - \phi_K(\alpha_K)}{\delta - \alpha_K} = \phi'_K(\alpha_K) \tag{48}$$

$$\frac{(s_V + r_V) - \phi_V(\alpha_V)}{\delta - \alpha_V} = \phi'_V(\alpha_V) \tag{49}$$

The determination of the optimum rates α_K and α_V is illustrated in Figs. 1.1 and 1.2.

For a given level $k = K/V$ of the ratio of private capital to social capital, the level $c = C/V$ of private consumption per unit of social capital can be determined so that the following equilibrium condition is satisfied:

$$c = f(k) - \phi_K(\alpha_K)\, k - \phi_V(\alpha_V) \tag{50}$$

together with conditions (44), (45), (47), (48) and (49). It is easy to see that, for a given k, the optimum rates of capital accumulation, α_K and α_V, and the level c of private consumption per unit of social capital, are all uniquely determined. It can also be shown that an increase in k implies an increase in either c or α_V, but also results in a decrease in α_K.

It has been thus shown that the optimum path can be approximated by the path for which the level of private consumption $c(t)$ and the rates of capital accumulation, $\alpha_K(t)$ and $\alpha_V(t)$, are determined for the capital ratio $k(t) = K(t)/V(t)$ at each moment of time t, by means of the above procedure. It follows that the imputed prices of private and social capital, p_K and p_V, can then be approximated by

$$p_K = p\phi_K(\alpha_K), \qquad p_V = p\phi_V(\alpha_V) \tag{51}$$

where p is the marginal utility defined by (46).

The above approximation theorem can be extended to the general case involving a number of producers and consumers, together with increasing marginal costs for the accumulation of both private and social capital. Equation (26), describing the marginal social costs associated with the use of social overhead capital, can be applied to this general case, except that the third term $\mu'(X/V)$ has to be multiplied by the imputed price p_V of social overhead capital evaluated at each point in time. The imputed price of social overhead capital p_V can be approximated by equation (51), and it is easy to see that an increase in the capital ratio k results in an increase in the optimum rate of accumulation α_V for social overhead capital and in an increase in the approximate imputed price p_V, thus increasing the level of marginal social costs.

1.7 *CONCLUDING REMARKS*

In this paper I have outlined the analysis of social overhead capital and have discussed how the optimum management of such capital can be achieved. It is easy to see that such an analysis can be extended to the more general situation where there are different kinds of social overhead capital performing various functions in a national economy; one can simply replace the stock of social overhead capital by

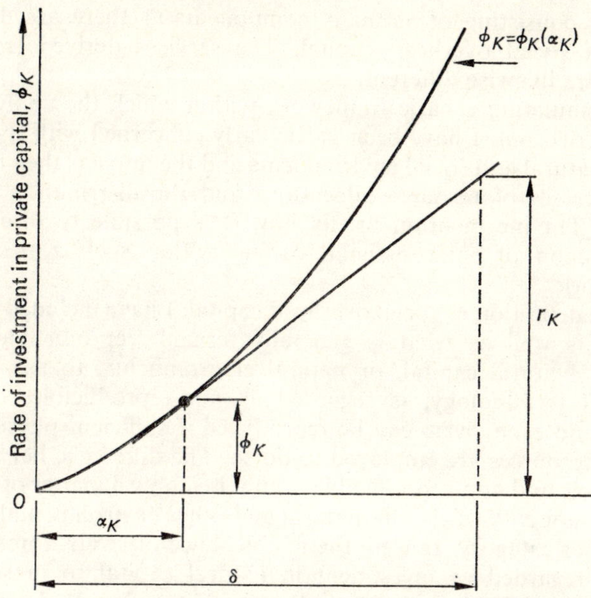

FIG. 1.1   The optimum rate of accumulation of private capital

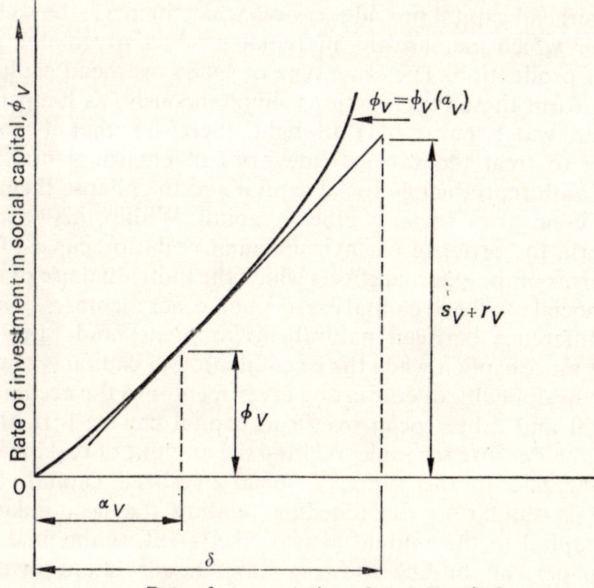

FIG. 1.2   The optimum rate of accumulation of social capital

a vector consisting of as many components as there are different kinds of social overhead capital. The services derived from the capital are likewise different.

In formulating a basic framework within which the analysis has been carried out, I have been particularly concerned with the problem of natural and social environments and the impacts they have on the processes of resource allocations and the distribution of real income. Let me mention briefly how it is possible to discuss the phenomenon of environmental capital within such a theoretical framework.

In the definition of social overhead capital, I have included natural capital as well as what is generally termed 'reproducible social capital'. Natural capital, or natural environments, to use a more standard terminology, is regarded as non-reproducible, but air, forests and even rivers can be reproduced if sufficient physical and human resources are employed to do so. The difference between reproducible and non-reproducible capital is simply a matter of degree, and it is not easy to classify natural and other environmental capital into either category. Indeed, the maintenance of rivers, forests, etc., can be regarded as investment in natural capital to preserve its functional efficiency and the public expenditure involved represents one of the most important functions a government can perform.

On the other hand, it is difficult to distinguish between the functions of natural environments and other social capital. Both types of social overhead capital provide services which increase the welfare of society or which increase the marginal productivity of the private means of production. The same type of social overhead capital may often perform these two functions simultaneously, as happens with roads, air, water, etc. I have thought, therefore, that it would be advisable to treat the various categories of environmental capital together with reproducible social capital and to collapse them into a uniform concept of social overhead capital. Within this conceptual framework, the problem of environmental pollution can be formalised in terms of the external effects which the individual use of the services of social overhead capital exerts upon other members of society.

The difference between natural environments and other social capital is more explicit when the accumulation of capital is examined. The ease or difficulty of converting investment into the accumulation of natural and other social overhead capital can be formalised in terms of the Penrose schedule, relating the amount of real investment to the increase in the stock of social overhead capital. Such a schedule is similar to the schedule relating the accumulation of private capital to the amount of real investment. In the final section of the paper, an outline showing how the optimum investment in social overhead capital can be determined in terms of the Penrose schedule has thus been presented, although the implications of such an analysis for the problems of environmental capital have not been spelled out. In fact, the usefulness of this analysis for the problems of environmental protection and pollution abatement are crucially dependent upon the possibility of verifying empirically basic concepts like the stock of social overhead capital and the Penrose schedule which, I must confess, cannot easily be resolved.

REFERENCES

[1] Ramsey, F. P., 'A Mathematical Theory of Saving', *Econ. J.*, vol. 38 (1928) pp. 543–59.
[2] Kneese, A. V., *Water Pollution: Economic Aspects and Research Models* (Resources for the Future, Inc., Washington, D.C., 1970).
[3] Uzawa, H., 'On the Economics of Social Overhead Capital' (unpublished, 1970).
[4] Samuelson, P. A., 'The Pure Theory of Public Expenditures', *Rev. Econ. & Stats.*, vol. 36 (1954) pp. 387–9.
[5] Rothenberg, J., 'The Economics of Congestion and Pollution: An Integrated View', *Amer. Econ. Rev.*, vol. 60 (1970) pp. 114–21.
[6] Uzawa, H., 'Time Preference and the Penrose Effect in a Two-Class Model of Economic Growth', *J. Polit. Econ.*, vol. 77 (1970) pp. 628–52.

Discussion of the Paper by H. Uzawa

Formal Discussant: Mills. Uzawa has presented an elegant and perceptive model of the optimum accumulation and use of social overhead capital. The salient characteristics of social overhead capital in his macro-growth model are: it can be accumulated by devoting scarce resources to its production; it contributes to the production of goods and utility; and it depreciates by an amount that depends upon its stock and on the extent to which it is congested by use. The most obvious example of the author's concept is probably roads.

I have doubts, however, whether the environment fits easily into the author's conceptual apparatus. Suppose a metropolitan area is on an estuary and the quality of the water and the ambient air over the metropolitan area are components of the area's environment. The quality of the environment can be impaired by congestion, i.e. by discharging residuals to it. And it can be improved to, or beyond, its natural condition by devoting resources to the task, e.g. by treating wastes. But I can see no natural interpretation to the notion that more, or a better quality of, environment can be produced with a production function that is homogeneous of degree one. Zero discharge implies a natural upper limit to environmental quality and I cannot interpret the concept of a homogeneous production process to produce more environment. In fact, recent research on environmental issues has emphasised the finiteness of the environment. The simplest and most natural axiomatisation of the finiteness of the environment implies that there must be upper limits to the population and production that the earth can support. Of course, popular writers make lots of mistakes on this matter, among the most important being their underestimation of the role of technical progress in shifting the limits or constraints. But that issue is not relevant to the author's model.

I conclude that the author's model is about just what he says it is about: the accumulation and use of social overhead capital. The previous paragraph is a warning against an interpretation he does not give it.

Machlup asked how the stock of social overhead capital, V, was measured or quantified, especially in what units it was expressed. He had no difficulty with the units in which the *services* of social capital were measured, even if their productivity or utility depended on the level of congestion (i.e. on the degree of use made by all users). He could not see, however, how the *stock* of social overhead capital, and increases in the stock, were measured. It could probably not be done in terms of 'capacity to render services', since these services were subject to so much variation. Had the author visualised that the measurement was in terms of inputs, at least with regard to produced and reproducible real capital? If so, we would run into problems whenever real capital was not reproducible. A physical description of real capital might be possible, but if we had different types of capital – roads and harbours, water and air, and so forth – we would confront the old problem of adding heterogeneous things. If we proposed measurement in terms of value, we would run into all the well-known problems set out in lectures on capital theory. We surely could not determine the value of a stock of capital by capitalising the value of the future flow of services, since the value of these services would depend on the unknown congestion resulting from total use. To get around the problem of valuation by assuming, as the author did, that there was only one kind of capital was to dodge a problem that we must solve if the author's model was to be at all useful.

Rothenberg thought that the concept of social overhead capital did capture a great deal of what was involved in describing environmental variables – particularly pollution and congestion. We were not really speaking about the reproducibility of air, or of water, or of nature. We were referring to their ability to produce certain services. What we meant by reproducing or increasing (e.g. the capacity of a river) was that we increased its assimilative capacity – its ability to render the services that it does. The item X/V was thus the level of the current total uses of this type of resource in relation to the total potential capacity of its use. In other words, it measured the degree of 'congestion' associated with this use.

Kneese thought that a deficiency of almost all theoretical models of the economic growth process was their failure to treat technological change in a meaningful way. We knew from empirical studies, like those of Barnett and Morse [1], that when long periods of time were under consideration, the factor of technological change became the dominant one. He had nevertheless not seen this factor treated in a satisfactory way in any of the models of economic growth that aspired to analyse the problems associated with the finite nature of resources, commodities or environmental assimilative capacity.

Beckerman claimed that there was a fallacy in the Mills discussion. The empirical argument that finite resources implied finite growth was only important if you were already fairly close to this limit. The idea that the world's resources will eventually be exhausted was trite: in so far as it was true, it was as true a thousand years ago as it was today. But if we were only going to reach that point in a million years' time, then why bother about it now.

Hicks also objected to Mills. He felt that he expressed an old materialist point of view. What we really meant by growth was growth in the means of satisfying wants, and this was a very delicate matter when wants changed over time. There was no reason to suppose, for example, that unlimited growth in some quite appropriate sense was impossible even while using an absolutely constant amount of physical resources.

Tulkens pointed out that the author intended to reach an optimal solution by charging a uniform price $\theta = -\left[\sum_{\alpha}(u^{\alpha}_{x}/u^{\alpha}_{c_{\alpha}}) + \sum_{\beta}F^{\beta}_{x}\right]$ for the amounts X_{α} and X_{β} of commodity X used by consumers and firms. He did not see, however, how this price fulfilled the author's optimality condition (12). Indeed, if p was the price of commodity C (denoted as Q for firms), individual profit maximisation at prices p and θ would lead the firms to an equilibrium implying the condition $\theta = p.F^{\beta}_{X_{\beta}}$, which differed from (12) by the factor p. Given that X played the role of a public good in the model, should not one call on the recent theorems on pseudo-equilibrium with public goods, and have the price θ individualised for each agent, i.e. defined as θ_α and θ_β? To be specific, if we defined the following price vector: $(p, \theta, \theta_\alpha, \theta_\beta, \kappa)$, and considered what happened when consumers maximised $u_\alpha(C_\alpha, X_\alpha, X)$, subject to $pC_\alpha + \theta X_\alpha + \theta_\alpha X = R_\alpha$ (when R_α was consumer α's income) and firms maximised $\pi_\beta = pQ_\beta - \theta X_\beta - \theta_\beta X - \kappa K_\beta$ (where π_β denoted firm β's profit) and subject to $Q_\beta = F_\beta(X_\beta, K_\beta, X)$. If the individualised prices θ_α and θ_β were chosen so that $\sum_\alpha \theta_\alpha - \sum_\beta \theta_\beta = \theta$, it was apparent that the first-order conditions of these individual maximising behaviours would indeed meet condition (12) in the author's paper. The charges for the use of social overhead capital should thus be discriminatory if they were to lead to a Pareto optimum.

Hicks, referring to Tulkens' argument, said that as soon as you introduced the assumption of constant returns to scale, then the discrimination was usually clear. However, as soon as this assumption was dropped, one's budgets no longer balanced and one had to levy taxes (or pay subsidies) where they were most effective to make up the difference.

He then asked what the distinction was between public and private capital in this model. If you were discussing optimum paths, what did this distinction have to do with it? Private property was simply one particular institution which sought to reach an optimum path. If we were trying to determine an optimum path, then this distinction did not belong in that part of the problem. Private and social capital were the same thing as far as the optimum path was concerned. One could simply take over the results of Ramsay [2] and others, and it then became simply a question of the institutional complications of the private and public sectors.

Rothenberg said that private capital could be thought of as a stock in which individual units could be appropriated for individual exclusive use. Social capital had common uses. It was used by groups of individuals without any dimunition of stock. One type of capital could be appropriated to a decentralised private use with no externalities in its use. The other could not help being used in a non-exclusive way by the users, and this gave rise to externalities.

Kneese observed that we often placed so much emphasis on the exclusion characteristics of public goods that we often forgot that the central question of resource allocation was that of jointness of production. In the case of a toll road, for example, you sacrificed some real efficiency for the sake of a privately administered arrangement. Exclusiveness was therefore only one aspect of public goods.

REFERENCES

[1] Barnett, H. J., and Morse, C., *Scarcity and Growth: The Economics of Natural Resource Availability* (Johns Hopkins Press, Baltimore, 1963, 1968).
[2] Ramsey, F. R., 'A Mathematical Theory of Saving', *Econ. J.* (1928); reprinted in Arrow, K., and Scitovsky, T. (eds.), *Readings in Welfare Economics* (Richard Irwin, Homewood, Ill., 1969).

2 Industrial Structure, Growth and Residual Flows*

Finn R. Førsund and Steinar Strøm

2.1 *BASIC CONSIDERATIONS*

The background for economic analyses of environmental pollution is excellently presented in several recent contributions (see, e.g., Kneese *et al.*, 1970; Russell and Spofford, 1972). It suffices here, then, to give but a brief introduction to the concepts used in the study.

The impact on the natural environment of man's production and consumption activities stems broadly from (a) occupying space, (b) extracting raw materials, and (c) discharging leftovers. The pervasive and extensive negative effects from the last activity are a significant new feature of environmental problems now experienced by the industrialised 'throughput' economies. This justifies concentrated attention to this aspect.

The 'materials balance' approach (cf. Kneese *et al.*, 1970) clearly underlines the generality of *residuals* as the normal outcome of the throughput of materials in the course of production and consumption activities. Residuals measured in weight are defined by the difference between the weight of the total material inputs to an activity and the weight of the products which are the objective of the activity, plus the weight of net accumulation of tangible assets in the activity (e.g. machinery, buildings, consumer durables, etc.).

Residuals can be classified as (a) material residuals (solid, liquid and gaseous) and (b) energy residuals (heat, noise and radiation).

The material residuals can, for example, be classified according to their chemical nature. (In Appendix 1 the 38 different types of

* The analysis of environmental problems presented in this study is a continuation of research on pollution problems carried out by the two authors in co-operation with the Norwegian Ministry of Finance. The data presented in the study have been published in a report from the Ministry of Finance to the Norwegian Parliament, 'Spesialanalyse 1, Forurensninger. Særskilt vedlegg til Langtidsprogrammet 1973–77'. The Planning Division of this Ministry has organised the collection of data. In this connection we are especially indebted to Mr. Kjell Slyngstad of the Ministry of Finance, who organised the data groups, and to the members of these groups, in particular Mr. Torstein Dale. We are further indebted to our colleagues at the Institute of Economics, Oslo, Mr. Sigbjørn Atle Berg and Mr. Eilev Jansen, for their assistance in making the calculations and preparing the paper. Finally, we thank the Ministry of Finance for their financial support.

material residuals considered are specified.) The two main categories of material residuals are organic and inorganic substances.

The natural environment is used as a sink for these residuals. This function will be called the *residual disposal service* of the environment. Pollution problems arise because the different geographically located receiving bodies, the *recipients*, also have other competing functions.

The recipients are also suppliers of raw materials and recreation services. The extraction of raw materials includes flows of timber, minerals, oil, fish, water and air.

The recreation services encompass activities like swimming, fishing, boating, skiing, enjoyment of aesthetic values from scenery and the intrinsic value of Nature.

Pollution problems are often stated in terms of different kinds of damages, e.g. health effects, wear and tear of buildings, etc. In this context, damages are conceived as reducing the levels of flows of raw materials or recreation services. Such reduction can take place at the same time as the residuals are discharged, or in the future.

A residual is here defined as a pollutant if the corresponding disposal service of recipients negatively affects, quantitatively or qualitatively, the raw materials or recreation services 'produced' by the recipients. The discharge of residuals does not of necessity generate pollution. The natural environment has an *assimilative capacity*. Owing to dilution, decay, decomposition, chemical transformations, etc., occurring in nature, there are certain *threshold values* of ambient residual concentrations that must be exceeded before harmful effects appear.

The policy problem posed by pollution is quite simple to state in general terms: it is just the classical economic problem of utilisation of scarce resources. The task is to strike a balance among the competing uses of the recipients, i.e. to determine an optimal compromise on the possibility frontier for residual disposal services and all other goods and services.

The nature of pollution problems necessitates policy measures both across and along the time axis. The time dimension is essential when the negative effects of residuals are due not to current flows, but to *stocks of accumulated flows* of residuals. There are two different categories. In the first place we have residuals which are accumulating in the environment due to their own chemical nature, i.e. they are persistent. This holds, e.g., for heavy metals, fluorine, pesticides, halogenated hydrocarbons and plastics. Some of these residuals may of course pass through different recipients in the environment before they reach the final recipient where accumulation takes place. It is a common characteristic of these residuals that they adversely affect the extraction of raw materials; that is, the quality and/or quantity of future flows of fish, meat, vegetables and so on is reduced, thereby, for instance, causing human health problems.

In the second place, the residuals can accumulate owing to discharges of residuals in excess of recipients' assimilative capacity.

Where negative effects result only from current *flows* of residuals, the environment regains its natural conditions in a comparatively short time after 'closing the tap'. Such situations can be dealt with 'across the time dimension', unlike where accumulated stocks matter. For control of the latter, the time pattern of discharge is essential. At a given point in time the negative externalities from the stocks of residuals cannot be optimally adjusted by myopic considerations about current discharge rates only (provided the time horizon of society is not negligible).

Recently there have been several articles published on this stock–flow problem of pollution. In what follows, we are using a growth model for calculating future discharges of residuals in Norway. Optimal growth and optimal accumulation of waste, as considered in d'Arge and Kogiku (1973), Keeler *et al.* (1971), Smith (1972) and Strøm (1972), will not be discussed in this paper. On the other hand, our growth model can be used to calculate the consequences of the current emission of waste on future stocks of residuals. This will be done for residuals which are accumulating due to their chemical nature and under alternative assumptions concerning public policy.

2.2 *FRAMEWORK FOR EMPIRICAL STUDIES*

The implementation of optimal control measures requires empirical information on several aspects of pollution problems. Empirical effort in the field of environmental pollution can be organised along the following main lines (cf. Russell and Spofford, 1970), taking the recipients as reference:

(a) The generation of residuals in production and consumption and discharge to recipients.
(b) The natural processes taking place in the recipients due to discharge of residuals, e.g. diluting, decaying, decomposing, transportation between recipients, transformation of residuals, etc.
(c) Defining the environmental services of the recipients and establishing the impact on these of ambient concentrations of residuals.
(d) Evaluating the preferences attached to changes in environmental services, including the time perspective (of the 'present generation').

In this study only (a) is considered. The data problem is not so severe here as for the other aspects.

As regards the generation of residuals and the accompanying use of residual disposal services, there are at least the following main possibilities of change:

  (i) The scale of the activities.
  (ii) The input mix in an activity.
  (iii) The location of generating sources.
  (iv) Modification of primary residuals in order to substitute for residual disposal services generating pollution with disposal services having less harmful effects. This is achieved by giving the primary residuals different physical forms and/or discharging them to other recipients. (Following Russell and Spofford, 1972, the concept of 'modification' is used instead of 'waste treatment' to underline the conservation of mass. The residuals do not physically disappear by waste treatment.)
  (v) Recycling the residuals to economic activities, thus reducing the need to extract new raw materials and relieving the environment of the residual disposal function – i.e. recycling means internalising the disposal function.

The possibilities stated so far can be realised without changing the production technique of the activities. Modification and recycling are conceived of as additive processes not interfering with the operation of the activities. But the change of basic techniques is an important possibility for changing the level and composition of discharged residuals. This is especially the case when considering the time dimension. All the different aspects of technique change are included in the last possibility of change:

  (vi) The technique of production/consumption, including changes in the products themselves.

In what follows, only (i) will be considered. As mentioned earlier, we are using a macro-economic model for calculating future discharges and stocks of residuals. The model, however, consists of several production sectors. Regarding each sector as representing one activity, the model therefore permits a study of the changes in the output mix in the economy. A main point of the macro-economic model we are using is that over time the output mix will change due to several factors which will be explained below.

2.3 *THE SCOPE OF THE STUDY*

As underlined in section 2.1, the pollution problems considered are conceived of as locally experienced problems in the sense that it is the environmental services of geographically located recipients that are negatively affected by the discharge of residuals. It follows, then, that regional empirical studies are called for to implement policy measures in the 'problem sheds' of a river basin, an estuary, the air over a city, etc.

This study, however, takes up pollution problems on a more aggregated level. The geographical unit is the whole nation and the units of production are production sectors.

What, then, can be obtained from such a macro-economic analysis?

For a government agency concerned with the development of the economy in a macro-economic perspective, it is essential to have a model framework that permits tracing the total overall effects on the economy of the various regional measures put forward, for instance, by municipalities and special agencies. It is a question of great importance for macro-economic policy whether large-scale action on serious pollution problems has any impact on macro-economic magnitudes. When applying policy measures to the generation of residuals, the *interdependences* of economic activities must be taken into consideration. Existing operational macro-economic models are based on data obtained from national accounts. Extending the models to cover the generation of residuals requires that various types of data gathered by different bodies be systematised and organised meaningfully to link the generation of residuals with economic activities. The task of incorporating information about residuals into the existing system of national accounts is, in fact, a main purpose of this study.

In the pollution debate, the question of the meaning of what is in fact measured by the national accounts has again appeared. This question will not be pursued here. It is not necessary to introduce new definitions of, say, G.N.P. in order to analyse tradeoffs – across and along the time axis – among production and consumption of goods generating polluting residuals and environmental services.

The controversy about economic growth as a conflict between G.N.P. and environmental qualities is clearly a macro-economic issue. Rapid economic growth has been an important goal of economic policy in most countries. The growth rate of G.N.P. has been used as a measuring-rod towards that goal. It is no wonder that allegations of conflict between growth and protection of the environment is a hotly disputed issue.

It is important to note that the economic growth model we are using in calculating future flows of residuals is a multi-sectoral model. An important feature of an economy like that of Norway is the real disparity in the growth rates of different sectors. The classic pattern is that the agricultural sector tends to expand less rapidly than other sectors. This and other tendencies are taken care of in the specification of our multi-sectoral growth model.

There are several factors which can explain the disparity in sectoral growth rates. The best-known factor is differences in the income elasticities of different commodities. The income elasticities of demand for agricultural and fishery products, for example, are low compared with those for industrial products, which in turn are lower than for services. Other factors are the rate of technological progress in the different sectors, the scale properties of long-run production functions, the input–output structure, etc. In Johansen (1960), detailed discussion of the reasons for non-proportional growth of the different sectors is given.

The non-proportional expansion of the different sectors means that a prediction of the residuals discharged in the future based on the amount discharged in an initial year and the growth rate of G.N.P. can be quite misleading. The residuals of the different types will also grow non-proportionally. Considering the numerous studies of future levels of pollution (see, for instance, Meadows *et al.*, 1972) where this non-proportionality is disregarded, an attempt to make predictions by means of a multi-sectoral growth model where non-proportionality is intrinsic may be of educational importance.

Input–output models provide a multi-sectoral framework for studying the generation of residuals from economic activities. In the studies by, e.g., Leontief and Ford (1971) and Victor (1971), final demand is exogenous. In our study, however, significant parts of final demand are *endogenous*. Moreover our model is dynamic.

2.4 *THE MULTI-SECTORAL GROWTH MODEL AND THE GENERATION OF RESIDUALS*

The analytical framework of the study is a model used by the Planning Division of the Norwegian Ministry of Finance for calculating long-term projections of general patterns of possible economic development of the Norwegian economy. This multi-sectoral growth model – M.S.G. for short – was first developed by Johansen (1960). An extended version is now used (see Schreiner, 1972). A brief outline of the main elements of the formal structure of the model is given in Appendix 3.

The interdependences among the twenty-six endogenous production sectors through intermediate deliveries are treated in an input–output framework, but in addition the gross production in each sector depends on inputs of labour and capital. These production functions are of the Cobb–Douglas type. A multiplicative exponential time function expresses the exogenous neutral technical change. The allocation of exogenously given *total* labour force and capital to the sectors is governed by the rule of marginal adjustments of the factors. The demand for sectoral products is partly exogenous, e.g. export and government demand, and partly endogenous. Private consumption functions for the sectoral products are specified.

The main outputs of the model are endogenously determined growth rates for sectoral production, product prices and distribution of labour and capital to the various sectors. The time changes in the variables are approximated by an iterative solution method (cf. Spurkeland, 1970) which ensures that market clearing conditions and definitional relationships are obeyed. We have chosen to calculate growth rates for three-year periods. The length of the period for which growth rates are constant can be parametrically varied.

In this study, data for residuals generated in production are gathered at the production sector level. The number of production sectors in the Norwegian national accounts is nearly 130, each sector being an aggregate of establishments. The data on residuals at the level of national accounts are aggregated to the M.S.G. level according to the sector code given in Appendix 2. As regards measuring the residuals, there are problems about the unit of measurement and the stage at which to measure the amount. The unit of measurement may be different from the unit most natural at the source of generation when considering different types of harmful effects. The amount of residuals can be measured at the point of primary generation, after waste treatment, or upon arrival at specific recipients. The only possibility for this study was to measure the residuals in technical units suitable for measuring at the source of generation, and to measure the amounts at the factory gate, i.e. *after* modification of residuals at the premises of the producer. Thus, information about modification possibilities was not obtained.

The generation of residuals in the production sectors is introduced as joint production (cf. Førsund, 1972). The functional forms are, however, not specified. The joint production aspect is quite simply represented by *discharge* coefficients, defined as the proportion between a residual from a sector and the gross production of that sector:

$$d_{ij}{}^{\alpha} = \frac{W_{ij}{}^{\alpha}}{X_j} \tag{1}$$

where $d_{ij}{}^{\alpha}$ is the discharge coefficient, $W_{ij}{}^{\alpha}$ is the discharge of residual of type i from production sector j to the recipient α (air, land and water), and X_j is the volume of gross production in sector j measured in kroner.

The generation of residuals through private consumption is introduced in the study by assuming a constant proportion between residuals and total private consumption. The reason for doing so is, of course, the lack of data.

The residuals from private consumption consist of sewage, solid waste and residuals from the use of private motor-cars. Heating is included as one input activity in the production sector for dwellings (no. 23). The demand for motor-cars is exogenously given in the M.S.G. model. The residuals generated (in the future) from the use of these cars can therefore not be calculated by the model. The data for the three categories of residuals mentioned above are calculated by applying the growth rate for total private consumption as determined by the run of the M.S.G. model to the observed amounts in the initial year.

There are two other activities in the economy which must be dealt with along more or less the same lines. The level of activity in military defence is exogenously given in the model. The residuals generated in the future through the use of aircraft and military vehicles are estimated by applying the growth rate for military defence used in the standard run of the M.S.G. model. Formally, the same procedure is followed in calculating the generation of asphalt dust in the future from the use of roads in Norway. The quantity of wear and tear of asphalt each year is exogenously given in the model.

At a point in time, the total amount R_i of residual i discharged into recipient α is:

$$R_i^\alpha = \sum_{j=1}^{26} d_{ij}^{\alpha} X_j + R_i^{C\alpha} + R_i^{G\alpha} + R_i^{f\alpha} \tag{2}$$

where $R_i^{C\alpha}$ is the amount of residual i generated through consumption and $R_i^{G\alpha}$ indicates residuals stemming from military defence and the use of the roads. As explained above, the development of $R_i^{C\alpha}$ and $R_i^{G\alpha}$ over time is to some extent predicted outside the model. The X_j's, however, are determined in the model. The d_{ij}'s are estimated according to equation (1) on the basis of data for an initial year.

A part of the residuals generated by activities in Norway is transferred from national recipients to foreign ones, and some of the Norwegian activities like ocean water transport are performed abroad. In order to get the right picture of the load on national recipients, it is necessary to specify the net import of residuals to Norway. Thus, in equation (2) $R_i^{f\alpha}$ is equal to the import of residual i to the national recipient minus the amount of residuals generated by Norwegian activities abroad, mostly by ocean water transport. The residuals transferred through air and water from locations in

Norway to other countries are not considered in our study, though, as mentioned above, in principle it should be. In the first place it is a question of data, and in the second place the impacts on the recipients of the discharges of residuals and the transport of residuals in general between recipients are not studied.

Due to the lack of reliable data on the transport of residuals from abroad, this also has been excluded from the study.

Thus, in the empirical section, the 'export' aspect of $R_i^{\alpha f}$ will reflect only the residuals generated by Norwegian activities abroad, and will exclude the residuals generated by ocean water transport.

As pointed out previously, *stocks* of some residuals are central to pollution analysis. In addition to residuals generation by sources, the depreciation of quality in the recipients should be known. In this study the depreciation aspect has unfortunately had to be disregarded because of lack of information. Three types of stocks of residuals will be calculated in the sequel. These persistent residuals are mercury, lead and pesticides. The rule for the calculation of these stocks is:

$$S_i(t) = \int_{\tau=t_0}^{t} \left[\sum_\alpha R_i^{\alpha}(\tau) \right] d\tau + S_i(t_0) \tag{3}$$

(3) implies that at each point in time the residuals discharged into the different recipients are added. We think this is a reasonable approach as long as most of the damages from the stocks of these residuals stem from the residuals being accumulated in vegetables, fish, animals, etc.

The stocks of these residuals could have been calculated for water and land separately, but at this stage of the study such a separation is of no practical interest.

2.5 *THE DATA*

The most important source of information is a questionnaire investigation by the Norwegian Industry Association among member firms about what was called 'problem waste'. This investigation provides cross-section data for 1970, covering about 70 per cent of total production. On the basis of this information, the expert groups organised by the Planning Division identified the thirty-eight categories of residuals listed in Appendix 1.

The response rate to the questionnaire varied between the two extremes, 70–80 per cent on the average. The data from the replying firms were blown up to cover the respective sectors, mostly by using residuals per employee, but residuals per unit of output was also used in some cases.

This procedure is, of course, liable to bias the data. If the more efficient firms have the highest response rate, then these 'observations' of residuals may be biased downwards when considering efficiency in utilisation of raw materials. On the other hand, higher capital–labour ratios in more modern firms might misleadingly show residuals *per employee* here that are *above* the average for the sector. Moreover, if the firms have limited knowledge about generation of residuals, they might be expected to understate the real amounts. To assume the opposite is somewhat unreasonable.

Another more important source of error is the fact that the different plants in a given sector will rarely produce the same composite of output. When the firms that reply produce a different composite than those that did not reply, it will be nearly impossible to arrive at the correct total quantity of each residual.

These estimates, based on data from the firms themselves, were supplemented by, and compared with, other available information. The sources were special studies, information given by the firms in connection with applications for permits to discharge residuals to water or air, data on the firms' use of inputs, etc. This supplementation process resulted in major changes in the original data together with a lot of additional information.

The procedure for quantifying the discharges was somewhat different from one sector group to the other.

Discharges from agriculture and forestry were quantified on the basis of data for production and use of inputs. These data were supplemented with data from special studies previously made at the Norwegian School of Agriculture.

Discharges from the burning of fuel for heating were estimated on the basis of data on the use of the various types of fuel in each sector.

Discharges from transportation activities were estimated from data on the use of fuels and other kinds of input.

Data for residuals from private consumption activities cover only garbage and sewage in addition to residuals from transportation activities. Estimates for the composite and quantity of garbage and sewage per household were made on the basis of existing special studies.

Even in the sectors where some discharges are quantified, there certainly may exist other discharges which are not. The reason for this is mainly technical problems of measurement.

The quantity and quality of the data are very variable, but no group of sectors can be said to have particularly unreliable data. Among the manufacturing sectors, reliability might be thought to be related to number of questionnaire replies. But other kinds of information played so great a part in the making of our data that not much weight should be attached to the number of replies.

Peskin (1972) has estimated the discharges from the Norwegian chemical pulp industry, based only on engineer's data. These estimates are very close to those of the present project. As our data have been constructed independently of Peskin's work, this may be taken as an indication that the quality of the data is relatively good. The data are shown in the tables in Appendix 4 on the M.S.G. sector level.

2.6 *THE UTILISATION OF THE M.S.G. MODEL*

The M.S.G. model provides a balanced, meaningful framework for the understanding of general patterns of possible economic development. When extending the model to cover residuals generation, the accompanying patterns of discharges of residuals are revealed.

The analysis of residuals generation is carried out in two steps:

(a) No feedback effects on the economy from policy measures concerning residuals are specified.
(b) Feedback effects on the scale and composition of production activities, etc., from environmental policy measures are specified.

The M.S.G. model can be run under alternative assumptions as regards the exogenous variables. In step (a) the development of the exogenous variables follows a pattern regarded as plausible by the Planning Division. Step (b) allows repercussions in the economy. This is, of course, the most interesting exercise from a policy point of view.

As regards formulating concrete environmental policy programmes, it must again be stressed that there is, of course, no unambiguous linkage between the discharge of residuals on a national level and the pollution of the environment. Information on the localisation of generation sources and the specific properties of recipients must be known in order to assess pollution damages, and the sector level must be disaggregated to process levels to assess the cost of changing technology, etc.

In the following sections the M.S.G. model is utilised to give projections on the flow of residual discharges when assuming fixed discharge coefficients, and to analyse the effects of introducing a form of charges on residuals. Other experiments of interest are to work out the necessary changes in the discharge coefficients required to keep both the flows and stocks of residuals within limits set by environmental standards; changing the amount and composition of exogenous government demand, thus reflecting increases in resources used for public waste-treatment installations; and changing exogenous total amount of capital available for the production sectors partly corresponding to an increase in public demand and partly simulating a policy of changing the saving ratio.

2.7 *THE PROJECTION EXPERIMENT*

The projections are based on a run of the M.S.G. model performed by the Planning Division in 1972 for internal use. This run is referred to as the 'reference run'. No explicit consideration of pollution control trol was introduced in that run.

By assuming constant discharge coefficients, the time path of residuals discharged is found by utilising equation (2). The stocks of persistent residuals are calculated according to equation (3). The initial discharges in 1970 and the discharges of residuals in the year 2000 are given in Appendix 4. Constant discharge coefficients are clearly – and happily – an unrealistic assumption as regards problem-causing residuals. Changes due to process substitutions, innovations, waste treatment, recycling, etc., surely imply reductions in the coefficients. Information about such changes can readily be included when calculating projections for residuals.

The projections of residuals provide environmental scientists with information enabling them to work out corresponding plausible changes in the environment. This 'look into the future' is especially useful when considering residuals which are accumulating in the environment.

When the environmental scientists are working out the plausible changes in the environment due to the projection of residuals given in the reference run, they must combine these projections with exogenous information on the localisation of economic activities and the spatial distribution of residual flows. In this way, the present approach can form a start towards an environmental programme for Norway.

As mentioned in section 2.4, a main reason for employing a multi-sectoral growth model is to take into consideration the non-proportional growth in the different sectors. In Table 2.1 below, the projections from the reference run of five selected residuals are compared with a naïve projection. In this naïve projection a uniform growth rate is assumed for the different sectors. This uniform rate of growth is taken to be the rate of growth of G.N.P. as implied by the reference run – 4·3 per cent.

The accuracy of the figures concerning both the observations for 1970 and the projections based on the M.S.G. model can of course be questioned. The experiment, however, stresses how serious errors may be from following a naïve projection. These errors will probably be more serious the more industrialised the economy is.

TABLE 2.1

DISCHARGES OF FIVE RESIDUALS, OBSERVATIONS IN 1970: PROJECTIONS FOR 2000 BASED ON THE M.S.G. MODEL AND ON UNIFORM SECTORAL GROWTH

Residuals	(1) *Observations, discharges of residuals in 1970* (tons)	(2) *Projections, discharges of residuals in 2000* (*the reference run*): *non-proportional growth* (tons)	(3) *Projections, discharges of residuals in 2000* (*naïve projection*): *uniform growth at a rate of 4·3%* (tons)	(4) *Overestimate, naïve projection* $\dfrac{(3)-(2)}{(2)}\times 100$ (%)
Sulphur oxides to air	167,150	531,740	625,976	18
Dust to air	331,240	1,031,075	1,248,508	16
Zinc to water	4,200	12,250	15,960	30
Phosphorus compounds to water	4,063	9,655	15,439	60
Biologically decomposable substances to water	762,565	1,645,295	2,897,747	76

Tables 2–4 in Appendix 4 show that both phosphorus compounds and biologically decomposable substances are, to a greater extent than other residuals, connected to agricultural activities, including the processing of agricultural products and the consumption of these products. These sectors are among those which expand at a lower rate than G.N.P. This is the main reason for the large overestimate in projecting the discharges of these residuals on the basis of the naïve approach.

Sulphur oxides and dust to air, however, are generated in nearly the whole spectrum of the economy. Nevertheless, the discharges of these two residuals are overestimated by nearly 20 per cent. For other residuals than the five specified above, the overestimate can be even larger than 75 per cent. Of course, some of the overestimates are smaller than 20 per cent.

The naïve projection can be improved by taking the disparity among growth rates into consideration without making use of a model like M.S.G. One procedure of this kind would be to observe the amount of residuals discharged in, say, the last five years and so calculate growth rates. On the basis of these rates a more reliable projection than the purely naïve projection can be made. We have done this by using the growth rates for the production sectors in the period 1966–70 and the discharge coefficient for the year 1970. The projection errors relative to M.S.G. predictions, however, still remain at a level of 10–20 per cent for most of the residuals. These figures refer to the year 2000. In earlier years the overestimate is, of course, of smaller size.

Nevertheless, one of the conclusions to be drawn from the preceding calculations is that to project future amounts of residuals, a macro-economic, but multi-sectoral, framework is necessary.

To reveal the impact of non-proportional growth, the projection results are rearranged by aggregating the M.S.G. sectors into six more general sectors. These sectors are labelled:

   I. *Consumption goods* (M.S.G. sectors 1, 3, 5, 6, 7, 8, 11, 12, household sewage).
  II. *Investment goods* and intermediate deliveries (M.S.G. sectors 2, 4, 9, 10, 13, 14, 16, 17, 18, 26).
 III. *Energy* (M.S.G. sectors 15 and 21).
 IV. *Transport* (M.S.G. sectors 24, military defence and part of households: the use of private cars).
  V. *Dwellings* (M.S.G. sector 23).
 VI. *Services*, trade and other manufacturing (M.S.G. sectors 20, 22, 25).

In Table 2.2 the discharge in 1970 and 2000 of some residuals from these six aggregate sectors are specified.

From these tables it follows that the activities belonging to 'consumption goods' contribute relatively less to the discharges of residuals in 2000 than in the initial year 1970. The main reason for this is the relatively low rate of expansion in these activities during the years 1970–2000. In consequence of the economic forces of growth in Norway, the contribution from transport, dwellings and services in the discharges of residuals increases during these thirty years.

Sulphur oxides are of special interest. The shares for both consumption goods and investment goods decrease during the period. In 1970 the activities belonging to 'consumption goods' generate 12 per cent of sulphur dioxide to air. In 2000 the share has dropped to 5 per cent. The greatest increase occurs in the sector labelled 'energy'. The main function of this sector is not only to cover the demand for energy in Norway, but in addition to exploit oil and gas reserves in the North Sea and to refine oil and liquid natural gas. Owing to the findings of oil and gas in the North Sea the expansion of this sector in Norway is underestimated in the present run of the M.S.G. model. The main contributor to the discharge of sulphur oxides is the industry group denoted 'investment goods'. This is the case in 2000 as it was in 1970.

For the remaining three residuals discharged to air, the main contributor is the sector called 'transport'. Note that in this sector the use

TABLE 2.2

DISCHARGE DISTRIBUTION OF SEVEN RESIDUALS TO AIR AND WATER IN 1970 AND 2000 PROJECTED BY THE REFERENCE RUN

(percentages)

I. *To Air*

Group of industries	Sulphur oxides 1970	2000	Nitrogen oxides 1970	2000	Lead 1970	2000	Dust 1970	2000
Consumption goods	12·01	4·94	18·24	7·65	1·46	0·70	1·55	0·60
Investment goods	61·94	56·60	34·17	35·60	0·24	0·33	47·82	42·40
Energy	4·13	12·03	1·92	6·04	0	0	0·92	2·70
Transport	7·61	8·33	41·29	44·75	97·56	98·59	48·31	52·52
Dwellings	13·70	17·21	4·08	5·51	0	0	1·38	1·74
Services	0·62	0·88	0·23	0·45	0·73	1·31	0·02	0·03
Total discharge in tons	167,150	531,740	88,340	261,145	410	1,065	326,840	1,031,075

II. *To Water*

Group of industries	Zinc 1970	2000	Phosphorus compounds 1970	2000	Biologically decomposable organic substances 1970	2000
Consumption goods	0	0	97·30	93·58	24·48	24·11
Investment goods	95·24	92·61	3·69	6·42	75·52	75·89
Energy	0	0	0	0	0	0
Transport	0	0	0	0	0	0
Dwellings	0	0	0	0	0	0
Services	4.76	7.39	0	0	0	0
Total discharge in tons	4,200	12,250	4,063	9,655	762,565	1,644,295

Note: Columns do not sum to 100 per cent because of rounding errors.

of private cars is included. The leading part played by this sector is strengthened in 2000 compared with the initial year 1970. This is especially so for dust. A large part of the amount of dust discharged to air is generated through the wear and tear of aphalt. This activity is included in the transport sector. The role played by 'investment goods' in the generation of dust to air is significantly less in 2000 than in 1970. In 1970 transport and investment goods shared approximately equally the amount of dust discharged to air. In 2000 transport has taken an unquestionable leading part.

During the years 1970–2000 the national income in the economy rises by approximately 4 per cent per year. The income elasticities for services provided by transport are certainly higher than the income elasticities for goods and services provided by the activities included in the sector 'investment goods'. There are of course many reasons behind the disparity of growth rates, but the differences in income elasticities are of great importance.

Turning to the three residuals discharged to water, the same phenomenon is demonstrated. In the case of zinc to air the share of discharges provided by 'services' is increased at the expense of the share for 'investment goods'. The share of phosphorus compound discharges contributed by 'investment goods' is increased at the expense of 'consumption goods'. In this last activity the sewage from households is included. Only in the case of biologically decomposable organic substances are the relative shares stable over the period. This is a result of the relatively low rate of expansion in the pulp and paper production sector. A large part of biologically decomposable organic substances is generated through this sector.

As stressed previously, the accumulation of persistent residuals is of major importance when analysing conflicts between growth and environmental quality. The following tables give some information on the accumulation of persistent residuals for the period 1970–2000. As mentioned in section 2.4, we do not distinguish among recipients for this category of residuals. Besides, the stocks in the initial year 1970 are unknown. They are therefore arbitrarily set equal to zero.

These dramatic figures should certainly stress the need for a radical change in the generation and hence in the accumulation of these three residuals.

TABLE 2.3

(a) ACCUMULATION OF MERCURY DURING THE YEARS 1970–2000

Source of generation (M.S.G. sector)	Discharge 1970 (tons)	Gross stock in 2000 implied by the reference run (tons)
Agriculture	0·9	31
Forestry	0·1	6
Fertiliser	2·0	120
Basic metal industries	1·0	48
Total	4·0	205

(b) ACCUMULATION OF PESTICIDES DURING THE YEARS 1970–2000

Source of generation (M.S.G. sector)	Discharge 1970 (tons)	Gross stock in 2000 implied by the reference run (tons)
Agriculture	1,300	44,922
Forestry	200	12,526
Total	1,500	57,448

(c) ACCUMULATION OF LEAD DURING THE YEARS
1970–2000

Source of generation (*M.S.G. sector*)	*Discharge 1970* (tons)	*Gross stock in 2000 implied by the reference run* (tons)
Agriculture	6	207
Forestry	1	63
Mining	1	57
Basic metal	1,700	81,286
Mechanical and electrical	5	360
Other manufacturing	3	216
Transport	78	5,042
Military defence	5	229
Households	438	21,812
Total	2,237	109,272

2.8 *THE RESIDUALS CHARGE EXPERIMENT*

The results of the previous section strongly suggest that reductions in the growth rates of residuals discharge are necessary.

One way of achieving this change is to levy effluent charges. These charges may, for example, cause changes in technology, including the degree of recycling of the raw materials in use and changes in the localisation of firms, as mentioned in section 2.2. The only change we are able to consider here is in the scale of the activities. Referring to the multi-sectoral character of our macro-economic framework, this means a change in the product mix in the economy.

The only way 'pollution charges' can be introduced in our framework is by hypothetically applying indirect taxes to the production of sector products. We must indicate in what way this procedure may be illuminating.

One reason might be that, owing to political constraints, the only way 'pollution charges' can be introduced is in the form of special taxes on the products themselves whose processing is offensive, rather than on the generation of harmful residuals within the production process. The tax per unit of production is then equal across the spatial distribution of firms.

A second reason might be that for some residuals it is only the total amount discharged that determines harmful effects, the localisation of generating sources being without significance. Examples here are some air pollutants.

A third reason might be high 'transaction costs'. In some cases it may be too difficult to levy charges on only some of several residuals discharged by firms, i.e. levying the charge on output is the only practical possibility. Another case is discharges from moving sources like cars and aeroplanes. It may be prohibitive to implement geographically differentiated charges.

Despite these, the way we are forced to deal with 'pollution charges' here is an inferior solution to the problem. The crucial linkage between any residual type i and the effects of this residual on the recipients is missing. In addition, the distribution of residuals in space is generally needed when formulating environmental policies.

Our approach, however, may serve as a first step towards formulating such policies. A second step would be to differentiate charges on the products according to geographical location. A third would be to differentiate charges with respect to both localisation in space and to the different types of residual. Until better information is obtained concerning the relation between certain residuals and environmental services provided by the recipients, the first and second steps could be used in formulating a concrete environmental policy programme.

The connection between residual charges and the indirect taxes on production used here is as follows:

Let $\tau_s{}^\alpha$ denote the effluent charges on residual type s discharged to recipient α. (The localisation in space is not considered.) Let $Z_{sj}{}^\alpha$ denote the discharge of this residual from sector j. It then follows from equation (2) that the total amount of effluent charges paid by sector j can be written:

$$\sum_\alpha \sum_i Z_{ij}{}^\alpha \tau_i{}^\alpha = \sum_\alpha \sum_i d_{ij}{}^\alpha X_j \tau_i{}^\alpha = \theta_j X_j \tag{4}$$

where

$$\theta_j = \sum_\alpha \sum_i d_{ij}{}^\alpha \tau_i{}^\alpha \tag{5}$$

The indirect taxes θ_j are the sum of the effluent charges per unit of gross production in sector j. They will be a part of the indirect taxes per unit of gross production otherwise present in the model. Varying these θ_j's, the impact on the product mix in the economy and hence the impact on the amount of the various residuals discharged can be studied.

In order to find the proper taxes it is necessary to estimate the polluting effect of an increase in the gross production of the different products. At the present stage of information on pollution damages in Norway or other countries, these estimates have had to be based on educated guesses.

In consultation with environmental scientists, five residual categories were selected. The intention was to select from the whole spectrum of polluting residuals in the most useful way. The five selected residuals were ranked according to their probable polluting effect in 1970, taking the 1970 discharges into consideration. In this educated guess on the polluting effect of residuals, the chemical nature of the residuals and the localisation of the discharges were considered.

The five selected residuals are, starting with the most serious:

1. Biologically decomposable organic substances (2.2) to water.
2. Sulphur oxides (1.3.1) to air.
3. Zinc (1.1.4) to water.
4. Phosphorus compounds (1.6.1) to water.
5. Dust (1.7) to air.

The sectors are now ranked according to their contribution of these residuals. The residuals are converted to the same unit of measurement by introducing 'residual prices', aggregate damage tradeoffs obtained from rank numbers. Thus, we assume that the total amount of biologically decomposable organic substances (rank 1) is five times more 'polluting' than the total amount of dust (rank 5), etc. Dividing the ranks (5, 4, 3, 2, 1) by the total amounts discharged in each in 1970 gives the residual prices for each type of residual. (The 'price' of dust is normalised to 1.) The ranking of pollution damage by the production sectors is given in Table 2.4 below.

TABLE 2.4

RANKING OF THE M.S.G. SECTOR WITH RESPECT TO POLLUTING EFFECTS

Rank	M.S.G. sector	M.S.G. sector no.	Rank score per million kroner
1	Pulp and paper	9	493
2	Mining	4	391
3	Basic metal production	17	311
4	Mechanical and electrotech. industry	19	197
5	Heavy non-metallic mineral production	16	84
6	Chemical products of fish	12	80
7	Various chemical industries	14	56
8	Iron and steel, metal mills, etc.	18	52
9	Refining of oil and coal	15	47
10	Business buildings and dwellings	23	45
11	Fertilisers	13	34
12	Fishery	3	26
13	Canned fish and meat, etc.	6	22
14	Agriculture and forestry	1·2	21
15	Wood and paper products	10	16
16	Internal transport	24	12
17	Processing of agricultural products	5	10
18	Textiles and clothing	8	9
19	Other manufacturing industries	20	8
20	Sweets, tobaccos, beverages	7	7
21	Printing and publishing	11	0·2
22	Electricity supply, trade, services, building	21, 22, 25, 26	0

The 'environmental costs' following from an increase of 1 million kroner in the gross production of each sector j are specified in the column on the right. These 'cost' figures are denoted rank score per million kroner and the numbers follow directly from the sum of the product of discharge coefficients and the 'prices' on the residuals.

The 'prices' and the 'cost' figures are of course not observable. They are imputed and fictitious magnitudes. The purpose they serve is only to suggest the size of difference among the rank scores of the sectors.

The ranking indicates the pulp and paper industry to be the worst polluter in Norway. This sector, together with the next three, are significantly different from the rest of the sectors.

It should be noted that in Norway electricity supply is provided by hydroelectric power stations. Owing to the findings of oil and gas in the North Sea this picture might be changed in the future. The ranking above does not therefore reflect future activities in Norway based on oil and gas.

To induce a change in the output mix of the economy, we now hypothetically change the indirect taxes on sector products in accordance with the ranking of the sectors. The tax structure is not changed at one point of time, but rather gradually from 1975 to 1981. The main reason for suggesting a gradual change in the tax structure is the high cost implied by a too sudden change in the economic structure of the economy. The economic structure may be non-optimal with respect to environmental externalities, but a too sudden correction of these externalities would create other social costs.

As regards export industries, the increase in taxes should imply a reduction in the amount of exports because of a consequent worsening of their competitive position. Because exports, and not prices, are exogenous in the model, the time development of exports has to be adjusted in correspondence with the new tax levels.

The assumption of full employment of a growing labour force and increasing amounts of total capital in the M.S.G. model means that the G.N.P. will continue to grow despite increasing the indirect taxes. The reduction in exports, however, does imply a reduced rate of growth in G.N.P. and an increasing import surplus in the economy.

In order to show what can be obtained in reducing discharges of pollutants by a mere change in the output mix of the economy without affecting the growth rate of G.N.P. and the balance of payments, the exports from the least polluting sectors are increased. The indirect taxes for these sectors are then *reduced* to make such a development more probable.

The changes in the indirect taxes and the exports are thus specified in such a way that the rate of growth in G.N.P. over the period from

1970 to 2000 is the same as implied by the reference run and the balance of payment is the same in the year 2000 as in the reference run. The overall changes in indirect taxes and exports will, however, improve the environmental services by reducing discharges of pollutants. Of course, the improvement could have been much higher by reducing the rate of growth in G.N.P.

The improvement of environmental services must not be interpreted as a reduction in all the residuals generated. A crucial point in the alternative run of the M.S.G. model is that while some of the residuals are reduced in quantity, others are increased. The new levels of taxes are given in Table 2.5 below. Here, the greatest increase occurs of course in the indirect tax per unit of output paid by pulp and paper. The indirect tax per unit of gross production in the pulp and paper sector is increased from 0·12 per cent in 1966 to 17·12 per cent in 1981. The other taxes are changed according to the rank scores given in Table 2.4. The 17 per cent increase for pulp and paper is of no intrinsic interest. The point is the relative changes in competitive position for the various sectors of the economy. An increase of 17 per cent may seem so high in the real world that one could fear a closedown of the whole sector. The model does not permit such a development, but even for the actual economy a 17 per cent increase distributed over the years from 1975 to 1981 could probably be tolerated.

In the alternative run, some of the indirect taxes are decreased. In fact, only the first six industries in the ranking of Table 2.4 have their taxes increased, the other twenty sectors experiencing a decrease in the indirect tax rate. This means subsidising. The reason for this is that the increase in the indirect taxes paid by the six top-polluting industries will raise the costs of production in other industries. Since the indirect taxes in the alternative run of the M.S.G. model are changed under the constraint that the growth rate in G.N.P. should not be affected, it is then necessary to subsidise some of the industries in order to counteract the higher cost of production caused by the higher indirect taxes paid by the six top-polluting industries. The rule for subsidising the industries is also based on the ranking given above. The least polluting industry gets the highest subsidy (3 per cent in 1981).

The exogenous export demand forms a significant part of the final demand in the six top-polluting sectors. The larger that exogenous demand, the more necessary it is to correct it in correspondence with the new competitive situation indicated by the new tax levels.

The corrections are accomplished by modifying exports in 2000. The top six polluting industries have their exports decreased in 2000. The greatest export drop is for pulp and paper products. The export of these products is reduced by 20 per cent in 2000. The industries which are subsidised have their exports increased. The numerical changes are given in Table 2.5.

The new export levels in each year are based on the levels in 2000 and the level of exports indicated by the reference run in 1975. The changes are distributed equally over the years in this period.

TABLE 2.5

EXPORTS AND INDIRECT TAXES IN THE REFERENCE RUN AND IN THE ALTERNATIVE RUN

Sectors	Exports in 2000 (1966 prices): the reference run (m. kroner)	Exports in 2000 (1966 prices): the alternative run (m. kroner)	Indirect taxes as percentages of gross production: the reference run, 1966	Indirect taxes as percentages of gross production: the alternative run, 1981
Agriculture	400	440	−4·8	−6·90
Forestry	100	110	−2·16	−4·26
Fishing	175	185	−1·67	−3·57
Mining	700	600	−10·17	2·63
Processing agr. prod.	650	740	−13·74	−16·34
Canned fish	1,600	1,740	−0·07	−2·17
Sweets, tobacco	1,000	1,170	49·20	46·50
Textiles	1,700	1,960	0·92	−1·68
Pulp and paper	2,800	2,240	0·12	17·12
Wood and paper prod.	1,650	1,850	1·38	−0·92
Print and publishing	170	200	1·88	−1·12
Chem. fish prod.	800	780	−3·31	−3·11
Fertiliser	2,950	3,160	−1·00	−2·60
Various chem. ind.	4,200	4,200	3·10	2·40
Coal and oil	5,100	5,300	0·16	−0·94
Cement, etc.	750	710	1·81	2·21
Basic metals	5,400	4,820	0	9·60
Iron, steel, etc.	5,000	5,100	0·50	−0·40
Mech. and elec.	14,200	13,070	2·47	7·47
Other manuf.	2,700	3,130	2·95	0·25
Electricity supply	150	180	2·76	−0·24
Trade	3,800	3,800	33·00	30·00
Buildings and dwellings	0	0	1·69	0·49
Internal transport	2,650	2,990	−0·64	−3·14
Services	850	1,020	−3·45	−6·45
Building and Construction	0	0	6·51	3·51
Investment machinery	250	250	—	—
Distribution account	2,300	2,300	—	—
Total	62,045	62,045	—	—

Furthermore, the changes are specified under the constraint that the balance of payment should be maintained.

There are many objections that can be raised against our approach. Rules of thumb and educated guesses are involved. Other weaknesses are the lack of geographical dimension, the lack of information on damage costs, the absence of the possibility of technological change in the discharge coefficients, etc. Moreover too little attention is paid to persistent residuals. None of the persistent residuals is among the five residuals chosen as most serious. If more attention were paid to them, then sectors like agriculture, forestry and transport would have ranked higher in pollution damage. This reservation is important to bear in mind when it comes to the interpretation of the results. A chief contributor to the generation of lead residuals is the basic metal industry. This industry is given a high polluting rank anyway, due to its heavy generation of the five key residuals selected. Last, but not least, a significant part of the overall pollution stems from sources whose levels of activity are exogenously given in the M.S.G. model. This holds good for the use of private cars and the use of roads. Despite such weaknesses, the present approach may serve as a first step towards the study of where to levy pollution charges.

Table 2.6 compares the results for the average growth rates in production calculated in the two runs (from 1966 to 2000) of the M.S.G. model. We find that the growth rates for the top four polluting industries are reduced from 2·6 per cent to 2·2 per cent (pulp and paper), from 3·8 per cent to 3·6 per cent (mining), from 2·7 per cent to 2·3 per cent (basic metal) and from 5·1 per cent to 5·0 per cent (mechanical and electrotechnical industry). In consequence of the reduced rate of growth in pulp and paper, the growth rate for forestry goes down from 4·3 per cent to 4·2 per cent.

Among the sectors which expand at a faster rate in the alternative run than in the reference run we find agriculture, processing of agricultural products, fishing, canned fish and meat, sweets, tobacco and beverages, fertilisers and textiles. Some of these sectors are not among the least polluting industries (see Table 2.4 above). The least polluting industries like printing and publishing, electricity supply, trade, services, and building and construction grow at the same rate as in the reference run.

The changes in the tax and export structure result in a reduction in the growth of industries producing investment goods and typically intermediate deliveries, while industries producing consumer goods or connected with these through input–output linkages grow at a higher rate. Services and trade expand at the same rate as before.

In interpreting this result it is necessary to remember the conditions for an untouched rate of growth in G.N.P. The changes studied are therefore caused by redistributions of real capital and labour. The level of total investment per year and the rate of growth in labour are the same in both runs of the model. The industries which suffer particularly from the increased effluent charges are capital-intensive sectors like pulp and paper, mining and basic metal industries. The redistribution of labour and capital investment caused by the changes in indirect taxes and exports is therefore mainly to the benefit of other capital-intensive sectors. This is the

TABLE 2.6

STRUCTURAL CHANGES OF THE ENVIRONMENTAL POLICY MEASURES

Production sectors	(1) *Average growth* rate, reference run*	(2) *Average growth* rate, alternative run*	*% change in employment, year 2000* $\frac{(2)-(1)}{(1)}$	*% change in real capital, year 2000* $\frac{(2)-(1)}{(1)}$
Agriculture	0·9	1·0	1·3	2·8
Forestry	4·3	4·2	−4·0	−2·9
Fishing	0·8	0·9	5·1	4·0
Mining	3·8	3·6	−6·8	−6·1
Processing agr. prod.	0·6	0·7	1·5	1·9
Canned fish	0·7	0·9	8·7	9·7
Sweets, tobacco	1·0	1·2	9·1	7·2
Textiles	2·8	2·9	3·2	3·6
Pulp and paper	2·6	2·2	−12·1	−12·5
Wood and paper prod.	4·4	4·4	1·3	1·7
Print and publishing	6·4	6·4	0·2	0·1
Chem. fish prod.	0·4	0·3	0·0	−1·3
Fertiliser	4·1	4·3	6·7	6·7
Various chem. ind.	5·7	5·7	1·5	0·3
Coal and oil	7·8	7·8	3·6	2·0
Cement, etc.	5·2	5·2	−1·6	−1·1
Basic metals	2·7	2·3	−8·1	−10·3
Iron, steel, etc.	5·8	5·8	−0·6	−0·4
Mech. and elec.	5·1	5·0	−3·6	−3·1
Other manuf.	5·1	5·4	9·8	9·9
Electricity supply	4·3	4·3	−2·9	−0·3
Trade	3·5	3·5	−0·5	−1·3
Buildings and dwellings	4·7	4·7	−4·6	0·1
Internal transport	4·5	4·6	0·9	1·5
Services	5·1	5·1	1·1	−0·2
Building and construction	5·3	5·3	0·2	−0·8

* In both runs the growth in total endogenous production is the same, 4·3%. Moreover the growth rate in the gross production in 'exogenous sectors' such as whaling (− 30·0%), ocean transport (2·0%), total public consumption (3·9%), public capital (4·0%) and military defence (2·5%), are the same in both runs. G.N.P. therefore grows at the same rate, 4·1%, in both runs. Total private consumption grows at a rate of 3% in the reference run. This is also the case in the alternative run.

case for sectors producing consumer goods, but not for sectors producing services.

In economic planning, forecasts about future price developments are important (see Johansen, 1967). The course of future prices can only be discovered within a macro-economic framework. Moreover, changes in, for instance, indirect taxes due to concern for the environment will have an impact on future prices. This impact can be calculated in the M.S.G. model. Thus, another purpose in using a macro-economic framework for analysing environmental problem can be ascertained. The prices in the M.S.G. model can certainly not be interpreted in a straightforward way. The sector product is not a single product. On the other hand, the information on changes in future sector prices can be of some help in the planning of a growing economy.

The new set of sector prices has been calculated. In this new set of prices, the rate of growth per year in the price of pulp and paper goes from 3·7 per cent in the reference run to 4·4 per cent in the alternative run. These growth rates are calculated for the whole period 1970–2000. The prices on products from basic metal industries grow at a rate of 2 per cent in the reference run, but at 2·4 per cent in the alternative run. The growth rate for the prices on consumer goods are on the whole *reduced* by 0·2–0·4 per cent.

Finally, it is time to consider the impact on the residuals. In Table 2.7 the differences between residuals generated through the alternative run and the reference run are specified.

The results given in this table indicate a reduction in the discharges of residuals to water, but an increase in the discharges of residuals to air and land. Exceptions to this rule are the discharges of cyanide, arsenic and plastic substances to water. These are *increased*. Discharges to air which are reduced are the discharges of sulphur oxides, fluorine and dust.

In accordance with the redistributions of labour and capital shown in Table 2.6, a redistribution of discharges of residuals occurs. Some discharges increase while others decrease. (The total discharge measured in tons, however, is reduced by 1·3 million.)

The most significant reductions appear for residuals like lead to water (a 9·07 per cent decrease in the discharge in 2000), soda lye to water (6·96 per cent), other bases to water (9·54 per cent), fluorine to air (5·84 per cent), bark to water (9·94 per cent) and biologically decomposable organic substances (8·68 per cent).

The greatest increase in the discharge of residuals occurs for plastic substances to water, which is increased by 8·70 per cent in 2000.

The impact on the gross stocks of the persistent residuals has also been calculated. The gross stock of lead in 2000 is *reduced* through

TABLE 2.7

CHANGES IN THE DISCHARGES OF RESIDUALS FOR THE YEAR 2000: DISCHARGES ALTERNATIVE RUN MINUS DISCHARGES REFERENCE RUN

Residuals	Air		Water		Land	
	tons	%	tons	%	tons	%
Mercury	0	0	0	0	—	—
Lead	+5	+0·47	−335	−9·07	0	0
Cadmium	—	—	0	0	—	—
Zinc	—	—	−550	−4·49	—	—
Copper	—	—	−90	−3·35	—	—
Iron	—	—	−810	−0·25	−25	+0·04
Chrome	—	—	−15	−2·78	—	—
Sulphur oxides	−11,450	−2·15	−15,740	−4·09	—	—
Hydrochloric acid	—	—	−400	−2·46	—	—
Nitrogen oxides	−225	−0·09	−25	−0·21	—	—
Other acids	0	0	−15	−0·18	—	—
Soda lye	—	—	−2,700	−6·96	—	—
Other bases	0	0	−9,715	−9·54	—	—
Fluorine	−435	−5·89	0	0	—	—
Cyanide	—	—	+25	+1·49	—	—
Arsenic	—	—	+55	+1·93	—	—
Carbon monoxide	+3,785	+0·42	—	—	—	—
Phosphorus compounds	—	—	−15	−0·16	—	—
Nitrogen compounds	+5	+0·02	+75	+0·10	—	—
Mine tailings, sludge, dust	−22,230	−2·16	−970,290	−4·14	−64,700	−1·25
Pesticides	—	—	—	—	0	0
Aliphetic halogenated hydrocarbons	0	0	0	0	—	—
Bark	—	—	−8,540	−9·94	−31,490	−6·62
Fibre	—	—	−14,040	−7·94	—	—
Wood chip	—	—	—	—	+5,820	+1·36
Plastic substances	—	—	10	+8·70	+1,685	+1·96
Biologically decomposable organic substances dissolved	0	0	−122,920	−8·68	—	—
Biologically decomposable organic substances suspended	—	—	+950	+0·41	0	0
Oil and oil products	+2,265	+0·78	−255	−0·50	+65	+0·20
Dispurging agent	—	—	−5	−0·05	0	0
Taste and smells substances	++30	+0·11	−55	−3·57	—	—
Organic solvents	+65	+0·23	+10	+0·21	+15	+0·28
Other organic substances	—	—	0	0	—	—
Solid waste	—	—	—	—	−27,050	−3·05

—=Category not applicable or data not available.   0=No change.

the reference run by 5235 tons, i.e. 5 per cent. The gross stocks of mercury and pesticides are *increased* by 3 tons and 1546 tons respectively.

If the aspects of environmental policy had been introduced into the model under a condition other than an unchanged rate of growth of G.N.P., the reductions in the discharge of residuals would have been larger. The scope of the study, however, was to study this partial problem. The results outlined above indicate the probable directions for environmental action in Norway. They also indicate the outcome of this policy with respect to which of the discharges would be decreased and which increased.

2.9 *TOWARDS ENVIRONMENTAL POLICY MEASURES*

There are two general ways of extending the analysis towards making it more relevant to concrete policy decisions. One is to incorporate more of the substitution possibilities mentioned in section 2.2 by further disaggregation of economic activities, the other to divide the discharge of residuals into a more disaggregated list of recipients corresponding to the real 'problem sheds'.

The central part of the national accounts comprises two matrices, one for the deliveries of commodities to sectors and one for the deliveries of commodities from sectors. This makes it possible to replace the sector-to-sector approach in input–output models with a commodity–activity–sector approach (see Bjerkholt and Longva, 1970). If information on residuals can be obtained at the activity level, more realistic substitution possibilities in the economy will be available for model studies. Waste treatment can be introduced as separate activities.

The introduction of the spatial pattern of discharge of residuals necessitates incorporation of the transport of residuals occurring in the environment. As regards the generating sources, it is not too difficult to establish an initial regional distribution of economic activity, although the problem sheds seldom correspond to existing regional administrative divisions. But to be of any real help for policy decisions, such a disaggregated model must contain mechanisms for *endogenous* redistribution of activities resulting from applying policy measures.

There is no way to by-pass some sectoral and regional detail if this analysis is to be changed from its present exploratory character into the more practical planning character demanded for concrete action.

REFERENCES

Bjerkholt, O., and Longva, S., 'Modis IV: The Basic Framework of an Input–Output Planning Model, with a Commodity–Activity–Sector Approach', Working Paper from the Central Bureau of Statistics of Norway (1970).

d'Arge, R. C., and Kogiku, K. C., 'Economic Growth and the Environment', *Review of Economic Studies*, vol. 12 (*Jan* 1973).

Førsund, F. R., 'Allocation in Space and Environmental Pollution', *Swedish Journal of Economics*, vol. 74, no. 1 (Mar 1972).

Johansen, L., *A Multi-Sectoral Study of Economic Growth* (North-Holland Publishing Company, Amsterdam, 1960).

——, 'Some Problems of Pricing and Optimal Choice of Factor Proportions in a Dynamic Setting', *Economica* (May 1967).

——, 'Explorations in long-term Projections for the Norwegian Economy', *Economics of Planning*, vol. 8, no. 1–2 (1968).

Keeler, E., Spence, M., and Zeckhauser, R., 'The Optimal Control of Pollution', *Journal of Economic Theory*, no. 4 (1971).

Kneese, A. V., Ayres, R. A., and d'Arge, R. C., *Economics and the Environment: A Materials Balance Approach* (Johns Hopkins Press, Baltimore and London, for Resources for the Future, Inc., 1970).

Leontief, W., 'Environmental Repercussions and the Economic Structure: An Input–Output Approach', *Review of Economics and Statistics*, vol. 3, no. 3 (1970).

—— and Ford, D., *Air Pollution and the Economic Structure: Empirical Results of Input–Output Computations* (Harvard Univ. Press, Cambridge, Mass., 1971).

Meadows, Do. and De., Randers, J., and Behrens III, W., *The Limits to Growth* (Pan Books Ltd., London, 1974).

Peskin, Henry M., 'National Accounting and the Environment', Article No. 50 (Statistical Central Bureau, Oslo, 1972).

Russell, C. S., and Spofford, W. O., Jr., 'A Quantitative Framework for Residuals Management Decisions' in Kneese, A. V., and Bower, B. (eds.), *Environmental Quality Analysis* (Johns Hopkins Press, Baltimore and London, 1972).

Schreiner, P., 'The Role of Input–Output in the Perspective Analysis of the Norwegian Economy', in Brody, A., and Carter, A. P. (eds.), *Input–Output Techniques* (North-Holland Publishing Company, Amsterdam, 1972).

Smith, V. L., 'Dynamics of Waste Accumulation: Disposals versus Recycling', *Quarterly Journal of Economics* (1972).

'Spesialanalyse 1, Forurensninger. Særskilt vedlegg til Langtidsprogrammet 1973–1977', Report from the Ministry of Finance to the Norwegian Parliament.

Spurkeland, S., 'M.S.G.: A Tool in Long-term Planning', paper presented at the First Seminar in Mathematical Methods and Computer Techniques, Norwegian Computing Centre (mimeographed, 1970).

Strøm, S., 'Dynamics of Pollution and Waste Treatment Activities', Memorandum from Institute of Economics, University of Oslo, 10 May 1972.

Victor, P. A., 'Input–Output Analysis and the Study of Economic and Environmental Interaction', Ph.D. thesis (University of British Columbia, Apr 1971).

APPENDIX 1

SPECIFICATION OF RESIDUALS

1. *Inorganic substances*

1.1	*Metals*
1.1.1	Mercury
1.1.2	Lead
1.1.3	Cadmium
1.1.4	Zinc
1.1.5	Copper
1.1.6	Iron
1.1.7	Chrome

1.2　*Radioactive substances*

1.3	*Acid substances*
1.3.1	Sulphur oxides
1.3.2	Hydrochloric acid
1.3.3	Nitrogen oxides
1.3.4	Other acids

1.4	*Bases*
1.4.1	Soda lye
1.4.2	Other bases

1.5	*Other poisonous substances*
1.5.1	Fluorine
1.5.2	Cyanide
1.5.3	Chlorine
1.5.4	Arsenic
1.5.5	Carbon monoxide

1.6	*Nutrients*
1.6.1	Phosphorus compounds
1.6.2	Nitrogen compounds

1.7　　*Mine tailings, inorganic sludge and dust*

2. *Organic substances*

2.1	*Slowly biologically decomposable organic substances*
2.1.1	*Dissolved*
2.1.1.1	Pesticides
2.1.1.2	Aliphetic halogenated hydrocarbons
2.1.1.3	Other halogenated hydrocarbons
2.1.2	*Suspended*
2.1.2.1	Bark

2.1.2.2   Fibre
2.1.2.3   Wood chip
2.1.2.4   Plastic substances

2.2   *Biologically decomposable organic substances*
2.2.1   Dissolved
2.2.2   Suspended

2.3   *Organic substance with special effects*
2.3.1   Oil and products of oil
2.3.2   Dispurging agents
2.3.3   Taste- and smell-producing substances
2.3.4   Substances with acute poisonous effect
2.3.5   Organic solvents
2.3.6   Other organic substances

3. *Waste*, unspecified

APPENDIX 2

NATIONAL ACCOUNTS AND M.S.G. SECTOR CODE

Sector Code		Name of Accounts	
M.S.G.	National accounts	National accounts	M.S.G.
1	11.1 11.2 11.3 13.0	Agriculture, production of crops Agriculture and livestock production Agriculture capital formation work done by the sector's own production factors Hunting, etc.	AGRICULTURE
2	12.1 12.2	Forestry Standing forests	FORESTRY
3	14.0	Fishing, etc.	FISHING
4	17.0 18.1 19.1 19.2	Coal mining Metal mining Non-metallic mineral products Quarrying and mining n.e.c.	MINING
5	20.1 20.2 20.3 20.6 20.7 20.9	Slaughtering and preparation of meat Dairy products Margarine Grain mill products and livestock food Bakery products Other food preparation	PROCESSING OF AGRICULTURAL PRODUCTS

Appendix 2—*contd.*

Sector Code		Name of Accounts	
M.S.G.	*National accounts*	*National accounts*	*M.S.G.*
6	20.4 20.5	Canning of fish and meat Fish processing	CANNED FISH AND MEAT AND OTHER FISH PRODUCTS
7	20.8 21.1 21.3 22.0	Cocoa, chocolate and sugar confectionery Distilling, rectifying and blending of spirits Breweries and soft drinks production Tobacco	SWEETS, TOBACCO AND BEVERAGES
8	23.0 23.1 23.2 23.3 23.9 24.1 24.3 24.4 24.9 29.0 30.0	Spinning and weaving of wool Textiles not elsewhere classified Knitting mills Cordage, rope and twine Impregnated and coated textiles, linoleum, etc. Footwear products and repairs to footwear Working clothes and other garments Fur goods, gloves Other made-up textile goods n.e.c. Leather and leather products Rubber products	TEXTILES AND AND CLOTHING
9	27.1 27.2 27.3	Mechanical pulp Chemical pulp Paper, paperboard and cardboard	PULP AND PAPER
10	25.1 25.9 26.1 26.2 27.4 27.5	Sawmills, planing mills and wood preserving Other wood and cork products Furniture Wooden fixtures Wallboards, etc. Paper and paperboard products	WOOD AND PAPER PRODUCTS
11	28.1 28.2	Publishing, etc. Printing, bookbinding, etc.	PRINTING AND PUBLISHING
12	31.6 31.7 31.9	Fish liver oil Herring oil and fish-meal Other oil refineries, etc.	CHEMICAL PRODUCTS OF FISH
13	31.1 31.2	Calcium carbide and cyanamide Other fertilisers	FERTILISERS

Appendix 2—*contd.*

Sector Code		Name of Accounts	
M.S.G.	*National accounts*	*National accounts*	*M.S.G.*
	31.3	Explosives	
	31.4	Synthetic fibres, resins and moulding powders	
	31.5	Other basic industrial chemicals	
14	31.8	Vegetable oil mills	VARIOUS CHEMICAL
	32.0	Paints, varnishes and lacquers	INDUSTRIES
	32.1	Pharmaceutical preparations	
	32.2	Soap, cosmetics, other toilet preparations and candles	
	32.3	Other chemical products	
15	32.9	Products of petroleum and coal	REFINING OF COAL
	51.0	Gas supply	AND OIL
	33.1	Structural clay products	
	33.4	Cement (hydraulic)	
	33.5	Cement products	
16	33.8	Limestone mills, lime and mortar works, grinding of other non-metallic mineral products, cut-stone and stone products	HEAVY NON-METALLIC MINERAL PRODUCTS
	33.9	Other non-metallic mineral products	
	34.0	Ferro alloys	BASIC METAL PRODUC-TION (FERRO ALLOYS,
17	34.3	Refining of aluminium	ALUMINIUM AND
	34.4	Crude metals not elsewhere classified	OTHER CRUDE METALS)
	34.1	Iron and steel works and rolling	
	34.2	Iron and steel foundries	IRON AND STEEL,
18	34.8	Rolling, refining and smelting of non-ferrous metals	METAL MILLS AND FOUNDRIES
	34.9	Non-ferrous metal foundries	
	35.0	Wire and wire products	
	35.1	Other metal building articles and steel structural parts	
	35.2	Metal packaging materials, etc., and metal household articles	
19	35.3	Arms, ammunition, hand tools and implements	MECHANICAL AND ELECTROTECHNICAL INDUSTRY
	35.4	Metal fittings	
	35.9	Other metal products	
	36.1	Mining and industrial machinery	
	36.9	Other machinery	
	37.0	Wires and cables	

Appendix 2—*contd.*

Sector Code		Name of Accounts	
M.S.G.	*National accounts*	*National accounts*	*M.S.G.*
	37.1	Transformers, generators and electric motors	
	37.2	Other distribution equipment	
	37.3	Signalling, radio and other tele-communication equipment	
	37.8	Accumulators and batteries, electric lamps and electrotechnical repair shops	
19	37.9	Other electrical products	MECHANICAL AND
	38.0	Building and repairing of steel ships	ELECTROTECHNICAL
	38.1	Building and repairing of wooden ships	INDUSTRY
	38.2	Other services for ships	
	38.3	Railroad equipment	
	38.4	Motor vehicles, parts and repairing of motor vehicles	
	38.5	Bicycles, motor-cycles and other transport equipment	
	38.6	Aircraft	
	33.2	Glass and glass products	
	33.3	Pottery, china and earthenware	
20	39.4	Jewellery and related articles	OTHER MANUFAC-TURING INDUSTRIES
	39.7	Plastic products not elsewhere classified, buttons, toys, etc.	
	39.8	Other products not elsewhere classified	
21	50.0	Electricity supply	ELECTRICITY SUPPLY
22	53.0	Trade	WHOLESALE AND RETAIL TRADE
	53.1	Customs duties (part of the trade industry)	
23	58.0	Commercial buildings	BUSINESS BUILDINGS AND DWELLINGS
	59.0	Dwellings	
	70.2	Coastal water transport	
	73.0*	Services related to water transport	
	74.0	Railway transport	
	75.0	Tramway and suburban railway transport	INTERNAL TRANSPORT AND COMMUNICA-TIONS
24	76.0	Land transport n.e.c.	
	77·0*	Air transport	
	78·0	Services related to transport and storage	
	79.0	Communications	

Appendix 2—*contd.*

Sector Code		*Name of Accounts*	
M.S.G.	*National accounts*	*National accounts*	*M.S.G.*

	52.0*	Water supply	
	54.1	Bank of Norway	
	54.2	State banks and loan associations	
	54.3	Commercial and savings banks, postal current account, post office savings banks, etc.	
	55.1*	Life insurance	
	55.2	Non-life insurance	
	55.3*	Social insurance	
25	82.0*	Educational services	SERVICES
	83.0*	Medical and veterinary services	
	84·0*	Religious and welfare activities	
	85.0*	Non-business organisations and institutions	
	86.0*	Legal, technical and business services	
	87.0	Recreation services	
	91.0*	Domestic services	
	92.0	Hotel and restaurant services	
	93.0*	Laundry, cleaning and other personal services	

	41.1	Building and repairing of agricultural buildings	
	41.2	Building and repairing of buildings in manufacturing	
	41.3	Building of dwellings	
26	41.4	Building of commercial and public buildings	BUILDING AND CONSTRUCTION
	41.5	Repairing of dwellings, commercial and public buildings	
	41.6	Other construction	
	41.8	Construction, *debit*	

27	All N.A. capital formation accounts except 70.0 (ocean and coastal transport). (All deliveries except buildings and construction.† † An approximate figure for investment in coastal transport is added.	OTHER INVESTMENT GOODS (EXCEPT FOR OCEAN WATER TRANSPORT)

28	05.0	Distribution account of commodities and services for administrative use (requisities, soap, transport services, telephone, etc.)	DISTRIBUTION ACCOUNT FOR VARIOUS GOODS
	02.0	Installation and repair work on machinery in establishments by their own workers	

Appendix 2—*contd.*

Sector Code M.S.G. National accounts		Name of Accounts National accounts	M.S.G.
	09.2	of central government military consumption	
	09.3	of central government civil consumption	
	09.4	of local government civil consumption	
29	09.6 Transfer account for deliveries	from private consumption to export	TRANSFER ACCOUNT
	09.7	from capital formation accounts to export	
	09.8	from capital formation accounts to use in produc-duction or to general government consumption	
	09.9	between capital formation accounts	
	80.0	Government administration	
	81.0	Military defence services	
Exogenous sectors, columns only	div.*	Part of wages and salaries in sectors denoted with * above	PUBLIC CONSUMPTION
	04.3	Central government depreciation on fixed real capital	
	04.4	Local government depreciation on fixed real capital	
	70.1	Ocean water transport	OCEAN TRANSPORT
	15.0	Whaling	WHALING

APPENDIX 3

THE MAIN ELEMENTS OF THE M.S.G. MODEL

$$(1) \quad X_i = A_i N_i^{\gamma_i} K_i^{\beta_i} e^{\epsilon_i t} \qquad \gamma_i > 0, \ \beta_i > 0; \ \gamma_i + \beta_i \le 1$$

$$(2) \quad X_{ji} = a_{ji} X_i \qquad a_{ji} > 0$$

$$(3) \quad M_i = \mu_i X_i \qquad \mu_i > 0$$

Production structure

$$(4) \quad \begin{cases} K_{Mi} = (1 - \kappa_i) K_i \\ K_{Bi} = \kappa_i K_i \end{cases} \qquad 1 \ge \kappa_i \ge 0$$

$$(5) \quad \begin{cases} D_{Bi} = \delta_{Bi} K_{Bi} \\ D_{Mi} = \delta_{Mi} K_{Mi} \end{cases} \qquad \begin{array}{l} \delta_{Bi} > 0 \\ \delta_{Mi} > 0 \end{array}$$

(6) $\gamma_i P_i^* X_i = W_i N_i$ ⎱ Marginal adjustment of
(7) $\beta_i P_i^* X_i = Q_i K_i$ ⎰ labour and capital

(8) $P_i^* = P_i - \Sigma_j a_{ji} P_j - P_0 \mu_i - \theta_i$ ⎱
(9) $Q_i = P_B(\delta_{Bi} + R_i)\kappa_i + P_M(\delta_{Mi} + R_i)(1 - \kappa_i)$ ⎰ Definitions

(10) $R_i = \rho_i R$

(11) $C_i = V g_i(P_0, P_1, \ldots, P_n, Y)$ ⎱ Demand functions

(12) $\quad X_i = \Sigma_j X_{ij} + C_i + Z_i$
$\quad X_{Bi} = \Sigma_j D_{Bj} + \dot{K}_B + Z_B$ ⎱ Equilibrium conditions
$\quad X_{Mi} = \Sigma_j D_{Mj} + \dot{K}_M + Z_M$

(13 $\Sigma_j N_j = N, \ \Sigma_j K_j = K$ } Definitions

$\quad X_i$ = gross production in sector i
$\quad X_{ji}$ = deliveries of intermediate goods from sector j to sector i
$\quad N_i$ = employment in sector i
$\quad K_i$ = total stock of fixed capital in sector i
$\quad N$ and K are totals for the economy
$\quad K_{Mi}$ and K_{Bi} are stocks of machines, etc., and building and plants
$\quad D_{Mi}$ and D_{Bi} are depreciation volumes of these stocks
$\quad M_i$ = non-competitive imports of raw materials
$\quad C_0$ = non-competitive imports of consumption goods
$\quad C_i$ = consumption of goods delivered by sector i
$\quad P_i$ = level of domestic seller's price of goods from sector i
$\quad P_0$ = price level of non-competitive imports
$\quad \theta_i$ = indirect taxes per unit of output
$\quad R_i$ = rate of return to capital in sector i
$\quad R$ = overall rate of return to capital in the economy
$\quad Z_i$ = net exogenous demand for good i (government demand + exports − competitive imports of type i)
$\quad Y_i$ = consumption expenditures per head, V = number of heads
$\quad Q_i$ = cost of using capital in sector i

The parameters represented by Greek letters and the ordinary input–output coefficients, a_{ij}, are constants. The consumption function is derived by assuming maximisation of individual utility functions subject to budget restrictions. The model therefore assumes perfect competition throughout. The wage rates, W_i, are constant over time. The labour force is a numeraire good.

The endogenous variables are:

$$\dot{N}_i/N_i, \ \dot{K}_i/K_i, \ \dot{X}_i/X_i, \ \dot{P}_i/P_i, \ \dot{R}/R \text{ and } \dot{Y}/Y.$$

The exogenous variables are:

$$\dot{K}/K, \ \dot{N}/N, \ \dot{V}/V, \ \dot{Z}_i, \ \epsilon_i, \ \dot{P}_0/P_0$$

and the unit taxes θ_i.

From the model several other growth rates may be derived, for instance the growth rate for the gross national product:

$$\frac{1}{GNP}\frac{dGNP}{dt}=\frac{1}{GNP}\left[\sum \dot{X_j}-\sum X_{ij}-\sum_{i,j}M_j+\dot{D}_{\text{gov}}+W_{\text{gov}}\dot{N}_{\text{gov}}\right]$$

Given the magnitudes of the exogenous variables, the model will generate the growth rates for the endogenous variables.

APPENDIX 4

Table 1.

THE DISCHARGE OF RESIDUALS TO AIR FROM NORWEGIAN SECTORS. OBSERVATIONS IN 1970. PROJECTIONS FOR THE YEAR 2000 CALCULATED IN THE REFERENCE RUN AND THE ALTERNATIVE RUN.

(tons)

Residuals Sectors, source of generation	Mercury 1970	Mercury 2000 Ref. run	Mercury 2000 Alt. run	Lead 1970	Lead 2000 Ref. run	Lead 2000 Alt. run	Sulphur oxides 1970	Sulphur oxides 2000 Ref. run	Sulphur oxides 2000 Alt. run	Nitrogen oxides 1970	Nitrogen oxides 2000 Ref. run	Nitrogen oxides 2000 Alt. run	Ammonia 1970	Ammonia 2000 Ref. run	Ammonia 2000 Alt. run	
1 Agriculture				6	7·5	7·5	2,800	3,530	3,575	700	885	895	13,000	16,395	16,590	1
2 Forestry				1	3·5	3·5	400	1,365	1,325	100	340	330	2,000	6,820	6,630	2
3 Fishing							3,550	4,290	4,420	12,000	14,495	14,940				3
4 Mining							1,150	3,495	3,335	310	940	900				4
5 Processing agr. prod.							2,775	3,215	3,260	660	765	775				5
6 Canned fish, etc.							1,140	1,330	1,435	290	340	365				6
7 Sweets, tobacco, etc.							630	795	850	150	190	200				7
8 Textiles, etc.							2,425	5,270	5,430	620	1,350	1,390	10	20	20	8
9 Pulp and paper							33,400	69,740	62,190	6,400	13,365	11,915				9
10 Wood and paper prod.							7,180	25,380	25,725	1,880	6,645	6,735				10
11 Print. and publ.							50	325	325	10	65	65				11
12 Chem. fish prod.							6,700	7,510	7,415	1,680	1,885	1,860				12
13 Fertilisers	1	3·5	3·5				4,000	13,675	14,505	6,500	22,225	23,570				13
14 Var. chem. ind.							3,995	21,605	21,655	1,320	7,140	7,155				14
15 Coal and oil							6,900	63,985	65,370	1,700	15,765	16,105				15
16 Cement, etc.							9,610	42,400	41,920	2,430	10,720	10,600	15	65	65	16
17 Basic metal							34,500	74,380	67,690	8,900	19,190	17,460				17
18 Iron, steel, metal mills							7,160	39,335	39,180	1,830	10,055	10,015				18
19 Mech. and electronic							2,135	9,610	9,330	520	2,340	2,275				19
20 Other manuf.				3	14	15	1,030	4,665	5,100	260	1,180	1,285				20
23 Build. and dwellings							22,900	91,530	91,555	3,600	14,390	14,395				23
24 Internal transport				58	221	225	10,020	38,255	38,645	20,920	79,870	80,685				24
M Military defence				4	9	9	1,500	3,145	3,145	2,300	4,825	4,825				M
H Household				338	820	820	1,200	2,910	2,910	13,260	32,180	32,180				H
R Roads, asphalt																R

58

TABLE 1 (*continued*)

	Residuals															
	Dust			Other acids			Other bases			Fluorine			Carbon monoxide			
Sectors, source of generation	1970	2000 Ref. run	2000 Alt. run	1970	2000 Ref. run	2000 Alt. run	1970	2000 Ref. run	2000 Alt. run	1970	2000 Ref. run	2000 Alt. run	1970	2000 Ref. run	2000 Alt. run	
1 Agriculture	220	275	280										4,800	6,055	6,125	1
2 Forestry	30	105	100										800	2,725	2,650	2
3 Fishing	4,200	5,075	5,230										10,400	12,565	12,950	3
4 Mining	530	1,610	1,535													4
5 Processing agr. prod.	120	140	140													5
6 Canned fish, etc.	60	70	75													6
7 Sweets, tobacco, etc.	30	40	40													7
8 Textiles, etc.	110	240	245				10	20	20							8
9 Pulp and paper	2,470	5,155	4,600													9
10 Wood and paper prod.	470	1,660	1,685													10
11 Print. and publ.																11
12 Chem. fish prod.	340	380	375													12
13 Fertilisers	4,500	15,385	16,315							450	1,540	1,630				13
14 Var. chem. ind.	3,040	16,440	16,475													14
15 Coal and oil	3,000	27,820	28,420													15
16 Cement, etc.	6,510	28,720	28,395													16
17 Basic metal	118,300	254,410	231,515							2,700	5,820	5,295				17
18 Iron, steel, metal mills	20,560	112,955	112,505													18
19 Mech. and electronic	170	765	745							5	25	25				19
20 Other manuf.	80	360	395	5	25	25										20
23 Build. and dwellings	4,500	17,985	17,990													23
24 Internal transport	6,580	25,120	25,380										61,520	234,880	237,285	24
M Military defence	870	1,825	1,825										4,740	9,945	9,945	M
H Household	450	1,090	1,090										266,000	645,580	645,580	H
R Roads, asphalt	150,000	513,450	513,450													R
Total	326,840	1,031,075	1,008,805	5	25	25	10	20	20	3,155	7,385	6,950	348,260	911,750	914,535	

TABLE 1 (*continued*)

Sectors, source of generation	Residuals			Al. halog. hydrocarb.			Decomp. dissolved org. subst.			Oil and oil products			Taste and smell subst.			Organic solvents			
	1970	2000 Ref. run	2000 Alt. run	1970	2000 Ref. run	2000 Alt. run	1970	2000 Ref. run	2000 Alt. run	1970	2000 Ref. run	2000 Alt. run	1970	2000 Ref. run	2000 Alt. run	1970	2000 Ref. run	2000 Alt. run	
1 Agriculture										1,400	1,765	1,785	95	120	120				1
2 Forestry										200	680	660	15	50	50				2
3 Fishing										20,300	24,520	25,275	2,560	3,090	3,185				3
4 Mining																			4
5 Processing agr. prod.																			5
6 Canned fish, etc.																			6
7 Sweets, tobacco, etc.																			7
8 Textiles, etc.	5	10	10													1,140	2,480	2,555	8
9 Pulp and paper													1,000	2,090	1,860	45	95	85	9
10 Wood and paper prod.																2,035	7,195	7,290	10
11 Print. and publ.																			11
12 Chem. fish prod.																390	435	430	12
13 Fertilisers																			13
14 Var. chem. ind.																110	595	595	14
15 Coal and oil										110	1,020	1,040							15
16 Cement, etc.							2	10	10										16
17 Basic metal																			17
18 Iron, steel, metal mills																			18
19 Mech. and electronic																3,110	13,995	13,595	19
20 Other manuf.																737	3,340	3,650	20
23 Build. and dwellings																			23
24 Internal transport										38,200	145,850	147,340	4,450	16,990	17,165				24
M Military defence										4,120	8,645	8,645	700	1,470	1,470				M
H Household										45,000	109,215	109,215	1,300	3,155	3,155				H
R Roads, asphalt																			R
Total	5	10	10	2	10	10				109,330	291,695	293,960	10,120	26,965	27,005	7,567	28,135	28,200	

THE DISCHARGE OF RESIDUALS TO LAND FROM NORWEGIAN SECTORS. OBSERVATIONS IN 1970. PROJECTIONS FOR THE YEAR 2,000 CALCULATED IN THE REFERENCE RUN AND THE ALTERNATIVE RUN

(tons)

Residuals — Sectors, source of generation	Mercury 1970	Mercury 2000 Ref. run	Mercury 2000 Alt. run	Lead 1970	Lead 2000 Ref. run	Lead 2000 Alt. run	Iron 1970	Iron 2000 Ref. run	Iron 2000 Alt. run	Sludge, dust, etc. 1970	Sludge, dust, etc. 2000 Ref. run	Sludge, dust, etc. 2000 Alt. run	Pesticides 1970	Pesticides 2000 Ref. run	Pesticides 2000 Alt. run	
1 Agriculture	0·9	1·1	1·2										1,300	1,640	1,660	1
2 Forestry	0·1	0·3	0·3										200	680	660	2
3 Fishing																3
4 Mining																4
5 Processing agr. prod.																5
6 Canned fish, etc.																6
7 Sweets, tobacco, etc.																7
8 Textiles, etc.																8
9 Pulp and paper																9
10 Wood and paper prod.																10
11 Print. and publ.																11
12 Chem. fish prod.										3,000	3,365	3,320				12
13 Fertilisers										1,000	3,420	3,625				13
14 Var. chem. ind.																14
15 Coal and oil										5,000	46,365	47,370				15
16 Cement, etc.										49,000	216,190	213,740				16
17 Basic metal										270,000	582,120	529,740				17
18 Iron, steel, metal mills										440,000	2,417,360	2,407,680				18
19 Mech. and electronic										17,000	76,500	74,305				19
20 Other manuf.							50	225	250	2,000	9,060	9,900				20
23 Build. and dwellings																23
24 Internal transport				20	75	75										24
M Military defence				1	2	2										M
H Household				100	245	245	28,000	67,955	67,955	53,000	128,630	128,630				H
R Roads, asphalt										500,000	1,711,500	1,711,500				R
Total	1	1·4	1·5	121	322	322	28,050	68,180	68,205	1,340,000	5,194,510	5,129,810	1,500	2,320	2,320	

TABLE 2 (continued)

Sectors, source of generation	Residuals Bark			Wood chip			Plastic substances			Decomp. susp. org. subst.			Oil and oil products			
	1970	2000 Ref. run	2000 Alt. run	1970	2000 Ref. run	2000 Alt. run	1970	2000 Ref. run	2000 Alt. run	1970	2000 Ref. run	2000 Alt. run	1970	2000 Ref. run	2000 Alt. run	
1 Agriculture	105,000	132,405	133,980										1,050	1,325	1,340	1
2 Forestry	15,000	51,135	49,710										150	510	495	2
3 Fishing																3
4 Mining													425	1,290	1,235	4
5 Processing agr. prod.													380	440	445	5
6 Canned fish, etc.													170	200	215	6
7 Sweets, tobacco, etc.													5	5	5	7
8 Textiles, etc.													180	390	405	8
9 Pulp and paper	140,000	292,320	260,680													9
10 Wood and paper prod.				121,200	428,440	434,260							50	175	180	10
11 Print. and publ.																11
12 Chem. fish prod.													800	895	885	12
13 Fertilisers																13
14 Var. chem. ind.																14
15 Coal and oil													60	555	570	15
16 Cement, etc.													420	1,855	1,830	16
17 Basic metal													230	495	450	17
18 Iron, steel, metal mills													555	3,050	3,050	18
19 Mech. and electronic																19
20 Other manuf.							4,000	18,120	19,805				110	500	545	20
23 Build. and dwellings																23
24 Internal transport													2,900	11,070	11,185	24
M Military defence													200	420	420	M
H Household							28,000	67,955	67,955	440,000	1,067,880	1,067,880	4,000	9,710	9,710	H
R Roads, Asphalt																R
Total	260,000	475,860	444,370	121,200	428,440	434,260	32,000	86,075	87,760	440,000	1,067,880	1,067,880	11,685	32,885	32,950	

TABLE 2 (*continued*)

Residuals Sectors, source of generation	Dispurging agencies			Organic solvents			Unspecified, solid waste			
	1970	2000 *Ref. run*	2000 *Alt. run*	1970	2000 *Ref. run*	2000 *Alt. run*	1970	2000 *Ref. run*	2000 *Alt. run*	
1 Agriculture										1
2 Forestry										2
3 Fishing										3
4 Mining										4
5 Processing agr. prod.							12,000	13,910	14,100	5
6 Canned fish, etc.							2,420	2,820	3,045	6
7 Sweets, tobacco, etc.							6,400	8,065	8,625	7
8 Textiles, etc.							13,980	30,395	31,315	8
9 Pulp and paper							51,700	107,950	96,265	9
10 Wood and paper prod.							18,100	63,985	64,850	10
11 Print. and publ.							5,000	32,610	32,625	11
12 Chem. fish prod.							1,150	1,290	1,275	12
13 Fertilisers										13
14 Var. chem. ind.				1,000	5,405	5,420	7,060	38,180	38,265	14
15 Coal and oil										15
16 Cement, etc.							13,700	60,445	59,760	16
17 Basic metal							50,000	107,800	98,100	17
18 Iron, steel, metal mills							2,000	10,990	10,945	18
19 Mech. and electronic							66,350	298,575	290,015	19
20 Other manuf.							1,850	8,380	9,160	20
23 Build. and dwellings										23
24 Internal transport	20	75	75							24
M Military defence	2	5	5							M
H Household	36	85	85				42,000	101,935	101,935	H
R Roads, asphalt										R
Total	58	165	165	1,000	5,405	5,420	293,710	887,330	860,280	

THE DISCHARGE OF RESIDUALS TO WATER FROM NORWEGIAN SECTORS. OBSERVATIONS IN 1970. PROJECTIONS FOR THE YEAR 2000 CALCULATED IN THE REFERENCE RUN AND IN THE ALTERNATIVE RUN.

(tons)

Residuals Sectors, source of generation	Mercury			Lead			Cadmium			Zinc			Copper		
	1970	2000 Ref. run	2000 Alt. run	1970	2000 Ref. run	20000 Alt. run	1970	2000 Ref. run	2000 Alt. run	1970	2000 Ref. run	2000 Alt. run	1970	2000 Ref. run	2000 Alt. run
1 Agriculture															
2 Forestry															
3 Fishing															
4 Mining				1	5	5	1	5	5	1,000	3,040	2,900	500	1,520	1,450
5 Processing agr. prod.															
6 Canned fish, etc.															
7 Sweets, tobacco, etc.															
8 Textiles, etc.															
9 Pulp and paper													5	10	10
10 Wood and paper prod.															
11 Print. and publ.															
12 Chem. fish prod.															
13 Fertilisers	1	3·5	3·5												
14 Var. chem. ind.															
15 Coal and oil															
16 Cement, etc.															
17 Basic metal	1	2	2	1,700	3,665	3,335	10	20	20	2,200	4,745	4,315	100	215	195
18 Iron, steel, metal mills															
19 Mech. and electronic				5	25	20	1	5	5	800	3,600	3,495	301	1,355	1,315
20 Other manuf.										200	905	990	40	180	200
23 Build. and dwellings															
24 Internal transport															
M Military defence															
H Household															
R Roads, asphalt															
Total	2	5·5	5·5	1,706	3,695	3,360	12	30	30	4,200	12,290	11,700	946	3,280	3,170

TABLE 3 (continued)

Residuals — Sectors, source of generation	Iron 1970	Iron 2000 Ref. run	Iron 2000 Alt. run	Chrome 1970	Chrome 2000 Ref. run	Chrome 2000 Alt. run	Sulphur oxides 1970	Sulphur oxides 2000 Ref. run	Sulphur oxides 2000 Alt. run	Hydrochloric acid 1970	Hydrochloric acid 2000 Ref. run	Hydrochloric acid 2000 Alt. run	Nitrogen oxides 1970	Nitrogen oxides 2000 Ref. run	Nitrogen oxides 2000 Alt. run	
1 Agriculture																1
2 Forestry																2
3 Fishing																3
4 Mining																4
5 Processing agr. prod.													190	220	225	5
6 Canned fish, etc.							25	30	30							6
7 Sweets, tobacco, etc.																7
8 Textiles, etc.										20	45	45				8
9 Pulp and paper				15	35	35	160	350	360							9
10 Wood and paper prod.							60,300	125,905	112,280							10
11 Print. and publ.																11
12 Chem. fish prod.																12
13 Fertilisers							780	875	865							13
14 Var. chem. ind.	51,000	275,810	276,420	10	55	55				160	865	865	1,705	9,220	9,240	14
15 Coal and oil							39,705	214,725	215,200							15
16 Cement, etc.																16
17 Basic metal																17
18 Iron, steel, metal mills							10,600	22,855	20,795	300	1,650	1,640	120	660	655	18
19 Mech. and electronic							280	1,540	1,530	3,040	13,680	13,290	360	1,620	1,575	19
20 Other manuf.	11,020	49,590	48,170	100	450	435	4,060	18,270	17,745							20
23 Build. and dwellings							15	70	75							23
24 Internal transport																24
M Military defence																M
H Household																H
R Roads, asphalt																R
Total	62,020	325,400	324,590	125	540	525	115,925	384,620	368,880	3,520	16,240	15,840	2,375	11,720	11,695	

65

TABLE 3 (*continued*)

Sectors, source of generation	Residuals			Other acids									Cyanide		
	1970	2000 Ref. run	2000 Alt. run	1970	2000 Ref. run	2000 Alt. run	1970	2000 Ref. run	2000 Alt. run	1970	2000 Ref. run	2000 Alt. run	1970	2000 Ref. run	2000 Alt. run
1 Agriculture				3,275	4,130	4,180									
2 Forestry				500	1,705	1,655									
3 Fishing															
4 Mining	90	275	260							90	275	260			
5 Processing agr. prod.	25	30	30	1,040	1,205	1,220									
6 Canned fish, etc.															
7 Sweets, tobacco, etc.				730	920	985									
8 Textiles, etc.	280	610	625				170	370	380						
9 Pulp and paper				12,000	25,055	22,345	44,000	91,870	81,930						
10 Wood and paper prod.															
11 Print. and publ.															
12 Chem. fish prod.	20	20	20	150	170	165									
13 Fertilisers				100	340	365				50	170	180			
14 Var. chem. ind.	1,200	6,490	6,505	450	2,435	2,440				500	2,705	2,710			
15 Coal and oil							1,050	9,735	9,950				150	1,390	1,420
16 Cement, etc.															
17 Basic metal				260	560	510									
18 Iron, steel, metal mills													20	110	110
19 Mech. and electronic	250	1,125	1,095	500	2,250	2,185							40	180	175
20 Other manuf.	2	10	10	5	25	25				2	10	10			
23 Build. and dwellings															
24 Internal transport															
M Military defence															
H Household															
R Roads, asphalt															
Total	1,867	8,560	8,545	19,010	38,795	36,075	45,220	101,985	92,260	642	3,160	3,160	210	1,680	1,705

TABLE 3 (*continued*)

Residuals Sectors, source of generation	Arsenic			Phosphorous compounds			Nitrogen compounds			Mine tailings, sludge			Al. halog. hydrocarb.		
	1970	2000 Ref. run	2000 Alt. run	1970	2000 Ref. run	2000 Alt. run	1970	2000 Ref. run	2000 Alt. run	1970	2000 Ref. run	20000 Alt. run	1970	2000 Ref. run	2000 Alt. run
1 Agriculture				325	410	415	15,700	19,800	20,035						
2 Forestry				50	170	165	2,500	8,525	8,285						
3 Fishing															
4 Mining										7,000,000	21,273,000	20,300,000			
5 Processing agr. prod.				40	45	45	40	45	45						
6 Canned fish, etc.															
7 Sweets, tobacco, etc.				11	15	15				720	905	970			
8 Textiles, etc.				27	60	60				60	130	135	1	2	2
9 Pulp and paper							400	835	745	10,500	21,925	19,550			
10 Wood and paper prod.										3,600	12,725	12,900			
11 Print. and publ.															
12 Chem. fish prod.				10	10	10				500	560	555			
13 Fertilisers										160,000	547,400	580,160			
14 Var. chem. ind.							570	3,085	3,090	65,150	352,330	353,115			
15 Coal and oil	300	2,780	2,840				850	7,880	8,055						
16 Cement, etc.										178,050	785,555	776,655			
17 Basic metal	30	65	60							101,000	217,755	198,160			
18 Iron, steel, metal mills										19,000	104,085	103,970			
19 Mech. and electronic				100	450	435	80	360	350	1,590	7,155	6,950			
20 Other manuf.										130	590	645			
23 Build. and dwellings															
24 Internal transport															
M Military defence															
H Household				3,500	8,495	8,495	15,000	36,405	36,405	40,000	97,080	97,080			
R Roads, asphalt															
Total	330	2,845	2,900	4,063	9,655	9,640	35,140	76,935	77,010	7,580,300	23,421,195	22,450,845	1	2	2

TABLE 3 (continued)

Residuals	Bark			Fibre			Plastic subst.			Decomp. dissolved org. subst.			Decomp. susp. org. subst.			
Sectors, source of generation	1970	2000 Ref. run	2000 Alt. run	1970	2000 Ref. run	2000 Alt. run	1970	2000 Ref. run	2000 Alt. run	1970	2000 Ref. run	2000 Alt. run	1970	2000 Ref. run	2000 Alt. run	
1 Agriculture										16,750	21,120	21,375				1
2 Forestry										2,625	8,950	8,700				2
3 Fishing																3
4 Mining																4
5 Processing agr. prod.										7,100	8,230	8,345	8,400	9,735	9,870	5
6 Canned fish, etc.										5,000	5,825	6,290	5,100	5,940	6,415	6
7 Sweets, tobacco, etc.										110	140	150	2,150	2,710	2,900	7
8 Textiles, etc.													1,600	3,480	3,585	8
9 Pulp and paper	38,000	79,345	70,755	64,700	135,095	120,470				550,360	1,149,150	1,024,770				9
10 Wood and paper prod.	1,000	3,535	3,585	11,300	39,945	40,490				18,140	64,125	64,995				10
11 Print. and publ.																11
12 Chem. fish prod.													460	515	510	12
13 Fertilisers																13
14 Var. chem. ind.										570	3,085	3,090	4,000	21,630	21,680	14
15 Coal and oil				200	1,855	1,895										15
16 Cement, etc.										200	880	870				16
17 Basic metal																17
18 Iron, steel, metal mills																18
19 Mech. and electronic																19
20 Other manuf.							25	115	125							20
23 Build. and dwellings																23
24 Internal transport																24
M Military defence																M
H Household										64,000	155,330	155,330	76,000	185,450	184,450	H
R Roads, asphalt																R
Total	39,000	82,800	74,340	76,200	176,895	162,855	25	115	125	664,855	1,416,835	1,293,915	97,710	228,460	229,410	

TABLE 3 (continued)

Sectors, source of generation	Oil and oil products			Dispurging agencies			Taste and smell subst.			Organic solvents			Other organic substances		
	1970	2000 Ref. run	2000 Alt. run	1970	2000 Ref. run	2000 Alt. run	1970	2000 Ref. run	2000 Alt. run	1970	2000 Ref. run	2000 Alt. run	1970	2000 Ref. run	2000 Alt. run
1 Agriculture	9	11	11	5	6·5	6·5									
2 Forestry	1	4	4	1	3·5	3·5									
3 Fishing	4,200	5,075	5,230												
4 Mining	3,302	11,555	11,025	240	730	695	362	1,100	1,050						
5 Processing agr. prod.				310	360	365									
6 Canned fish, etc.				5	5	5									
7 Sweets, tobacco, etc.				15	20	20									
8 Textiles, etc.	260	565	580	430	935	965							8	15	15
9 Pulp and paper													3	10	10
10 Wood and paper prod.															
11 Print. and publ.															
12 Chem. fish prod.	550	615	610												
13 Fertilisers	500	1,710	1,815												
14 Var. chem. ind.	20	110	110	100	540	540				840	4,545	4,545			
15 Coal and oil	100	925	945												
16 Cement, etc.							100	440	435						
17 Basic metal	330	710	645												
18 Iron, steel, metal mills	172	945	940												
19 Mech. and electronic	930	4,185	4,065	30	135	130				10	45	45			
20 Other manuf.										20	90	100			
23 Build. and dwellings															
24 Internal transport	4,600	17,565	17,740	40	155	155									
M Military defence	40	85	85	5	10	10									
H Household	3,000	7,280	7,280	3,300	8,010	8,010									
R Roads, asphalt															
Total	18.514	51,340	51,085	4,481	10,910	10,905	462	1,540	1,485	870	4,680	4,690	10	25	25

69

Discussion of the Paper by
Finn R. Førsund and Steinar Strøm

Formal Discussant: Kavanagh. I intend to confine my comments in this paper to the role of micro- versus macro-models in environmental problems. The authors state that 'pollution problems considered are conceived of as locally experienced problems in the sense that it is the environmental services of geographically located recipients that are negatively affected by the discharge of residuals. It follows, then, that regional empirical studies are called for to implement policy measures in the "problem sheds" of a river basin, an estuary, the air over a city, etc.' I leave it to others to comment on the interesting features of their macro-model and the advantages of their approach in the study of environmental problems. In particular, perhaps Hicks will comment on the implication of their paper for social accounting systems.

My own preference and experience is to favour the development of micro-models with a strong empirical base. We have been asked to relate the papers for different sessions of the conference. In this respect, I draw attention to Kneese's paper (Chapter 3) on water quality. The studies summarised in his paper illustrate the kind of approach I have in mind. The more case studies there are of the kind cited there, the more I think we can advance the development of the operational models we require for economics to become an applied science of use to policy-makers. Finally, as our environmental problems have an underlying technology, we must seek to develop models of river systems and the like in which technological and economic data can be logically combined and emiprically tested.

Hicks, replying to Kavanagh's comments about social accounting systems, said it was probable that if we calculated the social accounts in the normal way, then, if we introduced any pollution controls, the rate of growth of the national product would slow down. This was nevertheless a misleading result.However, to avoid it we not only needed to have an effective way of measuring the gains from the pollution controls, but we also needed some means of correcting the existing measures for the loss which was at present being incurred due to the absence of pollution controls. This aspect, i.e. that of measuring the disutility associated with having no pollution controls, had not really been discussed in the authors' paper.

Russell observed that the main advantage of input–output analysis was that the input–output tables were generally available and you could simply add on the coefficients for residuals generation. The disadvantage was that input–output tables were generally estimated by sector, rather than by establishment or plant, and this left out the spatial component from the production process which was a very important variable in pollution controls. Leontief had carried out a study in the Boston area in which the tables had been disaggregated to the plant level. However, the volume of work involved in doing this had been colossal.

Mills said that concentrating on small spatial areas increased the relevance of input–output models for pollution controls. However, he also suspected that the extra work involved in doing it this way far outweighed the increased accuracy of the input–output matrix.

Kneese noted that Kavanagh had indicated that he felt that microeconomic models of environmental questions were much to be preferred to the inter-industry and more aggregative types of models used by the authors and others. He thought the correct point of view on this was that models had to be designed to fit the problems they were intended to analyse. Certainly, 'problem shed' type models were of great importance in analysing detailed environmental problems in specific regions, but one could envisage other useful models as well. For example, an inter-industry model might be used to analyse some of the structural effects on the economy of large-scale restrictions on pollution. A case in point was the demanding conditions which were now being laid upon automobile emissions in the United States. For other problems, even more aggregative models might be indicated: for instance, if one was interested in the balance of payment effects and national income effects of environmental controls, one might wish to use an aggregate Keynesian model of the economy. The key consideration was to fit the model to the problem.

Hoch asked how one handled cost increases when the technical coefficients for residuals were changing. Did these show up as changes in technical coefficients or in final demand? In any event, one might try to build into the input–output framework some way of measuring the effects of different costs of reducing residuals on the production of various sectors.

Prud'homme agreed that we often needed macro-, or rather national, information on environmental quality. However, it did not follow that this information was best provided by macro- (national) models.

Macro- (national) models yield aggregated or *averaged* information that may well be useless and even meaningless. This was because environmental quality indicators, as opposed to production indicators, were essentially located, and could not be aggregated over space. The concept of a given level of steel production for a nation makes sense; the concept of a given level of biochemical oxygen demand (B.O.D.) for a nation does not. National environmental quality indicators should probably be *distributions* of local environmental quality indicators. And local environmental quality indicators can only be produced by micro- (local) models.

Rothenberg said that the relative location of different emissions of pollutant was very important. To evaluate the effect of pollution we needed to know to whom, and at what rate, each emission flowed. It might be possible to divide the emissions into spatial sectors in which both spatial and inter-industry interaction could take place.

In so much pollution control we were trying to bring about various kinds of substitutions. We were attempting, through pollution controls, to achieve a substitution of inputs, of outputs as intermediate goods, and of final demand. Any realistic model should thus allow such substitutions to take place, and the input–output mechanism might not be the easiest way of doing this.

Thoss disagreed with Rothenberg. He felt that one could meet this criticism in an input–output model by introducing different production processes in each sector which allowed for process substitution. This did, however, imply continuous production functions in the limit which could not be incorporated into an input–output model.

Rothenberg continued by observing that models of this sort made an important contribution towards public policy by examining the direct and indirect effects of different kinds of pollution controls. One of the important effects was that if controls were placed on activities A, B and C, they were not the only activities likely to be affected in either a positive or a negative way. The input–output format thus gave a clear portrayal of the kinds of adjustments required to support a continued level of growth in real income.

Kolm noted that the issue of the durability versus the non-durability of residuals was extremely complicated. Some effects were durable through their effects on man. For example, noise was a pollutant which was supposed to be non-accumulating. However, it could affect both hearing and mental health and thus might persist for the lifetime of the individuals involved.

Kneese pointed out that the possibility of a catastrophic breakdown of an ecosystem had important implications for models that had long time horizons. Many forms of pollution simply changed the ecosystem, e.g. the large-scale pollution of a lake may turn it into dry land, or produce effects that were reversible only over very long time horizons. The really dangerous pollutants, on the other hand, probably consisted of small concentrations of highly toxic products which did not disintegrate and which slowly accumulated. It was very difficult for growth models to deal with these latter pollutants.

3 The Application of Economic Analysis to the Management of Water Quality: Some Case Studies

Allen V. Kneese

3.1 *INTRODUCTION: THE CASES*

Efforts to apply economic analysis to environmental management problems are rather new – although not quite so new as the recent flurry of activity in this area might lead one to suppose. The first book presenting reasonably serious efforts to use economic analysis in connection with water quality problems appeared in 1961 [1]. Following this, during the 1960s, several case studies using an economics framework were done in important American river basins. As a matter of fact, that decade was a rich period for economic studies of water quality management and I have chosen to discuss a couple of these in some detail. They are the studies of the Delaware estuary and of the Potomac river basin. I have selected these because they are important and because I had some direct involvement in both of them. I shall comment on a few other studies more generally.

The pattern of the paper is as follows: first, I state the general conclusions that I feel are supported by the studies. Then, for those readers who are not already familiar with them, I describe briefly some models of the natural hydrological system which were central components of them, next I present the cases, and finally I discuss how and why they have been influential or failed to be so.

3.2 *WHAT ARE THE MAIN CONCLUSIONS SUPPORTED BY THE CASES?*

Several significant conclusions follow from consideration of the case studies. I shall just list them here and then develop them in connection with the actual discussion of the studies:

1. Economic analysis can be successfully combined with 'ecological' or 'natural system' models and applied to complex real cases.
2. The standard administrative approach to water pollution which tries to apply uniform effluent standards to all waste water discharges is extremely inefficient.

3. The studies tend to support a preference for effluent charges over effluent standards.
4. Even the few efforts at 'comprehensive' planning for water quality improvement have not sufficiently examined the 'production function' to permit a least-cost solution to be found.
5. The application of 'cookbook' benefit–cost analysis to environmental quality problems can be seriously misleading.
6. The stochastic characteristics of hydrological systems make it necessary to consider the probability of an ambient standard being violated as well as its level. Decisions on this are usually made in a 'rule of thumb' manner which can be very costly in economic terms.

3.3 *SOME MODELS OF THE NATURAL SYSTEM*

To move the application of economic analysis much beyond the point of general statements about equating margins, it is necessary to bring it into conjunction with other models of the real world, drawn from other disciplines, and with technological information. In the case of water quality, the most pertinent models are drawn from sanitary engineering and from hydrology.

3.3.1 THE STREETER–PHELPS MODEL

Roughly speaking, polluting substances discharged to watercourses can be divided into two classes – degradable and non-degradable. Sodium chloride or ordinary table salt is a good illustration of the latter. Its behaviour in water is extremely simple since all that happens to it is dilution. So, calculating the concentration that results when a given amount of it is discharged to a watercourse is a simple matter of division. Degradable substances are different since they are transformed by biological or chemical action, and the transformation process itself often works undesirable changes on the water body. The most important example is degradable organic matter which is contained in household sewage and in many industrial effluents. Bacteria feed on these effluents and convert them to plant nutrients – nitrates and phosphates – and carbon. In the process these 'aerobic' bacteria use the dissolved oxygen (D.O.) in the water for respiration. This results in what is termed an 'oxygen sag'. Oxygen-demanding material is measured in terms of pounds of biochemical oxygen demand, or B.O.D.

One way of visualising the oxygen sag process is to think of a certain amount of organic waste – this is usually measured in terms of pounds of biochemical oxygen demand (B.O.D.) – being placed in a container of well-aerated water with an appropriate bacteria culture. As the biological reactions proceed, the unsatisfied B.O.D is reduced at a constant rate per unit time, and D.O. is depleted. At the same time some oxygen is restored to the water through the air–water interface. As time elapses, the oxygen-demanding waste is gradually consumed (that is, converted into inorganic substances) and the re-aeration process continues, with the level of dissolved oxygen again reaching the saturation point for water of that temperature when the oxygen-demanding waste is completely consumed. By a well-known law of gases, saturation level varies inversely with temperature.

Alternatively, one can think of a flowing stream with uniform characteristics over a certain reach in which there is a constant input of oxygen-demanding waste at a certain point. We now can regard distances downstream as being exactly equivalent to time in the former way of looking at this situation. Thus, as we proceed downstream, we first find a decrease in the dissolved oxygen content as oxygen is consumed in the biochemical reactions on the oxygen-demanding waste, and finally a rise in the dissolved oxygen content to the saturation level still further on downstream. This process is the 'oxygen sag' in a watercourse.

While the subsequent formulas are written in terms of time, they could just as well be written in terms of distance, provided the equivalences stipulated above are met. These formulas are commonly known as the Streeter–Phelps equations.

Biochemical oxidation is indicated as a first-order differential equation of the form

$$\frac{dL}{dt} = -k_1 L_t$$

where L_t is the unsatisfied B.O.D. in parts per million (p.p.m.), t is time (days), and k_1 is a rate constant which is a function of the characteristics of the waste and the water temperature. If L_a is the initial first-stage B.O.D. and is interpreted as the constant of integration, the result of integrating the above equation is

$$L_t - L_a e^{-k_1 t}$$

Re-aeration is also indicated as a first-order process. It is a function of the difference between actual D.O. concentration and saturation concentration, as follows:

$$\frac{dC}{dt} = k_2 (C_s - C_t)$$

where C_t is the concentration (p.p.m.) of D.O. at time t and C_s is the saturation concentration. The item k_2 is once again a rate constant which is primarily a function of temperature.

When the two reactions are combined and the resulting equation written in terms of the D.O. deficit ($D_t = C_s - C_t$), we obtain

$$\frac{dD}{dt} = k_1 L_t - k_2 D_t \ [2]$$

After substituting $L_a e^{-k_1 t}$ for L_t, we have

$$\frac{dD}{dt} = k_1 L_a e^{-k_1 t} - k_2 D_t$$

This is a first-order differential equation of the general form

$$\frac{dy}{dx} + P_y = Q$$

with

$$Q = k_1 L_a e^{-k_1 t}$$

The solution of this equation is

$$D_t = \frac{k_1 L_a}{k_2 - k_1}\left(e^{-k_1 t} - e^{-k_2 t}\right) + D_a e^{-k_1 t}$$

where D_a is the deficit and L_a is the B.O.D. concentration, both at time $t = 0$. The time at which the maximum deficit occurs, say t_c, can be found by taking the derivative of the above equation with respect to time, setting the result to zero, and solving for t_c. The resulting expression is

$$t_c = \frac{1}{k_2 - k_1} ln \left[\frac{k_2}{k_1}\left(1 - \frac{(k_2 - k_1)D_a}{k_1 L_a}\right)\right]$$

At the location corresponding to this time the deficit is

$$D_c = \frac{k_1 L_a}{k_2} e^{-k_2 t_c}$$

3.3.2 MULTIPLE SOURCES AND RECEPTORS UNDER STEADY-
       STATE CONDITIONS

The equations just discussed transform a number of pounds of B.O.D. discharged at a particular location into concentrations of dissolved oxygen (or, given the saturation level, dissolved oxygen deficits) at other (receptor) locations downstream. This is for 'steady-state conditions', i.e. the rate of stream flow, the rate of re-aeration, and temperature are all fixed. Even if one accepts the steady-state assumptions, in any even moderately developed river basin this will be too simple a model. There will be multiple points of discharge as well as multiple receptors. Before the oxygen sag will have fully recovered, another discharge will enter the system, and so on down the line. Thus the stream consists of a number of interconnected segments. When the Streeter–Phelps oxygen balance equations are applied to these, a system of linear first-order differential equations results. But the transfer functions, which are found by solving these equations and which relate the change in D.O. in segment i to an input of B.O.D. in segment j, fortunately simplify to a set of linear relationships if we continue to assume steady-state conditions. In matrix notation we can write the system as follows:

$$A\mathbf{x} = \mathbf{r}$$

where A is the matrix of transfer coefficients, $\mathbf{x}$ is the vector of B.O.D. discharges in pounds, and $\mathbf{r}$ is a vector of D.O. concentration at various specified (receptor) locations. This form is extremely convenient because, as we shall see below, it lends itself easily to incorporation into linear economic models. Sets of equations such as this are also used in several of the other papers at this conference [3].

3.3.3 STOCHASTIC HYDROLOGY

As everyone knows, the flow of rivers (one of the major determinants of their capacity to assimilate waste materials) is not constant with time. It varies seasonally in somewhat regular patterns but with a large random compartment. Traditionally, in designing flow-regulating structures (reservoirs) a device called a 'mass curve' is used to determine the yield that can be sustained from the reservoir during drought periods. Underlying this technique is the assumption that future flows (measured on a daily or monthly basis) will be an exact replication of flows observed historically (usually there are thirty years or so of record).

An alternative, which is more defensible on statistical grounds, is to assume that the historical record is a sample from a much larger population and that what will remain invariant in the future are only the lower moments (mean, standard deviation, and possibly skewness) of the distribution of observed flows. Based on the latter assumption, stochastic hydrology generators have been devised which can simulate long hydrologic sequences incorporating extreme values and patterns of events not in the historic record while maintaining moments of the frequency distribution of that record. There are difficult problems associated with such generators, but many of them have been overcome. The problems involve such things as serial correlation in the record of flows and maintaining cross- and serial correlations for separate gauging stations in the same system. I shall not attempt to treat these since my intent is just to acquaint the reader with the recursive relationships used in the generation.

In these Markovian models the basic recursive relation used can be represented by the following equation [4].

$$x_{i+1} = \mu + \beta(x_i - \mu) + t_{i+1}\sigma(1 - \rho^2)^{\frac{1}{2}}$$

In this model, x_{i+1}, the flow in the $(i+1)$st interval, is a linear function of x_i, the flow in the ith interval; of a random standardised deviate t_{i+1}; and of the population parameters μ (the population mean), σ (the population standard deviation), β (the regression coefficient of flows in the $(i+1)$st interval on values in the ith interval), and ρ (the correlation coefficient between flows in successive time periods). The standardised random deviate t_{i+1} has zero mean and unit variance.

It can be shown that if the distribution of flows is normal and the regression function of x_i on x_{i-1} is linear and homoscedastic (of constant variance), the conditional expectation of x_i, given x_{i-1}, is given by

$$E(x_i|x_{i-1}) = \mu + \beta(x_{i-1} - \mu)$$

and the conditional variance of x_i, given x_{i-1}, is

$$\mathrm{var}(x_i|x_{i-1}) = \sigma^2(1 - \rho^2)$$

Thus, we can see in the recursive equation that the first two terms on the right-hand side are the expected value of x_{i+1}, given x_i has occurred (February's flow, January having occurred), and the last term is the random component consisting of a randomly selected normal deviate which, when multiplied by the expected variance of x_{i+1}, given x_i, brings the result back into the proper dimension comparable with the first two terms of the right side. It can also be shown that to treat non-normal distributions it is sufficient to alter the distribution of the random additive component and thus maintain moments of observed data ([4] chap. 2).

It is readily seen that a recursive model of this type could be used to generate indefinitely long sequences of hydrological record to be used as inputs to a simulation model. As long a generated sequence of flows as is wished (say, several thousand years) can then be used in analysing the probabilistic performance of a reservoir, or other water quantity and quality management system elements. Stochastic hydrology is used in connection with the Potomac case study discussed below.

3.4 *DISCUSSION OF THE CASES*

3.4.1 INTRODUCTION

Armed with this quick review of a few important models from other disciplines, we can proceed to a discussion of the cases. First, let us consider the study of the Delaware estuary.

3.4.2 THE DELAWARE STUDY

(a) *The area.* The Delaware river basin, though small by the standards of the great American river basins and draining an area of only 12,765 square miles, holds a population of over 6 million. Portions of the basin, especially the Lehigh sub-basin and the Delaware estuary area, are among the most highly industrialised and densely populated regions in the world, and it is in these areas that the main water quality problems are encountered. The Delaware estuary, an 86-mile reach of the Delaware river from Trenton, New Jersey, to Liston Point, Delaware (see Fig. 3.1), is most important in terms of the quantity of water impacted, the area involved, the extent of industrial activity, and the number of people affected.

Despite early industrial and municipal development in the basin, water quality problems were neglected until the last few decades. The Interstate Commission on the Delaware River Basin (INCODEL) was formed in 1936, and under its auspices the states in the basin signed a reciprocal agreement on water quality control. This provided the legal basis for construction of treatment plants by municipalities after the Second World War. The standards of treatment achieved were not particularly high (on the average not much more than removal of the grosser solids), and the residual waste load from the plants, together with industrial discharges, continued to place very heavy oxygen demands on the estuary. Especially during the warm summer months, D.O. fell, and still falls, to low levels or becomes exhausted in a few portions of the reach of the estuary from Philadelphia to the Pennsylvania–Delaware state line.

There are many water quality characteristics which affect the value of the various services of a watercourse. As we have already seen, one of the most central is D.O., which is affected by meteorological and hydrological variables and by the discharge of organic wastes. D.O. is also something of a surrogate measure in other quality characteristics. The performance of waste water treatment plants is usually measured in terms of their ability to remove B.O.D. from waste waters. Thus, my discussion of the analysis of the Delaware estuary is focused primarily (but not exclusively) on oxygen conditions.

(b) *Background of the study.* In 1957–8, at the request of the Corps of Engineers (the Federal agency responsible for developing 'comprehensive' plans in the river basins of the eastern United States), the U.S. Public Health Service made a preliminary study of water quality in the Delaware estuary. The data it produced regarding the quality of the estuary led state and interstate agencies, concerned with water quality, to request a comprehensive study of the estuary under the provisions of the Federal Water Pollution Control Act.

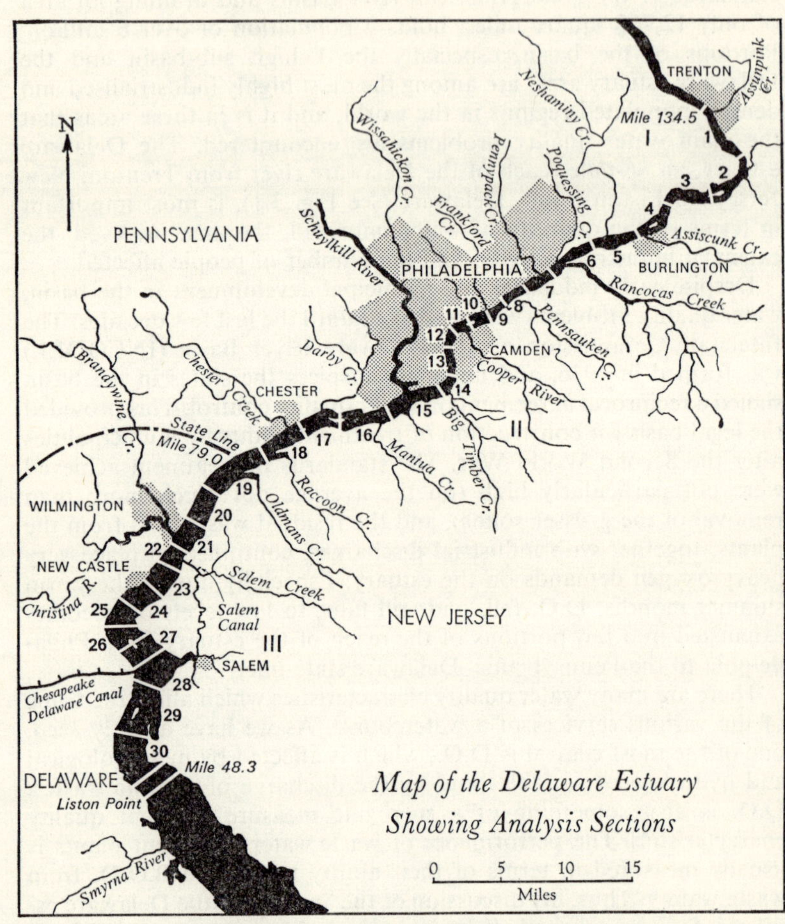

FIG. 3.1   Map of the Delaware estuary showing analysis sections
From: Federal Water Pollution Control Administration,
Delaware Estuary Comprehensive Study (1966)

The study was begun in 1961, and in the summer of 1966 a report was issued by the Federal Water Pollution Control Administration – *Delaware Estuary Comprehensive Study: Preliminary Report and Findings*. The study made an effort to measure external costs as well as costs of control associated with various policy alternatives. One of its main contributions was to link the model of waste degradation and re-aeration for multiple points of discharge, which I described in the previous section, to an economic optimisation model.

(c) *The model.* Assume that a watercourse consists of m homogeneous segments (thirty segments were used in the Delaware estuary study) and c_i represents the improvement in water quality required to meet a D.O. target in segment i. The target vector $\mathbf{c}$ of m elements can be obtained by changes of inputs to the watercourse from combinations of the m segments. Define another vector $\mathbf{x} = (x_1, \ldots, x_j, \ldots, x_n)$ in which the values of x refer to the volume of waste discharges in each of the segments. In a feasible solution, these values represent the waste discharges at the various points which meet the target vector $\mathbf{c}$. This vector generates D.O. changes through the mechanism of the constant coefficients of the linear system already described in section 3.2 above. Item $a_{ij} = \text{D.O.}$ improvement in segment i per unit reduction of x_j, $i = 1, \ldots, m$; $j = 1, \ldots, n$; and, of course, $x_j > 0$.

If we let A be the $(m \times n)$ matrix of transfer coefficients, then $A\mathbf{x}$ is the vector of D.O. changes corresponding to $\mathbf{x}$.

Now, recalling that $\mathbf{c}$ is the vector of target improvements, we have two restrictions on $\mathbf{x}$, namely, $A\mathbf{x} \geq \mathbf{c}$ and $\mathbf{x} \geq 0$. The reader will have noticed that mathematically these are sets of linear constraints such as those found in a standard linear programme. All we need is an objective function to complete the problem. Let $\mathbf{d}$ be a row vector where $d_j = \text{unit cost of } x_j$, $j = 1, \ldots, n$. Notice that this assumes linear cost functions, although programmes with linear constraints and non-linear objective functions can usually be solved if the non-linear function is not too complicated. So this condition would not necessarily have to hold. We can now write the problem as a standard linear programme. (The actual programmes needed to solve the problem encountered in the Delaware estuary were somewhat more complicated. The reader interested in the details should consult Sobel [3].)

$$\text{Minimise } \mathbf{dx}$$
$$\text{subject to } A\mathbf{x} \geq \mathbf{c}$$
$$\mathbf{x} \geq 0$$

Of course, the transfer coefficients (a_{ij}), as already explained, relate to a steady-state condition of specified conditions of stream flow and temperature. Thus the model turns out to be totally deterministic, and the variability of conditions is handled in this analysis by assuming extreme conditions usually associated with substantial declines in water quality. This is a weakness in programming-type models, and an alternative mode of analysis which can handle some stochastic elements (but with unfortunately its own set of weaknesses) is discussed in connection with the Potomac case.

A linear programming model of the general type just described was constructed for the Delaware estuary. In addition to D.O., it included other non-degradable types of material. Computation of the a_{ij}'s is much easier for these. Once done, the model provided an extremely flexible tool for the analysis of alternative policies.

(d) *Analysis of objectives*. A major part of the strategy of the Delaware study was to use the model to analyse the total and incremental costs of achieving five 'objective sets', each representing a different package and spatial distribution of water quality characteristics, with the level of quality increasing from set 5 (representing 1964 water quality) to set 1. In other runs, overall costs were minimised by the programming model. In others, additional constraints were added to represent more usual administrative approaches to the problem. The water quality characteristics and associated levels and the areas to which they apply are shown for objective set 2 in Table 3.1. The 30 sections referred to in the table are shown on the map in Fig. 3.1. An effort was then made to measure benefits associated with the improvement in water quality indicated by the successive objective sets. Before turning to the benefit estimation, it will be useful to describe the objective sets a bit further.

(e) *Water quality objectives*. Table 3.2 shows several water quality parameters with the associated levels for the five objective sets. The general nature of the objective sets is as follows:

Objective set 1. This is the highest set. It makes provision for large increases in water-contact recreation in the estuary. It also makes special provision for 6·5 p.p.m. levels of dissolved oxygen to provide safe passage for anadromous fish during the spring and autumn migration periods. Thus, this objective set should produce conditions in which water quality is basically no obstacle to the migration of shad and other anadromous (migratory) fishes.

Objective set 2. Under this set the area available for water-contact recreation is constricted somewhat. Some reduction in sport and commercial fishing would also be expected because of the somewhat lower dissolved oxygen objective. This set, like objective set 1, makes special provision for high dissolved oxygen during periods of anadromous fish passage.

Objective set 3. This set is similar to set 2. Although there is no specific provision for raising dissolved oxygen during periods of

Water Quality Parameter [a,b]	1	2	3	4	5	6	7	8	9	10	11	12	13	14	15	16	17	18	19	20	21	22	23	24	25	26	27	28	29	30
	Trenton			Bristol		Torresdale	Camden				Philadelphia							Chester		Wilmington				New Castle						Liston Point
Dissolved oxygen[c]	5·5					5·5	4·0											4·0		5·0								5·0		6·5
0·0.[i] 4/1–6/15 and 9/16–12/31	6·5																													6·5
Chlorides[d]														50		250														
Coliforms (=/100 ml.)	5,000[e]					5,000[e]	5,000[f]																							5,000[f]
Coliforms 5/30–9/15	4,000[e]			4,000[e]		5,000[e]	5,000[f]																					5,000[f]		4,000[e]
Turbidity (Tu)	N.L.+30																													N.L.+30
Turbidity 5/30–9/15	N.L.			N.L.		N.L.+30																		N.L.+30				N.L.		N.L.
pH[g] (pH units)	6.5-8.5																													6.5-8.5
pH[g] 5/30–9/15				7·0-8·5		6·5-8·5																						6·5-8·5		7·0-8·5
Alkalinity[g]	20-50					20-50	20-120																							20-120
Hardness[h]	95					95	150									150														
Temperature[g] (°F)	Present levels																													Present levels
Phenols[h]	·001					·001	·005									·005	·01													·01
Syndets[h]	·5					·5	1·0																							1·0
Oil and grease, floating debris	Negligible																													Negligible
Toxic substances	Negligible																													Negligible
Section	1	2	3	4	5	6	7	8	9	10	11	12	13	14	15	16	17	18	19	20	21	22	23	24	25	26	27	28	29	30

[a] mg/l unless specified. [b] Not less stringent than present levels. [c] Summer average. [d] Maximum 15-day mean. [e] Maximum level. [f] Monthly geometric mean. [g] Desirable range. [h] Monthly mean. [i] Average during period stated. N.L. = Natural levels.

Based on: A. V. Kneese and B. T. Bower, *Managing Water Quality: Economics, Technology, Institutions* (Johns Hopkins Press, Baltimore, 1968).

TABLE 3.2

COMPARISON OF WATER QUALITY GOALS FOR OBJECTIVE SETS 1–5
(Set 5 represents conditions in 1964)

Section

Water Quality Parameter	Set	1	2	3	4	5	6	7	8	9	10	11	12	13	14	15	16	17	18	19	20	21	22	23	24	25	26	27	28	29	30
		Trenton			Bristol		Torresdale	Philadelphia (Camden) →										Chester		Wilmington					New Castle						Liston Point
Dissolved oxygen, mg/l, summer average	1	6·5						6·5		5·5		4·5						4·5	5·5		5·5	6·5						6·5	7·5		7·5
	2	5·5						5·5	4·0											4·0	5·0							5·0		6·5	
	3	5·5						5·5	3·0												3·0	4·5						4·5	6·5		6·5
	4	4·0						4·0	2·5													2·5		3·5						5·5	
	5		7·0		5·1			5·8							1·0					1·0			4·2								7·1
Chlorides, mg/l, max. 15-day mean	1														50		250														
	2													50		250															
	3													50		250															
	4												50		250																
	5									50		100		250	400		1,340			2,400											
Coliforms, #/100 ml, 5/30–9/15, Monthly geometric mean	1	4,000[a]						4,000[a]	5,000[a]									5,000[a]			5,000[b]				4,000[a]						4,000[a]
	2	4,000[a]			4,000[a]			5,000[a]	5,000[b]																	5,000[b]					4,000[a]
	3			4,000[a]			5,000[b]																			5,000[b]					4,000[a]
	4	5,000[a]																								5,000[b]					4,000[a]
	→ 5		2,600		2,700		6,800		25,000						63,000		66,000		51,000		22,000		7,000					1,900			700
Turbidity, turbidity units, 5/30–9/15	1	N.L.						N.L.		N.L.+30										N.L.+30	N.L.										N.L.
	2	N.L.				N.L.		N.L.+30																		N.L.+30				N.L.	
	3	N.L.						N.L.+30																		N.L.+30				N.L.	
	4	N.L.+30																								N.L.+30				N.L.	

	Set	1	2	3	4	5	6	7	8	9	10	11	12	13	14	15	16	17	18	19	20	21	22	23	24	25	26	27	28	29	30
*p*H, *p*H units, desirable range, 5/30–9/15	1				7·0-8·5				6·5-8·5												6·5-8·5						7·0-8·5				
	2		7·0-8·5				6·5-8·5																	6·5-8·5					7·0-8·5		
	3		7·0-8·5				6·5-8·5																				6·5-8·5			7·0-8·5	
	4	6·5-8·5					6·5-8·5			Present levels								Present levels									6·5-8·5			7·0-8·5	
Present range →	5			7·0-8·7					6·9-7·6						6·6-7·3					6·4-7·0						5·6-7·6					6·1-7·8
Alkalinity, mg/l, desirable range	1	20-50							20-50			20-120																	20-120		
	2	20-50							20-50			20-120																	20-120		
	3	20-50							20-50			20-120																		20-120	
	4	20-50							20-50			Present levels																		Present levels	
Present range →	5			25·51					33-46						34-50					13-41						4-25					10·49
Hardness, mg/l, monthly mean	1	95							95	150									150												
	2	95							95	150					150																
	3	95							95	150					150																
	4	95							95	150		150																			
	5			83								122							467												
Phenols, mg/l, monthly mean	1	·001														·001		·01													·01
	2	·001					·001		·005			·005	·01																		·01
	3	·001					·001		·005			·005	·01																		·01
	4	·005										·005	·01																		
	5		·01		·02		·03		·04			·03		·05			·05		·06												
Section		1	2	3	4	5	6	7	8	9	10	11	12	13	14	15	16	17	18	19	20	21	22	23	24	25	26	27	28	29	30

[a] Maximum level.   [b] Monthly geometric mean.   *N.L.* = Natural levels.

Based on: Kneese and Bower, *Managing Water Quality: Economics, Technology, Institutions*.

anadromous fish migrations, there is comparatively little difference in the survival probability under objective sets 2 and 3. Under the waste-loading conditions envisioned for objective set 3, the estimated survival 24 years out of 25 would be at least 80 per cent – compared with 90 per cent for set 2.

Objective set 4. This provides for a slight increase over 1964 levels in water-contact recreation and fishing in the lower sections of the portion of the estuary studies. Generally, water quality is improved slightly over 1964 conditions and the probabilitiy of anaerobic conditions occurring is greatly reduced.

Objective set 5. This would maintain 1964 conditions in the estuary. It would provide for no more than a prevention of further water quality deterioration.

(f) *Costs of alternative programmes*. The costs of achieving objective sets 1 through 4 by various combinations of waste discharge reduction at particular outfalls for the waste-load conditions expected to prevail in 1975–80 are shown in Table 3.3. The range in costs as between the various treatment strategies is discussed further below.

TABLE 3.3

SUMMARY OF TOTAL COSTS OF ACHIEVING OBJECTIVE SETS 1, 2, 3 AND 4

(Costs include cost of maintaining present (1964) conditions and reflect
waste-load conditions projected for 1975–80)
Flow at Trenton = 3,000 cu. ft per sec.
($m. 1968)

Objective set	Uniform treatment			Zoned treatment			Cost minimisation		
	Capital costs	O&M costs[a]	Total costs	Capital costs	O&M costs[a]	Total costs	Capital costs	O&M costs[a]	Total costs
1	180	280 (19·0)	460[b]	180	280 (19·0)	460[b]	180	280 (19·0)	460[b]
2	135	180 (12·0)	315[c]	105	145 (10·0)	250[c]	115	100 (7·0)	215[c]
3	75	80 (5·5)	155[c]	50	70 (4·5)	120[c]	50	35 (2·5)	85[c]
4	55	75 (5·0)	130	40	40 (2·5)	80	40	25 (1·5)	65

[a] Operation and maintenance costs, discounted at 3 per cent, twenty-year time horizon; figures in parentheses are equivalent annual operation and maintenance costs in millions of dollars per year.

[b] High-rate secondary to tertiary (92–98 per cent removal) for all waste sources for all programmes. Includes in-stream aeration cost of $20 million.

[c] Includes $1–$2 million for either sludge removal or aeration to meet goals in river sections 3 and 4.

Based on: Kneese and Bower, *Managing Water Quality: Economics, Technology, Institutions*.

(g) *Benefits of improved water quality.* The Delaware estuary comprehensive study pioneered by broadening the range of benefits considered in the water quality planning process and by introducing quantitative estimates of recreation benefits (reduced external costs) into the process. While the benefit figures were necessarily rather rough, they appear to be sufficiently accurate to comprise a general guide to the decision-making process.

Three general categories of recreation benefits were considered: (1) swimming, (2) boating, (3) sport fishing. Analyses conducted at the University of Pennsylvania, and based on a highly simplified model of recreation participation, indicated a large latent recreation demand in the estuary region. Another study, separately sponsored, tended to confirm the order of magnitude of the estimates [5]. In computing the monetary values associated with recreation demand under each objective set, a number of factors were considered, including recreation-bearing capacity of the estuary as influenced by improved quality. A range of benefits was calculated by the application of alternative monetary unit values to the total use projected for the estuary. The analyses indicated that the increase in the present value of direct quantifiable recreation benefits for set 1 would range between $160 million and $350 million, for set 2 between $140 million and $320 million, for set 3 between $130 million and $310 million, and for set 4 between $120 million and $280 million. Since municipal and industrial benefits were deemed to be small and to some extent

TABLE 3.4

COSTS AND BENEFITS OF WATER QUALITY
IMPROVEMENT IN THE DELAWARE ESTUARY AREA[a]
($m.)

Objective set	Estimated total cost	Estimated recreation benefits	Estimated incremental cost		Estimated incremental benefits	
			minimum[b]	maximum[c]	minimum[b]	maximum[c]
1	460	160–350				
			245	145	20	30
2	215–315	140–320				
			130	160	10	10
3	85–155	130–310				
			20	25	10	30
4	65–130	120–280				

[a] All costs and benefits are present values calculated with 3 per cent discount rate and twenty-year time horizon.

[b] Difference between adjacent minima.

[c] Difference between adjacent maxima.

Based on: Kneese and Bower, *Managing Water Quality: Economics, Technology, Institutions.*

cancelled by negative features in regard to industrial water use (higher D.O. causes corrosion in cooling equipment), these ranges were taken to be rough estimates of the total benefits from improved water quality in the estuary.

A comparison of the recreation benefits with the cost estimates (Table 3.4) shows that objective set 4 appears to be justified, even when the lowest estimate of benefit is compared with the highest estimate of cost. The incremental costs of going from set 4 to set 3 suggest that the justifiability of set 3 is marginal. On the assumption that some of the more widely distributed benefits of water quality improvement may not have been appropriately taken into account, it can probably be justified. Clearly, however, the incremental benefits of going to sets 2 and 1 are vastly outweighed by the incremental costs. The reader will have noticed in both Tables 3.3 and 3.4 that the cost estimates cover a tremendous range. I shall comment further on that shortly.

(h) *Effluent charges on the Delaware estuary*. Following the planning study just described, another important study was undertaken using the same models and data. It concerned itself with the possible use of effluent taxes or charges as an economic incentive for controlling waste discharge. The study was done in connection with the work of a special interdepartmental committee on water quality control headed by Gardner Ackley, at the time Chairman of the President's Council of Economic Advisers. The written reports were prepared by the Federal Water Pollution Control Administration headquarters staff, primarily by Edwin Johnson [6, 7]. An excellent discussion of the effluent charges study is also found in Schaumberg [8].

Assuming that direct controls would be effective and that waste dischargers would respond rationally to economic incentives, the study analysed four programmes for achieving alternative dissolved oxygen objectives in the estuary.

The first, and in a sense a standard of comparison for the others, is the Least-Cost Linear Programming Solution (L.C.) – the low-cost figures in Tables 3.3 and 3.4 correspond to it. This solution uses the mathematical programming technique described earlier to obtain the minimum-cost distribution of waste removals. To implement this programme with the usual direct regulation as a control policy would require precise information on waste-treatment costs at all outfalls and direct controls on all waste discharges. It would result in radically different levels of treatment and treatment costs at different outfalls. The reason is simply that it would concentrate treatment at those points where the critical oxygen sag can be reduced most inexpensively.

The second is the Uniform Treatment Solution (U.T.). In this solution each waste discharger is required to remove a given percentage of the wastes previously discharged before discharging the remainder to the stream. The percentage is the minimum needed to achieve the D.O. standard in the stream and is the same at each point of discharge. This solution may be considered typical of the conventional administrative effluent standards approach to the problem of achieving a stream quality standard – the high-cost estimates in Table 3.4 correspond to it.

The third is the Single Effluent Charge Solution (S.E.C.H.). This solution involves charging each waste discharger in the estuary the same price per unit of waste discharge. The solution examines responses of individual waste dischargers and identifies the minimum single charge which will induce sufficient reduction in waste discharge to achieve the standard. Unfortunately, only treatment was permitted as a response to the charge in this study. Had process changes and by-product recovery been admitted, the costs of obtaining the objective would have been reduced. Still, there is no reason to think that the relative costs of the alternative strategies would be changed much.

The fourth is the Zone Effluent Charge Solution (Z.E.C.H.), which used a uniform effluent charge in each zone instead of a uniform charge over all reaches of the estuary. In Fig. 3.1 the zones are designated by roman numerals.

Among the general conclusions which the F.W.P.C.A. staff drew from this study were the following:

1. Effluent charges should be seriously considered as a method for attaining water quality improvement.
2. Costs of waste treatment induced by an appropriate charge level will approach the least costly treatment plan.
3. A charge level of 8 to 10 cents per pound of oxygen-demanding material discharged appears to produce relatively large increases in critical dissolved oxygen levels.
4. Major regional economic readjustments from a charge of that level are not anticipated to occur in the study area. In all but a few cases the total cost (cost of treatment plus effluent charge) was less than 1 per cent of the value of output. In most cases it was a small fraction of 1 per cent ([7] Table 8). Another study [9] shows roughly similar results with costs of waste treatment averaging approximately 1 per cent of value added in several industries.
5. Compared with a conventional method of improving water quality, the charge method attains the same goal at lower costs

of treatment, with a more equitable impact on waste dischargers. The F.W.P.C.A. staff did not state what criteria of
equity they were using, but presumably they regarded a charge
proportioned to the use of assimilative capacity by a waste
discharger as equitable. Also, the charge furnishes a continuing
incentive on the discharger to reduce his waste discharges and
provides a guide to public investment decisions.

(i) *Efficiency considerations.* Table 3.5 indicates the economic
costs associated with the programme for two levels of water quality.
The 3–4 p.p.m. standard approximately coincides with objective set
3 in Table 3.1 – the highest objective set which appears to be justifiable on economic efficiency grounds.

TABLE 3.5

COST OF TREATMENT UNDER ALTERNATIVE
PROGRAMMES

D.O. objective (p.p.m.)	L.C.	Programme U.T. ($m. per year)	S.E.C.H.	Z.E.C.H.
2	1·6	5·0	2·4	2·4
3–4	7·0	20·0	12·0	8·6

The analysis indicates that the effluent charges system would
produce the specified quality levels at about one-half the social cost
of the uniform treatment method. Especially at the higher quality
level, the cost saving is of a highly significant magnitude. The
present value of the cost stream saved is of the order of $150 million.

The least-cost system is capable of reducing costs somewhat
further than either the uniform or the zoned charge, since it programmes waste discharges at each point specifically in relation to the
cost of improving quality in the critical reach, but this comes at the
cost of detailed information on treatment costs at each point and a
distribution of costs such that some waste dischargers experience
heavy costs and others virtually none. The least-cost system is
closely approached by Z.E.C.H. at the higher quality level. In effect,
this zone charge procedure 'credits' waste dischargers at locations
remote from the critical point with degradation of their wastes in the
intervening reach of a stream before they arrive at the critical reach.
This is a necessary condition for full efficiency when effluent charges
are used to achieve a standard at a critical reach in a stream. The
reason that the Z.E.C.H. does not achieve quite the same efficiency
as L.C. is that the 'credit' is not specific to the individual waste
discharger but is awarded in blocks – three in this case.

Another way of putting this is that equalising marginal waste
water reduction cost at all outfalls is strictly speaking a necessary

condition for cost minimisation only when a homogeneous 'lump' of assimilative capacity is being allocated – or, more formally, when all the coefficients in the transfer matrix are identical. When they are not, thoroughgoing cost minimisation requires that prices be 'tailored' for each outfall. This explains why the solution based on a single charge only approaches but does not reach the programmed cost-minimisation solution. How closely it will approach is an empirical question relating to the magnitude of the a_{ij}'s. To see this, assume two industrial dischargers with the following cost functions for reducing waste discharge:

$$c_1 = f(x_1) \tag{1}$$
$$c_2 = f(x_2) \tag{2}$$

where x_1 = waste discharged from plant 1
x_2 = waste discharged from plant 2
c_1 and c_2 are costs of reducing waste discharge.

Assuming reach 6 is the 'critical' reach and recalling the meaning of the elements in the transfer matrix:

$$R_6 = a_{61}x_1 + a_{62}x_2 \text{ (i.e. 'binding' constraint)}$$

where

$$R_6 = \text{the 'standard'} \tag{3}$$

Form the Lagrangean:

$$L = c_1 + c_2 + \lambda(R_6 - a_{61}x_1 - a_{62}x_2) \tag{4}$$

At optimum:

$$\frac{\partial L}{\partial x_1} = \frac{dc_1}{dx_1} + \lambda(-a_{61}) = 0 \tag{5}$$

$$\frac{\partial L}{\partial x_2} = \frac{dc_2}{dx_2} + \lambda(-a_{62}) = 0 \tag{6}$$

If we include the constraint $R_6 = a_{61}x_1 + a_{62}x_2$, we have three equations and three unknowns (x_1 x_2 and λ). Solving for λ, we get:

$$\lambda = \frac{1}{a_{61}} \frac{dc_1}{dx_1} \tag{7}$$

and

$$\lambda = \frac{1}{a_{62}} \frac{dc_2}{dx_2} \tag{8}$$

Note that:

$$\frac{1}{a_{61}} \frac{dc_1}{dx_1} = \frac{1}{a_{62}} \frac{dc_2}{dx_2}$$

or

$$\frac{dc_1}{dx_1} = \frac{a_{61}}{a_{62}} \frac{dc_2}{dx_2}$$

Note also that $\lambda \neq 0$ unless either dc_1/dx_1 or $dc_2/dx_2 = 0$. Because both equations (5) and (6) are equal to zero, they may be set equal to each other:

$$\frac{dc_1}{dx_1} = \frac{dc_2}{dx_2} + \lambda(a_{61} - a_{62})*$$

(j) *Further comments on the effluent charge approach.* At an average effluent charge of 10 cents per pound of B.O.D., which the staff estimated would be needed for the zoned effluent charge programme ([6] Table 6), the funds collected by the administrative agency would amount to about \$7 million per year. Nevertheless, for the 3–4 p.p.m. D.O. objective, the total cost to industry and municipalities as a whole – effluent charge plus cost of treatment – is about the same as the cost of treatment only under the uniform treatment programme. About half this outlay does not represent an actual resources cost but rather a rental-type payment for the use of assimilative capacity.

It should be noted that an important efficiency advantage of the effluent charges programmes as contrasted with the L.C. programme is their relatively lesser demand for information and analytical refinement. A study of the type already performed for the Delaware estuary could serve as the basis for an effluent charge scheme. An order of magnitude estimate of the required charge is provided and changes could be made if necessary as responses to the charge reveal themselves. Actually, since the costs do not take account of the possibility of process change in industry, which is often cheaper than 'end of pipe' treatment, the 10 cents per pound of B.O.D. charge is probably too high and could be adjusted downward at a later point. Also, the charge provides a continuing incentive for the discharger to reduce his waste load by placing him under the continuing pressure of monetary penalties. He is induced to develop new technology and, as it develops, to implement it. As new technology develops, the effluent charge could be gradually reduced while the stream standard is maintained, or the standard could be allowed to rise if this is deemed desirable. The process of induced technology may produce particularly striking results in this field since the waste-assimilative services of watercourses have heretofore been completely unpriced.

The direct control measures implicit in the L.C. programme, on the other hand (as well as the effluent standard of the uniform treatment programme), provide only a more limited incentive to improved technology. Moreover, the minimum-cost programme would require not only detailed information on current cost levels at each individual outfall, but also information on changes in cost with changing

* I am indebted to my associate Walter Spofford for developing this demonstration.

technology in regard to industrial processes, product mix, treatment cost, etc., and would be extremely inequitable in its cost distribution.

(k) *Concluding statement on the Delaware study.* The Delaware estuary survey was a pioneering study of water quality management and is of continuing importance. Some of the tools it developed are in use by the Delaware River Basin Commission, and it is still frequently referred to in discussions of policy in the United States. It was the first study to embody, at least in a rudimentary form, an ecological model into an economic optimisation framework, and it provided an illuminating analysis of several policy options including effluent charges. It did have some major deficiencies, however. Among them are that it examined only a very limited range of technological alternatives for managing water quality (i.e. it did not explore the production function extensively) and that it did not incorporate an analysis of the stochastic aspects of water quality. These are among the matters which were addressed in the Potomac study to which I now turn.

3.4.3 THE POTOMAC STUDY

(a) *Introduction.* A 1961 amendment to the Federal Water Pollution Control Act of 1948, which, with minor exceptions, was the first step by the Federal Government into what had been an exclusive area of state sovereignty, is our starting-point for discussion of the Potomac case. The feature of the 1961 Act which is salient to our purposes is that it provided for the inclusion of storage in Federal multi-purpose reservoirs to augment low flows for the purpose of improving water quality. This opened the door to the possibility of including technological options other than treatment in the Federally sponsored water quality planning process.

In general, the costs of such storage were to be considered non-reimbursable – that is, no cost assessments or prices for the service are to be levied on the beneficiaries. Referring back to the Streeter–Phelps equations discussed above, it is clear that low-flow augmentation is capable of affecting both the dilution of waste and re-aeration in such a way as to improve water quality.

Pursuant to a resolution of Congress in 1936, the U.S. Army Corps of Engineers submitted the *Potomac River Basin Report.* This report was the first one in which a Federal water resource agency submitted a 'comprehensive' plan in which water quality management was the major consideration. The plan took a quality objective stated in physical terms (parts per million of dissolved oxygen) as given and recommended a programme of waste treatment and low-flow regulation to meet this objective in the future. While it was a pioneering effort, the benefit evaluation technique, based on the 'alternative cost procedure', was grossly deficient, and the range of alternatives considered for water quality improvement was still very narrow. Basically, the Corps of Engineers limited its planning to consideration of those quality improvement facilities which could clearly be implemented through existing governmental institutions. As we shall see, this constrained the range of choice greatly and led to the recommendation of a set of facilities which was far from the least costly which could have been devised to achieve the stated water quality objective.

A later Resources for the Future study, a full report of which can be found in Davis [10], which forms the basis for my further discussion, used the Corps of Engineers data plus considerable additional information to define further the range of alternatives for water quality management in the Potomac estuary in the neighbourhood of Washington, D.C., the locus of most of the water quality problems.

(b) *The basin and its problems.* The watershed of the Potomac, an area of about 14,000 square miles, lies in portions of four states and the District of Columbia. About three-fourths of the population in the basin is found in the Washington metropolitan area. This area, which extends beyond the District of Columbia into Maryland and Virginia, already has a population of over 2 million persons and is one of the most rapidly growing metropolitan areas in the nation. The Washington area lies near the head of the Potomac estuary, which is heavily used for recreation. Water supply for the area is taken from the Potomac river above the estuary. The estuary periodically experiences low dissolved oxygen – a condition which could get much worse as waste loads from the metropolitan area and upstream sources mount. The Corps of Engineers, as part of its planning effort, projected water demands and waste loads to the year 2010. One of the central objectives of the plan was to control the effects of waste loads expected to prevail then. Among the planning assumptions was that the maximum feasible control of waste loads would result from conventional secondary sewage treatment (about 90 per cent B.O.D. removal).

The plan made public in 1963 recommended the construction by the year 2010 of 16 major reservoirs in the Potomac basin and more than 400 headwater structures (see Fig. 3.2). These were meant to meet projected water supply, water quality and flood control objectives. Of the 16 major reservoirs, 10 were planned to meet projected upstream objectives for the low-flow regulation for water supply and quality control. At the same time, this group of reservoirs would provide a sufficiently higher sustained flow at Washington to meet

the projected municipal water diversions there. The remaining group
of six reservoirs (providing 60 per cent of the proposed yield – 2340
cu. ft. per sec. out of 3931 cu. ft. per sec.) was designed to augment
flows into the estuary sufficiently to maintain 4 p.p.m. of dissolved
oxygen (24-hour monthly mean for the minimum month). The

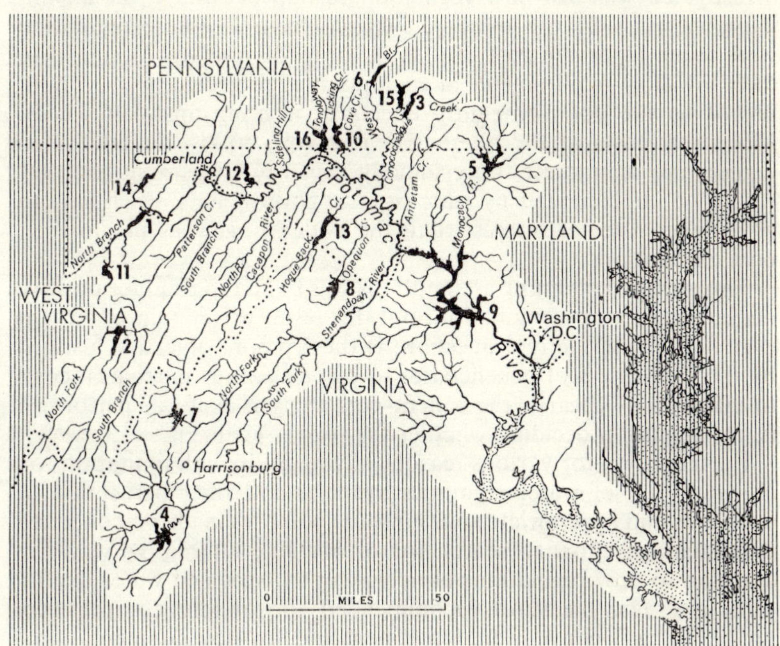

Fig. 3.2   Major reservoirs in recommended plan for Potomac basin
From: R. K. Davis, *The Range of Choice in Water Management*
(Johns Hopkins Press, Baltimore, 1968)

projected storage was based on counteracting residual 2010 waste
loads and an assumed replication of the historical record of flows
into the estuary (i.e. the mass curve approach was used to determine
required storage to sustain flows calculated to be needed to offset
residual waste loads to the extent of maintaining 4 p.p.m. D.O.).
From a statistical point of view, the mass curve approach has great
deficiences which were discussed in connection with our consideration
of stochastic hydrology in section 3.3.3 above.

A benefit–cost analysis was presented which indicated that the
benefits from flow regulation would outweigh its costs. But this
analysis did not do what benefit–cost analysis is intended to accomplish – assist in deciding whether a proposed investment is socially

worth while. The pre-set physical quality objective of 4 p.p.m. of D.O. was taken as given in the planning. It was assumed that treatment of the sewage from the Washington metropolitan area could not succeed in removing more than 90 per cent of the waste load. This is insufficient to meet the objective. The only alternative seriously investigated was low-flow regulation to improve the waste-assimilative capacity of the estuary. Benefits from low-flow regulation were taken to be the cost of a single-purpose reservoir designed to meet low-flow requirements at each point of projected need without regard to the complementary effects of meeting upstream needs. Moreover, costs of the alternative reservoirs were calculated at a higher rate of interest – presumably because it was assumed they would be implemented by state or local governments which have to pay higher rates of interest than the Federal agencies use in their own calculations.

This string of planning assumptions was bound to produce positive net benefits for flow regulation. But this result was obtained without even addressing the real question: is the 4 p.p.m. objective justified by the willingness of beneficiaries to pay at least as much as it costs to maintain it? Clearly, the benefit–cost analysis as it was performed is not helpful in deciding whether the plan is justified.

Furthermore, for various reasons having to do with limitations on the authority of water resources agencies and their perception of problems and appropriate solutions, the planners made no concerted or systematic effort to search for or evaluate alternative ways (in addition to flow regulation) of achieving the specified water quality objective. The measures recommended for quality control were limited to basic conventional treatment and low-flow augmentation – measures which could clearly be implemented by the Federal and local government agencies which are the traditional purveyors of water services in the United States. To have implemented a programme embodying the less conventional measures which our later study demonstrated would have entered into a cost-minimising solution that would have required institutional change. The possibility of such change was not contemplated in the planning process.

(c) *Searching for additional alternatives.* The subsequent Resources for the Future study showed that including in the plan certain collective measures which no existing agency had a clear authority or incentive to finance, construct and operate would have entered into a least-cost system for meeting the oxygen objective. Had such a system served as the alternative for benefit analysis, net benefits for flow regulation, at least for the larger-scale reservoir systems, would have been grossly negative. But, in general, the alternative cost-benefit estimation procedures are inappropriate in cases like this.

In going through the analysis, the Corps assumption that 90 per cent treatment would be given to wastes in the Washington metropolitan area was (somewhat arbitrarily) used as a baseline. The costs of other alternatives were then weighed off against the incremental cost of flow regulation for counteracting the residual oxygen deficit.

Costs were obtained for various levels of low-flow regulation by scaling down by various amounts the Corps's proposed low-flow regulation system. The scaled-down systems were roughly optimised by using a computer simulation program which permitted the historical trace of hydrology to be regulated by the various systems. In the initial instance, historical hydrology was used to maintain close comparability with the Corps results. In computing costs for the successively smaller reservoir systems, account was taken of the difference in flood-damage reduction and recreation services realised by any scaled-down system in comparison with the full proposed *Corps of Engineers* system.

With the assistance of consultants, costs of several alternative ways of equivalently offsetting the waste load were developed. These included processes for further treatment of the waste load (microstraining, step aeration, chemical polymers, powdered carbon and granular carbon); costs for effluent distribution via pipeline along the estuary to use its naturally occurring assimilative capacity better; and reoxygenation of the estuary which, like low-flow regulation, improves assimilative capacity.

Computer simulation of the effects of these processes in view of variations of river flow (using the historical hydrologic trace) into the estuary show that they need to be operated on the average only 3·5 months per year in order to meet the D.O. objective. Because the alternative systems are high in operating cost and low in capital cost relative to low-flow regulation, they can be comparatively efficient if operated only as needed, but would not be competitive if operated continuously, thereby overshooting quality goals most of the time. Accordingly, they could only enter efficiently into the quality management system if institutional means existed for carefully articulated design and operation in conjunction with other elements of the system.

Establishing combinations which would meet the standard required that if one process was reduced, another had to be equivalently increased. It was possible to use computer simulation to exhaust all possible combinations of the feasible and sufficient processes, given the relatively large increments defined for them. The computer program gave a complete listing and cost ranking of all systems – some 300 in all.

A sampling of alternative feasible and sufficient systems is shown in Table 3.6.

TABLE 3.6

SYSTEM COSTS BY GENERAL CLASS OF PROCESS
COMBINATIONS, 3·5 MONTHS OPERATION

(Present worth, 50-year period, 4 per cent discount)

($m.)

Alternative systems	*Present worth*
1. Reoxygenation	20
2. Chemical polymers and reoxygenation	22
3. Step aeration and reoxygenation	25
4. Microstrainers and reoxygenation	28
5. Diversion and reoxygenation	33
6. Diversion, waste treatment and reoxygenation	45
7. Low-flow augmentation and reoxygenation	60
8. Low-flow augmentation, reoxygenation and waste treatment	60
9. Low-flow regulation	140

Adapted from: Davis, *The Range of Choice in Water Management.*

This analysis shows that many combinations of processes could achieve the objective at less cost than the proposed system based upon conventional treatment and flow regulation. It is notable that all of them except the flow regulation alternative would require the construction and closely articulated operation of facilities which have not traditionally been in the purview of either the Federal or local government (particularly reoxygenation and regional effluent distribution works). Another salient point is that while low-flow regulation is vastly more costly than reoxygenation or some of the other alternatives from the point of view of the people in the basin, it costs much less. Low-flow regulation for water quality improvement has been, as previously mentioned, a fully non-reimbursable purpose of Federal water development in the United States, while no subsidy at all is available for measures like reoxygenation and waste diversion. Thus, the fact that Federal water development policy is such that certain measures for development confer large subsidies on a region while others do not can also contribute to choices among alternatives which are distorted from a broader economic point of view.

Both these factors were undoubtedly implicit considerations in the plan recommended by the Corps. But by means of economic systems analysis which does not operate only within the existing institutional and policy restraints, it is possible to examine them and thus provide information on the desirability of institutional change. Such examination of institutional constraints should clearly be part of the planning process. In the case of the Potomac it appears that much could be gained by institutional arrangements permitting the planning design and operation of quality management systems embodying a wide range of alternatives.

(d) *Stochastic aspects*. So far, in my discussion of cases, I dealt with deterministic models. These, quite imperfectly, recognise the variability of river flow by specifying some low flow below which flow is likely to drop only with low probability. The Delaware estuary study took this approach and so did the Corps study of the Potomac, as well as much of the Resources for the Future follow-on study. But the availability of the Potomac reservoir simulation model made it possible to study some aspects of the probability question in a more explicit way. It is one of the major disadvantages of optimisation models (which otherwise have great advantages) such as that used on the Delaware that it is very difficult to incorporate stochastic aspects into them.

The Corps of Engineers based its design on the specification that D.O. concentrations in the estuary would not fall below 4 p.p.m. based on the 24-hour monthly mean for the minimum month. A standard kind of mass curve analysis was used to check that the yield of the proposed reservoirs system would be sufficient to meet the objective. As has already been noted, this analysis makes the statistically untenable assumption that future stream flow will be a replication of the past.

To help illuminate the probability aspect of the water quality standards conventionally used in planning, a stochastic hydrology was generated for long periods of time and applied to the reservoir simulation programme.

Table 3.7 presents some results on the implication of different probabilities of violating 2 and 4 p.p.m. D.O. levels when different systems of reservoirs are operated to achieve D.O. targets. In this presentation the Corps assumption that low-flow regulation is the only means used to counteract the residual deficit from 90 per cent

TABLE 3.7

PERCENTAGE OF TIME MONTHLY MEAN D.O. IS LESS THAN 2·0 p.p.m. FOR 500-YEAR TRIALS AT DIFFERENT D.O. TARGET AND SYSTEM CAPACITIES

Storage capacity (acre-ft)	D.O. target	% time < 2 p.p.m.	% time < target
82,000	2 p.p.m.	0·25	0·25
140,000	2 p.p.m.	0·03	0·03
600,000	4 p.p.m.	0·35	3·30
770,000	4 p.p.m.	0·22	1·03
970,000	4 p.p.m.	0	0·33

Based on: Davis, *The Range of Choice in Water Management*.

treatment is used. This is done to spell out clearly the implication for reservoir storage, even though in an optimised or least-cost system the incremental costs of achieving lower probabilities of violation would be less. It is interesting to note two main points emerging from this analysis:

1. Reducing the probability of violating the 4 p.p.m. objective from 3·30 per cent to 0·33 per cent costs about 370,000 acre-ft in storage and around $50 million. The 0·33 per cent level is about equivalent to the objective used in the Corps study.
2. If a system is operated to avoid violation of a 2 p.p.m. target, about the same low level of violation of the 2 p.p.m. level can be achieved with 82,000 acre-ft of capacity as with 770,000 acre-ft of capacity in a system operated to achieve a 4 p.p.m. target. Thus, what the level of the standard is meant to accomplish becomes a profoundly important question when reservoirs are used in a water quality management system. Put in another way, how much security concerning oxygen not falling to very low levels are we willing to sacrifice to keep it at higher levels more of the time? It is not necessarily true that a system operated to achieve high levels as much of the time as possible will provide greater security against extreme failure than a smaller system operated to achieve lower levels as much of the time as possible. In fact, in the Potomac instance quite the opposite was true. Water was released for the higher objective, and when it was exhausted the target was missed by large margins.

What this analysis has shown is that the probability aspect of standard-setting, which is usually treated by simple engineering rules of thumb, actually contains important valuation problems. The probability statement is just as important as the level in terms of its cost–benefit implications.

(e) *Conclusions from the Potomac study*. The Resources for the Future Potomac study accomplished three main things:

1. For the first time it revealed how wide a range of technical options is available for water quality management and how constraining it is to confine consideration of alternatives to those historically included in the missions of specific agencies. This conclusion has since been confirmed by a number of other regional studies, including those of the Miami basin in Ohio and the Wisconsin river, as well as by the historical experience of the Ruhr area *Genossenschaften* which is discussed in another paper at this conference.

2. It showed that the alternative cost–benefit evaluation procedure, justifiable in some specific situations, can be more deceptive than helpful when applied to water quality planning.
3. It illustrated the importance of the stochastic aspects of the problem.

As a result of this study and other objections to the Corps plan, it has been abandoned. A planning process is now going on in the basin which is much less limited in its consideration of alternatives.

3.5 *CONCLUDING DISCUSSION*

Work on water quality management during the 1960s, including several careful case studies, has made a compelling case for the efficiency of regional water quality management incorporating a wide array of technological options. This could probably best be done by regional river basin agencies having continuous management responsibilities and an intimate knowledge of their watercourses, including their stochastic aspects. This research has also pointed towards the efficiency (and equity) of effluent charges as a device for controlling the generation and discharge of waste waters at individual points and (implicitly) for financing the implementation of other alternatives.

How useful and influential has this research been? Here the picture is decidedly mixed. Some successes can be reported. The national water legislation of France is based directly on it and that of Canada was visibly influenced by it. Its results are discussed in connection with developing policy and management institutions in England, Germany and Sweden that I know of. Several international agencies such as the O.E.C.D. and E.C.E. are cognisant of it and it is helping to shape their environmental programmes. There has been some noticeable impact on planning processes. Its results are well known to academic researchers and their work has been influenced by them. Education in the traditional field of sanitary engineering has been significantly altered by this research. In the United States some rudimentary efforts at regional water quality management are under way in particular regions, at least in part, because of this research. Perhaps this is a larger than usual return for a few million dollars of research expenditure.

But one blatant failure of this research to be influential is at the level of national policy in the United States, the very country where it was done. The national legislation which we have and, especially, additional legislation that is pending neglect the regional approach almost altogether. Also, it relies on arbitrary – at least from an economic point of view – direct restrictions or prohibitions to reduce waste discharge. In addition, it contains major subsidies to municipalities and industries for the use of a very specific technology – waste water treatment plants. From an economic efficiency point of view, it is quite simply a nightmare.

Why should this research have so little honour in its own country – at least at the national level? That is a very difficult question for me to answer. It is not for want of communication. The results have been made available to, and interpreted for, members of Congress and their legislative staffs on numerous occasions, and they have been debated on the floor of Congress – the latter without shedding much light on why the legislative process is so impenetrable to them.

Truly, I can only speculate on this question. In part I think it is simply historical. The traditional view of both lawyers and sanitary engineers is that pollution should be subjected to direct regulation by existing governments – and they dominated the process of forming pollution control law in its early stages. The idea of viewing a river basin as an integral hydrological system has not been part of the folklore of sanitary engineers in the United States – as contrasted, for example, with the Ruhr area of Germany (see Chapter 4). And lawyers – well, lawyers are lawyers.

At a later stage in the development of policy and its associated debate, a strange alliance was formed to oppose the economists' approach. On the one hand, 'conservationists' who really wanted to see all waste water discharges stopped (a noble but unrealistic goal) were afraid that effluent charges would be a 'licence to pollute' – that they would not provide the positive control of direct regulation. Since their goal was halting waste discharges entirely, they were also not much interested in regional management. The conservationists were joined in their opposition to effluent charges by representatives of industry, who also argued, with a somewhat hollow ring, that effluent charges would be a licence to pollute. Their motive, I speculate, was to head off an approach which would be really effective and unavoidable, in contrast to the regulatory approach which, as it operated in practice rather than in theory, provided almost endless opportunities for manipulation, delay, and even total avoidance.

But the debate goes on and the story is not yet all told. Senator Proxmire has a bill in the Congress based squarely upon the results of the water quality management research of the 1960s, and Congressman Hamilton has sponsored a similar one in the House. Although these men have been joined by a number of prominent senators and congressmen in sponsorship, their bills have very little immediate chance for success, at least in part because of the peculiarities of the committee structure in the U.S. Congress. But the bills are providing an important educational function in the congress and are a focal point for outside supporters of the economic approach. Interestingly, these now include a number of conservation groups who have learned that the direct regulation approach is anything but positive and sure in its impact and who have come to believe that economic incentives will be more effective. Also, there is a chance that the economic incentive approach will get a boost from the related field of air pollution management where the national policy approach is still much more fluid. The President has proposed a national sulphur oxide emissions charge or tax as part of his environment programme. In addition, Senator Proxmire has introduced a bill to impose a stiffer charge than proposed by the Administration and has been joined in sponsorship by a large group of senators. So, the debate goes on.

REFERENCES

[1] Kneese, A. V., *Water Pollution: Economic Aspects and Research Needs*, 4th ed. (Resources for the Future, Washington, D.C., 1970).

[2] Fair, G. M., Geyer, J. C., and Okun, D. A., *Water and Wastewater Engineering* (Wiley, New York, 1968) vol. 2.

[3] Sobel, M. J., 'Water Quality Improvement Programming Problems', *Water Resources Research*, vol. 1, no. 4 (1965) pp. 477–87.

[4] Fiering, M. B., *Streamflow Synthesis* (Harvard Univ. Press, Cambridge, Mass., 1967).

[5] Davidson, P., Adams, F. G., and Seneca, J., 'The Social Value of Water Recreation Facilities Resulting from an Improvement in Water Quality: The Delaware Estuary', in Kneese, A. V., and Smith, S. C. (eds.), *Water Research* (Johns Hopkins Press, Baltimore, 1966).

[6] Federal Water Pollution Control Administration, *Report on the Effluent Charge Study* (F.W.P.C.A., mimeo, 1966).

[7] Johnson, E. L., 'A Study in the Economics of Water Quality Management', *Water Resources Research*, vol. 3, no. 2 (1967) pp. 291–305.

[8] Schaumberg, G. W., Jr., *Water Pollution Control in the Delaware Estuary* (Harvard University Water Program, mimeo, 1967).

[9] Bramhall, D. F., and Mills, E. S., *Future Water Supply and Demand* (Maryland State Planning Dept., 1965).

[10] Davis, R. K., *The Range of Choice in Water Management* (Johns Hopkins Press, Baltimore, 1968).

4 An Empirical Linear Model for Water Quality Management: Pilot Study for Four Regions in the Ruhr Basin

R. Thoss and K. Wiik

4.1 *OBJECTIVES*

The last few years have seen a lively theoretical discussion on problems of the environment and on quantitative economics during which time important advances have been made. The authors have included some of the most important results of this work in the following model in an attempt to formulate practical guide-lines for the planning of the environment. Since we wish to *apply* the results of this debate to practical planning problems, we have omitted many of the refinements that have been suggested in the theoretical literature because we cannot yet meet the data and computing requirements which would be involved if we included all (the nevertheless relevant) aspects of environmental problems.

The model described here considers only one partial aspect of environmental planning – the management of the quality of the common property resource 'Surface-Water' within an isolated river basin. Our partial approach with respect to one environmental medium, and one isolated area, must not be taken to imply that we purposely wish to neglect the fact 'that the quantities of residuals discharged to air, water, or land are highly inter-dependent so that treating one environmental medium, such as water, in isolation from others can lead to undesirable secondary effects [1]. We are likewise aware of the fact that geographical regions are never completely isolated. There are important linkages between activity levels in different regions, which mean that all decisions affecting regional production levels and production processes invariably lead to repercussions (e.g. multiplier effects) in other parts of the economic system.

On the contrary, we would like to emphasise that the model presented here is only one part of an integrated multi-medium, multi-sectoral, multi-regional model designed to produce guidelines for environmental planning in the Federal Republic of Germany. The total model includes sub-models of agricultural and recreational land use, air pollution and waste materials. We understand [1] that Russell and Spofford have already constructed a similar model for application to the Delaware estuary [2].

Our model is static and refers to 1970, the year of the last German census. All production data refer to the whole year; the ecological data are daily averages.

The four subregions we have selected for this present pilot study are all situated in the watershed of the Ruhr. Our decision to choose part of the Ruhr area as the subject of a water quality management study was taken for the following reasons:

(1) As Fig. 4.1 shows, this big industrial area has been divided into several water territories, in each of which the quality of water supply is managed and controlled by a different water resources association. The borders of those associations are the same as those of the respective river basins.

(2) The political problem of organising water management agencies, which seems to be of primary concern in other parts of the world, has been already solved in the territory discussed in this paper. Kneese and Bower have given a detailed description of the functions and institutional arrangements of the water resources associations in this part of the Federal Republic [3]. Since these associations levy effluent charges, they also produce a large stock of statistical data on water quantities and qualities which can be used in formulating the quantitative ecological relationships for a decision model.

It must, however, be mentioned, that no relevant *economic* data are available for the river basins, since these data have traditionally been organised on a county (Kreis) level. This means that some arbitrariness cannot be avoided when combining economic and ecological data sets, i.e. in assigning counties to river regions. But since the counties in the selected area are relatively small, this causes no major difficulties.

Fig. 4.2 shows that our subdivision of the Ruhr basin is a compromise, resulting from the need to use economic information – available only at the county level – as an indicator of economic activity in each reach of the watercourse and, on the other hand, that we have to take the topographical and hydrological reality into consideration. The regions – each consisting of several counties – are situated along the river and its tributaries. Roughly speaking, they may be defined as:

the upper reach of the Ruhr (region 1),
the tributary Lenne (region 2),
the tributaries Ennepe and Volme (region 3),
the lower reach of the Ruhr (region 4).

The tributaries in regions 1, 2 and 3 are not interconnected. This means that each region discharges its waste water into its own part of the river (or tributary) only. All three drain downstream into

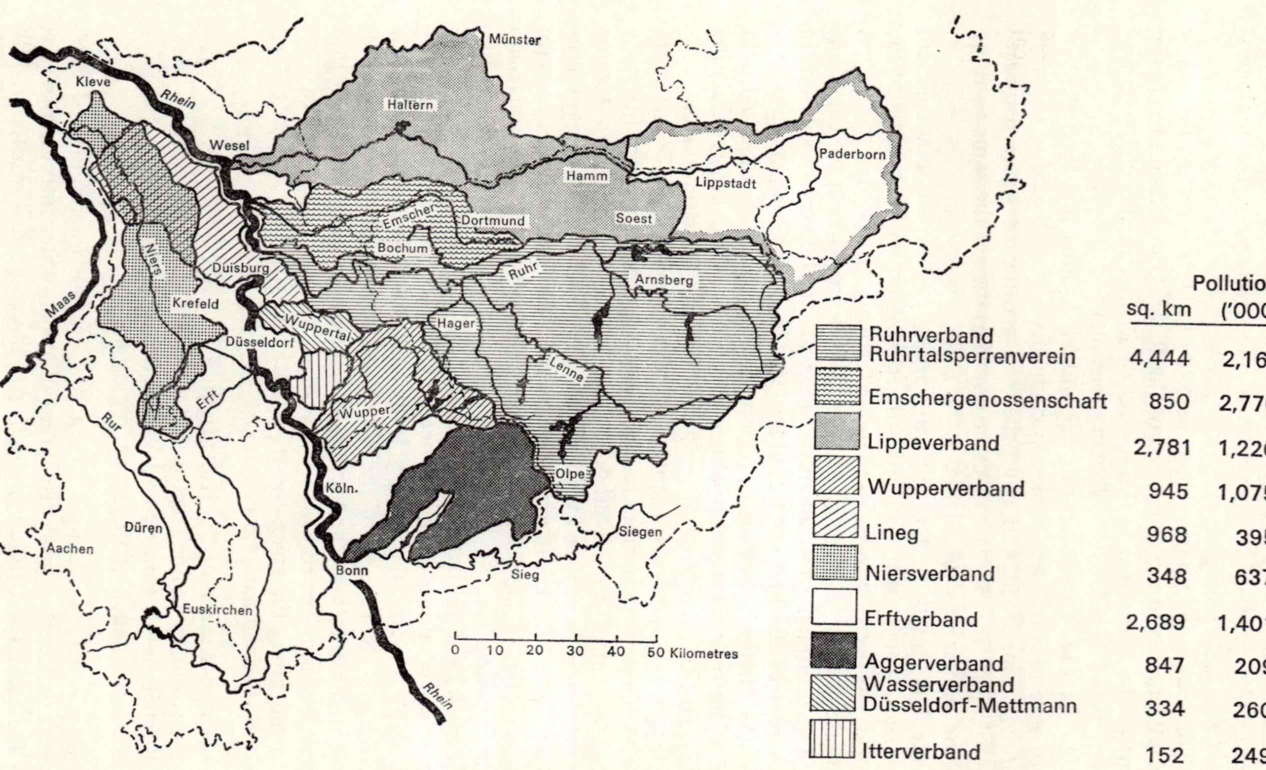

FIG. 4.1 Water resources associations of the Rhine–Ruhr industrial area

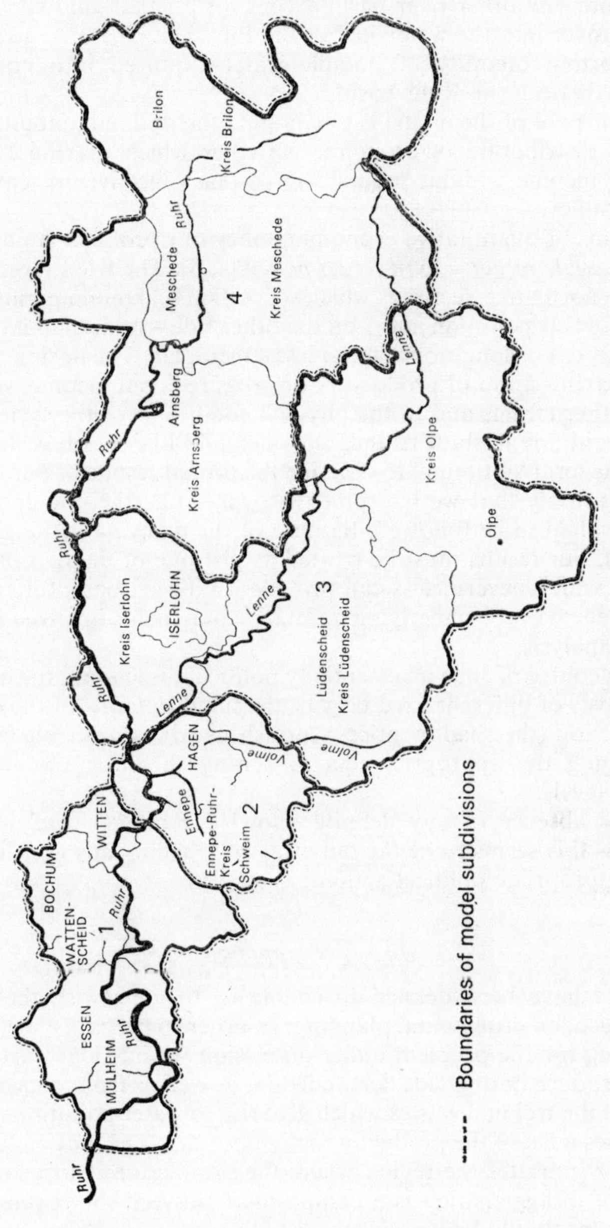

FIG. 4.2   Regional zoning of the investigation area

region 4, rom where all water enters the Rhine. Virtually no sewage passes into any other river basin except for a small amount that is pumped over into the watershed of the Emscher. There is, however, an important quantity of potable water exported into northern watersheds from the Ruhr basin.

The purpose of the model is to compute the optimal amounts and regional distribution of economic activities which maximise gross regional income without prejudicing certain objectives of environmental policy.

In terms of quantitative economic policy our problem is one of a mixed *flexible target – fixed target policy* [4, 5]. The fixed targets are made by normative decisions which set certain maximum permissible levels of water pollution and – on the other side – minimum required standards of consumption, capital investment, etc. The flexible target is the maximisation of production or gross regional income, subject to the other targets and to the physical conditions of the system.

To avoid any misunderstanding, we should like to stress that the use of the term 'optimum' to describe the primal results of our model does not imply that we interpret the solution as the one and only possible plain to be followed. Because of the many value judgments involved, our results must be treated as just one of many solutions, but one which nevertheless can provide a rational basis for further public debate of the underlying normative judgments involved in this type of analysis.

In the course of such an essentially political discussion, the opportunity costs of different fixed targets measured in terms of the objective function (the shadow prices) furnish us with invaluable criteria for judging the appropriateness of setting these targets at their original levels.

In the absence of any definite knowledge about 'true' welfare functions this seems to be the only way of reaching any conclusions about appropriate political action.

4.2 *ACTIVITIES*

Activities have been defined to enable us to cope with the main problems of environmental planning. In order to prepare a meaningful setting for the problem under discussion we introduce activities which produce both goods demanded in, as well as from outside, the area and the residual wastes which give rise to water pollution. Table 4.1 defines a list of the productive activities ($X_i{}^r$). The index i denotes sectors; r indicates the region where the production is located. The degree of disaggregation is a compromise between the requirement that sectors should be homogeneous with respect to the two outputs produced, i.e. goods and sewage, and the need to respect the statistical delineations used by German sewage statistics and in the Input–Output Study which we use to describe the economic interdependence between the various sectors of production.

TABLE 4.1

LIST OF SECTORS OF PRODUCTION

1. Agriculture	21. Iron, plate and metal products
2. Energy	22. Musical instruments, sport equipment, toys and jewellery
3. Coal mining	
4. Oil and gas extraction	23. Timber
5. Iron ore mining	24. Timber products
6. Mining of other materials	25. Pulp and paper
7. Chemicals	26. Publishing
8. Mineral oil processing	27. Leather
9. Artificial fibres	28. Textiles
10. Rubber and asbestos	29. Clothing
11. Quarrying	30. Milling
12. Glass and pottery	31. Edible oils and margarine
13. Iron and steel production	32. Sugar
14. Iron and steel founding	33. Brewing and malting
15. Non-iron metal	34. Food and tobacco
16. Sheet metal and cables	35. Craft and handwork
17. Steel manufacturing	36. Building
18. Light engineering, ship building, motors and aero space	37. Commerce
	38. Transport
19. Electrical and electronics	39. Services
20. Fine mechanical and optical	40. Government

Other economic activities are private consumption (C^r_{pr}); public consumption (C^r_{St}); public and private investment (ΔK^r); foreign imports $(M_A{}^r)$; foreign exports $(E_A{}^r)$; imports from the rest of the Federal Republic of Germany $(M^*_{I,i})$; exports to the rest of the Federal Republic of Germany $(E_I{}^*_{,i})$; capital demand for treatment plants (G^r); increase or decrease in stocks $(\Delta V_i{}^r)$ and regional income (Y^r).

One of the major sources of water pollution is domestic waste. With the exception of consumer durables, most of the materials consumed by households are returned to the environment after final consumption. A large part of it is disposed of in the form of water pollution. Therefore the number of persons (P^r) living in each of the regions is included as one of the model variables.

From our sewage statistics we can identify three types of industrial waste water: cooling water $({}_1O_i{}^r)$, processing water $({}_2O_i{}^r)$, and personnel sewage $({}_3O_i{}^r)$. In addition we can also obtain data on household sewage $({}_0O^r)$.

It seems appropriate to discriminate among these different types with respect to obligatory treatment. Whereas personnel industrial waste water (as well as household sewage) must receive obligatory treatment, water that has only been used as a cooling device is, in our model, discharged into the streams without treatment. This is possible because waste heat has not yet posed any serious problems in the area. Processing water may or may not be treated.

We indicate the type of treatment by a left superscript, e.g. $(_2{}^1O_i{}^r)$ and $(_2{}^2O_i{}^r)$, describing the quantities of untreated and treated processing water respectively. Total amount of treated water is described by $(^2O^r)$, and the total amount of residuals discharged into the streams is denoted by (R^r).

The capital demand (G^r) for installing the necessary capacity of treatment plants is a nonlinear function of the quantity of water treated. This nonlinear function has been piecewise approximated in four steps, which leads to the inclusion of four auxiliary activities in each region $(T_k{}^r)$. Land requirement for treatment plants is denoted by (F^r).

Other primal variables are the streamflow immediately before waste water discharge (S^r); streamflow immediately after waste water discharge (Q^r); water withdrawal (W^r); and water inflow from natural sources (Z^r).

4.3 *A GRAPHICAL DESCRIPTION OF THE MODEL*

We shall now turn to the assumptions and functional relationships of the model.

In order to simplify the presentation, the general form of the model is shown (somewhat simplified) schematically in Fig. 4.3 in matrix and vector notation for one region.

The concept of the multi-regional model, consisting of the four regions linked together by waterflow equations and export/import relations, is schematically shown in Fig. 4.4 for two regions.

We shall first concentrate attention on the simple one region model, this being the easiest way to clarify our concept. Afterwards it is demonstrated how the single regions are linked together.

Not counting the objective (flexible target) function, which is to be maximised, the matrix shown in Fig. 4.1 may be thought of as divided into three basic sets of constraints (or rows) describing:

(a) production of and demand for goods; creation of income; population; and input availability (labour);
(b) production and handling of residuals (wastewater balances and *BOD*-balance);
(c) the impact of residuals on water-quality.

The activities, defined above, have been entered into the heading of the tableau. The activity vectors (columns) may now be divided into five major groups: (a) production and demand; (b) regional income (Y); (c) population; (d) treatment of residuals; and (e) discharge of residuals. The righthand side (RHS) is a vector of constraint levels. The coefficients of the model (written in matrix, vector or scalar notation) are placed in boxes, so that one can easily identify with which activities they are to be multiplied.

A complete list of all parameters and primal variables for the programming model is given in the appendix.

Generally a small roman letter denotes a column vector; a large roman letter a matrix; a slash denotes transposition; a circumflex accent over a vector denotes a diagonal matrix formed from the vector and a bar denotes an average value.

Further we denote by I an identity matrix and by i a unit vector.

4.4 *THE OBJECTIVE FUNCTION*

The flexible target is the maximisation of regional income, with due regard for the balance of trade and the balance of production to demand, more precisely:

$$Z = Y - (E_A - M_A) - \sum_i \Delta V_i \longrightarrow \max! \quad i = 1, \ldots, 40 \qquad (1)$$

The choice of this objective function implies a growth goal for the region, and emphasises regional investment and capacity expansion rather than consumption. By introducing empirically established relations concerning production factors, investment, consumption and foreign trade, it can, however, be ensured that the model will reflect a realistic and politically defendable behaviour of the system under discussion.

4.5 *THE FIXED TARGETS (CONSTRAINTS) OF THE ONE REGION MODEL*

4.5.1 PRODUCT BALANCES AND FINAL DEMAND

In order to model the feature of interdependence among industries (i.e. industries which exchange their products) we incorporated a 40 by 40 Input-Output Matrix based on a 1966 national matrix of 56 sectors prepared by the Deutsches Institut für Wirtschaftsforschung in Berlin [6]. Thus the restrictions of our one region model include a set of 40 product balances requiring that the total production of each sector i is equal to product flows to itself, other sectors, final demands and to changes in stock investments, i.e.:

$$X_i = \sum_j a_{ij}X_j + d_{pr,i}C_{pr} + d_{St,i}C_{St} + b_i\Delta K + e_i E_A$$
$$\pm \Delta V_i \qquad (2)$$
$$i,j = 1,\ldots,40$$

The constants d_i, b_i and e_i indicate the proportions in which the sector i contributes to final demand. They are subject to the condition

$$\sum_i d_i = \sum_i b_i = \sum_i e_i = 1; \quad i=1,\ldots,40$$

For the input coefficients a_{ij}, together with the amount of value added per unit of output u_j and the rate of foreign imports m_j (in equation (5)):

$$\sum_i a_{ij} + u_j + m_j = 1; \quad i,j = 1,\ldots,40$$

Outputs and intermediate inputs are measured at constant (1966) prices.

The equation (Y) defines gross regional income, which is connected to production by:

$$Y = \Sigma u_j X_j; \quad j=1,\ldots,40 \qquad (3)$$

The constraint $(Y_{\min})$ ensures that the gross regional income per head achieves at least a certain minimum ω, i.e.

$$Y - \bar{\omega}P \geq 0 \qquad (4)$$

The foreign imports (M_A) required for intermediate and final demand are defined by:

$$(E_A - M_A) = \sum_j m_j X_j - m_{pr}C_{pr} - m_k\Delta K + (1 - m_E)E_A \qquad (5)$$
$$j = 1,\ldots,40$$

That part of private consumption which results from imports is given by the coefficient m_{pr}; m_k stands analogously for the import share of private and public investment, and m_E shows the proportion of imports in total exports (transit trade).

The share of regional product flowing to the different demand sectors is determined by relations (6) to (9). (6), (7) and (8) are normative relationships securing minimum shares of the regional product for consumption and total investment. (9) ensures that there there be a slight positive balance of trade.

$$\tau_1 Y \leq C_{pr} \leq \tau_2 Y \qquad (6)$$
$$v_1 Y \leq C_{St} \leq v_2 Y \qquad (7)$$
$$\varphi_1 Y < \Delta K \leq \varphi_2 Y \qquad (8)$$
$$\psi_1 Y \leq (E_A - M_A) \leq \psi_2 Y \qquad (9)$$

The parameters τ_1, ν_1, φ_1, ψ_1 and τ_2, ν_2, φ_2, ψ_2 define the lower and upper limits for (C_{pr}), (C_{St}), (ΔK) and $(E_A - M_A)$ subject to the condition that:

$$\tau + \nu + \varphi + \psi = 1.$$

4.5.2 POPULATION AND EMPLOYMENT STANDARDS

The population of a region is closely related to industrial production by sectoral labour coefficients l_j and the employment rate κ, thus leading to:

$$P = \frac{1}{\kappa}\sum_j l_j X_j \equiv \sum_j g_j X_j; \quad j = 1,\dots,40 \tag{10}$$

Since real capital is provided for in the block 'product balances', only the supply of labour remains to be introduced into the economic part of the model. This is done by defining fixed output per unit labour input for each sector. In addition, flexibility constraints are imposed for total regional labour (L_j) in each sector j requiring that employment in any sector (and therefore also production) may only vary by a maximum of 20 per cent relative to the employment levels of the reference period (1970), i.e.

$$0.8L_j \le l_j X_j \le 1.2L_j; \quad j = 1,\dots,40 \tag{11}$$

4.5.3 WASTE WATER PRODUCTION

The generation of household sewage is defined as a certain amount of sewage per head per day. The generation of waste water in the production sectors is represented by discharge coefficients, defined as a constant proportion between the output of waste water from a sector i and gross production in the sector. This does not imply that we are unaware that the amounts of sewage produced per unit of output have been continuously decreasing, both absolutely and relatively, over the last few years, and that important reductions have been achieved by in-plant treatment and recirculation of waste water. Of course, such a development leads to reductions of sewage output coefficients. It should be mentioned, however, that on average, progress in this field has not been very rapid in Germany. For our aggregate sectors, significant changes have not occurred in the last ten years. Sensitivity analysis can furnish information about strategic coefficients, i.e. coefficients where reductions would lead to the largest improvement of the value of the objective function. This could serve as a basis for political discussion about the strategy for further research on how and where to reduce sewage output most efficiently, without cutting down on production.

We therefore feel that the use of fixed proportions for sewage output best serves our purpose.

Having three types of waste water from each sector and two means of disposal – secondary treatment in sewage plants, or direct discharge into 'the' river of the region – the following three waste water balances are introduced into the model for each sector i:

$$_1u_iX_i = {}_1^1O_i \tag{12}$$

$$_2u_iX_i = {}_2^1O_i + {}_2^2O_i \tag{13}$$

$$_3u_iX_i = {}_3^2O_i \tag{14}$$

The amount of household sewage is given by

$$_0uP = {}_0^2O \tag{15}$$

In these balances (X_i) and (P) stand for industrial production of sector i and population respectively; $({}_0^2O)$ for (treated) household sewage; $({}_1^1O_i)$ for (untreated) cooling water; $({}_2^1O_i)$ for untreated processing water; $({}_2^2O_i)$ for treated processing water; $({}_3^2O_i)$ for (treated) personnel water. The fixed coefficients $_1u_i$, $_2u_i$ and $_3u_i$ give the amounts of waste water per unit of output per day [7]. $_0u$ denotes sewage per head per day.

4.5.4 WATER QUALITY INDICATORS

The supply of dissolved oxygen from atmospheric re-aeration and the self-purification capacity of streams are mainly functions of the load of organic material, which determines the total metabolism of the river. We thus use the load of organic material per unit of river water as a water quality criterion. We consider its impact on dissolved oxygen (D.O.) but also require that the relative load of organic pollution be kept below a certain limit. The reason for having two types of constraints on river quality in connection with the organic waste load is that there are at least three different uses of the Ruhr river. It serves not only as a waste transport medium, but also as a drinking water reservoir. Further, it is also important for recreation in the Ruhr area. It is not enough – in our view – to provide for sufficient oxygen in the river water, which can be achieved by artificial aeration but leaves the water in a very poor condition for many recreation purposes.

One would imagine that by restricting the model to the regulation of organic wastes one automatically excludes the consideration of other types of pollution (e.g. grease, acidity, dissolved solids, detergents) in the waste water produced by each sector of production. Methods have been developed, however, which are capable of integrating the effects of organic and toxic wastes on water quality into a single index number. This index – a special form of the population-equivalent biochemical oxygen demand (B.O.D.) – has been computed for the waste waters of different industries [8, 9, 10]. We thus use it for measuring the quality of waste water as well as river water.

Denoting by $_v^f h_i$ and $_0^2 h$ the relative contents of biochemical oxygen demand in industrial and household sewage with and without treatment respectively, we may define the total amount of B.O.D. discharged into the river in one region by:

$$R = \sum_f \sum_v \sum_i {}_v^f h_{iv}{}^f 0_i + {}_0^2 h_0{}^2 0 \qquad \left\{ \begin{array}{l} i = 1,\dots,40 \\ f = 1,2 \\ v = 1,2,3 \end{array} \right. \qquad (16)$$

The solution of the model provides information about the quantity of waste water to be treated. This amount depends on the B.O.D.-standard for the river water, on the desired amount of dissolved oxygen (D.O.), on the water flow, on the degree of treatment in the sewage plants, on the type and quantity of the waste water, and on the quality of the river water running into the region. Further, the self-purification capacity of the stream plays a very important role.

In order to provide for acceptable water qualities throughout the whole year, we need only concern ourselves with the critical season of the year. In our case, short periods of drought can be offset by maintaining the flow of the Ruhr river from upland reservoirs. In measuring the quantities of surface water available for dilution, we calculate those flows which can be maintained during the summer after two consecutive years of dry summers with little rain during the wintertime.

For calculating the relationships between the B.O.D.-pollution and the amount of dissolved oxygen in the stream, we introduce the well-known Streeter–Phelps equations into the model [11]. These equations allow us, under steady state conditions, to represent the interaction between biochemical degradation and re-aeration as a linear function of the B.O.D.-discharges, and thereby to calculate the D.O.-deficit, i.e. the difference between the D.O.-saturation value and the actual D.O.-level, that results from different levels of B.O.D.-inputs.

Since the B.O.D.-load of the river water is greatest immediately after the discharge of the B.O.D. into the river, the restriction concerning the B.O.D.-concentration in a certain region can be written as

$$R \leq \text{B.O.D.-standard} \qquad (17)$$

As time elapses, the oxygen-demanding waste is gradually reduced by chemical and biological processes in the stream. This however leads to a reduction of the dissolved oxygen in the stream, even if at the same time some oxygen is restored to the water. Therefore, proceeding downstream the D.O.-deficit will increase until a certain point (oxygen-sag), after which re-oxygenation (or re-aeration) dominates the process leading to a rise in the dissolved oxygen level.

We denote by the D.O.-standard a certain tolerable D.O.-deficit at the most critical point in the stream below the discharge point and, further, by γ the D.O.-deficit that occurs per unit of organic material during the flow time from the discharge point in a certain region to the above mentioned critical point.

Our restriction concerning the D.O.-deficit may be written as:

$$\gamma R \leq \text{D.O.-standard} \tag{18}$$

4.5.5 SEWAGE TREATMENT COST AND LAND REQUIREMENT

Modern sewage plant operation consists of two parts: primary and secondary treatment. The primary section utilises mechanical processes and is thus able to remove only the larger floating, suspended and undissolved particles. The secondary treatment stage is biochemical in nature, i.e. bacteria react with the degradable organic material by absorption, digestion, oxidation or decomposition. The reactions that take place in a biological treatment plant can be described as an accelerated self-purification process.

In the model we consider only one type of secondary treatment plant, namely the activated sludge plant. Compared to other secondary treatment methods, e.g. trickling filters, the activated sludge plant is characterised by greater compactness. It thus requires less land, which in the highly industrialised Ruhr area is extremely scarce. This type of plant can be operated efficiently at a very large size [14]. At the mouth of the river Emscher an activated sludge plant is presently being constructed with a treatment capacity of more than 100,000 cubic metres per hour. As we allow only one treatment plant in each region, all waste water purification has to be carried out by four plants only. Waste water throughput thus easily reaches a scale where this type of plant can be operated most efficiently.

The purification effect in activated sludge plants may reach levels above 90 per cent which we take to be constant, not variable as in Günther [12, 13] and the F.W.P.C.A. [15] study. As we only consider one type of sewage treatment plant with a fixed degree of purification of 90 per cent, and since operational costs only vary slightly with sewage throughput, it is not necessary to consider this type of cost explicitly. In the model, therefore, only total construction costs were introduced as an activity. Investigations have shown this type of cost to be dependent on the quantity of waste water to be treated per time unit [16, 17, 18].

Using the same notation as before, the total quantity of waste water per hour to be treated in one region is

$$^2O = \frac{_0{}^2O + \sum_i ({}_2{}^2O_i + {}_3{}^2O_i)}{n}; \quad i=1,\ldots,40 \tag{19}$$

where n denotes the average operating time in hours per day.

The function describing the relationship between construction costs and the amount of sewage water is taken from Schmidt [16], p. 43. It has the form:

$$G = 127 \cdot 665({}^2O)^{0.49} \tag{20}$$

In introducing this non-linear function into the model we had to linearise it. As (G) is replaceable by a polygonal approximation, this could be achieved by using a separable programming technique [19]. A graphical mapping showed that a piecewise function with four linear segments would give an appropriate approximation of equation (20).

Defining by ϕ_k, $k=1,\ldots,4$, the amount of sewage treated in the k-th segment, we have:

$$\sum_k \phi_k = {}^2O_{\max}; \quad k=1,\ldots,4$$

where ${}^2O_{\max}$ denotes the maximum capacity for a treatment plant, i.e. the upper limit of $({}^2O)$, for which we consider the separable function to be active.

In this model ${}^2O_{\max}$ was set equal to 200,000 cubic metres per hour, i.e. 55 cubic metres per second. For each segment k, construction costs Γ_k were calculated as shown in Table 4.2.

TABLE 4.2

SEGMENT SIZE AND CORRESPONDING CONSTRUCTION COSTS

k	1	2	3	4	
Φ_k	4	6	10	180	(1000 m³/hour)
$\Sigma\Phi_k$	4	10	20	200	(1000 m³/hour)
Γ_k	7·7	4·4	4·9	26·0	(million DM)
$\Sigma\Gamma_k$	7·7	12·1	17	53·0	(million DM)

Introducing a special activity (T_k), restricted so that $0 \le T_k \le 1$, we write the constraints describing the construction costs for sewage treatment in a certain region as:

$$^2O = \sum_k \phi_k T_k \tag{21}$$

$$G = \sum_k \Gamma_k T_k \tag{22}$$

$$0 \le T_k \le 1 \qquad k=1,\ldots,4 \tag{23}$$

Land for treatment plants is a scarce factor in our model, therefore:

$$\epsilon^2 0 \leq \text{available land} \qquad (24)$$

where ϵ denotes hectare land per cubic metre treated waste water per hour.

4.6 *THE MULTIREGIONAL MODEL*

On familiarity with the single region model, it is easy to recognise the structure of the multiregional model. Actually it consists of several identical regional models. By defining the exports from one region to another as the imports of the second region from the first, our four regions in the Ruhr area are tied together, so that they occur as *one* region set against the rest of the Federal Republic of Germany and the 'rest of the world'. Thus, analogous to the single region model, the objective function may be written as:

$$Z = \sum_r Y^r - \sum_r (E_A{}^r - M_A{}^r) - \sum_r \sum_i \Delta V_i{}^r \longrightarrow \text{max}! \qquad (25)$$

$$\begin{cases} r = 1,\dots,40 \\ i = 1,\dots,4 \end{cases}$$

i.e. we stipulate the maximisation of regional product net of the foreign export surplus and growth in stocks.

For the planning area as a whole, we require that the balance of trade be non-negative (see equation $E^* - M^*$ in Fig. 4.4):

$$\sum_r (E_A{}^r - M_A{}^r) + \sum_r \sum_i E_I{}^r{}_{,i} - \sum_r \sum_i M^r{}_{I,i} \geq 0 \qquad (26)$$

$$\begin{cases} i = 1,\dots,40 \\ r = 1,\dots,4 \end{cases}$$

As in the single region model the imports and exports to the 'rest of the world' are restricted, requiring a slight positive balance of trade. The imports from and the exports to the rest of the Federal Republic of Germany remain unrestricted.

The model, therefore, may be used to answer the questions of which and how many goods should be procured from outside the system to satisfy the pollution limits. This may have unfavourable consequences for some of the production sectors. However, the maximisation of regional income guarantees full employment of scarce factors. The system produces as many goods and services as are permitted by factor endowments and water quality standards.

An important part of the multi-regional model are the equations describing the water flows and the water quality. As we have already mentioned, we take the first three regions to be independent with respect to surface water. We assume that the waste water discharge takes place at the downstream regional border, either directly into the Ruhr or into one of the tributaries, and that there is only one treatment plant and one common point of discharge of treated and untreated waste water in each region.

Denoting by (S^r) the volume of water in the river immediately before the discharge of waste water in region r $(r=1,...,4)$, and by (Q^r) the total volume of water in the river immediately after the discharge of waste water in region r, we may write:

$$Q^r = \sum_i [_1^1O_i^r + _2^1O_i^r + _2^2O_i^r + _3^2O_i^r] + _0^2O^r + S^r \tag{27}$$

(W^r) denotes the volume of withdrawn water, and (Z^r) the inflow of water from natural sources in region r. We may then define (S^r) as:

$$S^r = \begin{cases} Z^r - W^r & \text{for } r=1 \\ Q^{r-1} + Z^r - W^r & \text{for } r=2,3,4 \end{cases}$$

In Fig. 4.2 these relationships are given by the equations Q^1, S^2 and Q^2 for regions 1 and 2.

As mentioned above, the B.O.D. concentration in river water is measured immediately after the discharges, whereas the oxygen deficits are measured at critical downstream points situated at certain distances from the respective discharge points.

In order to calculate the self-purification capacity of the river, we have divided the streams into several reaches. In particular, this is necessitated by the artificial re-oxygenation, the use of oxidation lakes, etc. in the Ruhr river, which results in very different rates of degradation of B.O.D. and different rates of re-aeration.

In all, we have four points of waste water discharge $(r=1,...,4)$, four points of measurement of the B.O.D. load $(s=1,...,4)$ and – owing to the dilutional effects of natural water inflow and the cumulative effects of discharges in different regions – six points of measurement for dissolved oxygen $(p=1,...,6)$.

We denote by $^s\pi^r$ a coefficient describing the speed of degradation of B.O.D. during the time from a discharge point r to a control point s (for $s=r$, $^s\pi^r$ equals one); by (R^r) the total amount of B.O.D. discharged in region r, and β^s the tolerable upper limit (standard) for B.O.D. at control point s.

Assuming that the B.O.D.-load and the D.O.-deficit of the river water (S^1) before the discharge point in region one and also that of the natural inflow (Z^r) are negligible, we may write the B.O.D.-constraint for the first region as:

$$R^1 - \beta^1 Q^1 \leq 0 \tag{28}$$

For the second measurement point immediately after discharge in region 2, and under consideration of the withdrawal of water (W^2) during the time of flow from the first measurement point to the second discharge point, we have:

$$R^1\,{}^2\pi^1 - \zeta^2 W^2 + R^2 - \beta^2 Q^2 \leq 0 \qquad (29)$$

where ζ^2 is the concentration of B.O.D. in the river water withdrawn in region 2.

Generally for each B.O.D. measurement point s we have the restriction:

$$\sum_{r=1}^{s} R^{rs}\pi^r - \sum_{r=2}^{s} \zeta^r W^r - \beta^s Q^s \leq 0; \quad s=1,...,4 \qquad (30)$$

The coefficient $^p\gamma^r$ denotes the oxygen deficit per unit of organic material which occurs during the time of flow from the discharge point r ($r=1,..,4$) to the oxygen measurement point p ($p=1,...,6$).

Note that it is assumed that the oxygen deficit is zero before the first sewage discharge.

At the first oxygen measurement point which lies immediately upstream from the discharge point in region 2, we have the following restriction:

$$R^1\,{}^1\gamma^1 - \delta^1 Q^1 \leq 0 \qquad (31)$$

where δ^p denotes the highest allowed oxygen deficit of river water at the oxygen measurement point p (see the restriction 'diss. oxyg. st.' in Fig. 4.4).

At the second measurement point (after the second discharge) we have:

$$R^1\,{}^2\gamma^1 - \epsilon^2 W^2 + R^2\,{}^2\gamma^2 - \delta^2 Q^2 \leq 0 \qquad (32)$$

where ζ^2 denotes the oxygen deficit of water withdrawn in region 2.

The oxygen constraints for the next four measurement points are similarly developed. Note that the coefficients $^s\pi^r$ and $^p\gamma^r$ may be different for each r, s and p. The parameters required to establish the value of these coefficients are obtained from values measured along the river.

Inclusion of the above restrictions in the model introduces a relationship among the water regions which makes it possible to test, on the one hand, the present structure of industry and housing with respect to the load they impose on the river water, and, on the other, to compute the desired structure and optimal allocation of industry and residential population.

4.7 DISCUSSION OF THE RESULTS

Before entering into a discussion of our results, let us point out that the planning proposals which may be derived from our model must not be interpreted in the sense that one should shut down existing industries at one place and construct them anew at another location. Rather, we are concerned to discover the optimal direction to follow when taking decisions about the location of *new plants*. New factories should be located where the optimal solution of the model indicates that there is room for industrial growth. Old factories at sub-optimal locations should not be bribed to stay in their old community *if they want to move*.

In the discussion of the results we use labour input as a measure of the level of economic activity.

In Table 4.3 the optimal change in sectoral labour input – under the present conditions of the system – is set out for each region and the total planning area. This table gives information on an aggregated sectoral level. Looking at the optimal change in total labour input, we note that in all regions production should be expanded – by about 10 per cent in region 1 (lowest) and by about 18 per cent in region 3 (highest).

Looking at the sectors in which this expansion should take place, we see that agriculture, the electricity sector and the tertiary sectors should be extended to the maximum in all regions. This result is not unexpected, for the following reasons: The first sector mentioned, agriculture, is specialised on products which contribute very little to water pollution. Agriculture in regions 1 and 2 concentrates on forestry with a relatively small amount of dairy farming. Agriculture in regions 3 and 4 concentrates on arable farming. There is only little concentrated life stock breeding in the whole area. The agricultural sector, therefore, contributes little to water pollution in the Ruhr basin.

The electricity sector consists mainly of water electricity works. Their contribution to water pollution is negligible. The contribution of the tertiary sectors to water pollution is also very small compared with most manufacturing industries.

Turning to sectors 6–35, the manufacturing industries, which naturally matter most in our context, we see that there should be significant changes in the regional distribution of these sectors taken as a whole. For instance, a decrease of 1·5 per cent in region 1 and an increase of 16 per cent in region 3 would be optimal. In order to explain this fact, we must look at the hydrology of the Ruhr basin and the performance of the Ruhr water authority.

The Ruhr river and its tributaries (Lenne, Ennepe and Volme) are all relatively small rivers. During the critical period of the year, the Ruhr river has a water flow of about 10 cubic metres per second at the downstream border of region 1. At the downstream border of region 2 the tributary Lenne and in the middle of region 3 the

TABLE 4.3

LABOUR (1970) AND OPTIMAL CHANGE (1970)

Sectors		Region 1		Region 2		Region 3		Region 4		Total Planning Area	
		Labour	Optimal Change	Labour	Optimal Change	Labour	Optimal Change	Labour	Optimal Change	Labour	Optimal Change
		(Pers.)	(%)	(Pers.)	(%)	(Pers.)	(%)	(Pers.)	(%)	(Pers.)	(%)
1	Agriculture	8,459	20·0	5,070	20·0	1,253	20·0	2,541	20·0	17,323	20·0
2	Electricity	780	20·0	1,464	20·0	1,223	20·0	3,635	20·0	7,102	20·0
3–5	Coal mining, oil and gas, iron ore mining	None in regions 1, 2 and 3						6,397	−11·6	6,397	−11·6
6–35	Manufacturing industries	52,045	−1·5	143,013	4·2	42,102	16·0	152,261	11·3	389,421	7·5
36–40	Tertiary sectors	49,437	20·0	97,632	20·0	53,835	20·0	130,191	20·0	331,095	20·0
1–40	All sectors	110,721	9·9	247,179	10·8	98,413	18·3	295,025	14·9	751,338	13·3

tributaries Ennepe and Volme enter the Ruhr river. At the downstream border of region 3 the river has a water flow of about 28 cubic metres per second during the critical period. Of course, where water flow is highest, dilution capacity is also highest.

The general objective of the water authority in the Ruhr basin is to maintain water quality suitable for household water supply and recreation. More than a thousand million cubic metres per year are withdrawn, and of this amount about four hundred million cubic metres are exported to other river basins. This is only possible because of large upland reservoirs which regulate the river's flow during the course of the year. The water works are situated along the Ruhr river. The biggest water works are located at the upper border of region 3, and therefore the heaviest withdrawals take place in this area. Large amounts of water are also withdrawn in regions 1 and 2. High water quality is to be maintained for this purpose.

In regions 1 and 2 the rivers are rather fast running. It is no problem to keep a sufficient level of dissolved oxygen in these reaches, owing to the turbulent flow and the short flow time during which only a little amount of the discharged B.O.D. is oxidated. In regions 3 and 4 a series of shallow lakes in the Ruhr itself has been constructed for neutralisation, precipitation and oxidation of the river's water. In order to cope with the heavy consumption of dissolved oxygen in these reaches, the water authority has built several weirs and has installed many devices for mechanical re-aeration in the reach between the downstream border of region 2 and the Rhine river.

Concerning region 1 and region 2, the water authority in the Ruhr basin has stated its policy as follows: 'In particular, it is sought to prevent the location of heavy water polluting industries in the upper watershed and to preserve suitable areas in this part of the river basin for water collection and water storage.' [20]

In order to achieve this goal the water authority has set high water quality standards in the water courses running through regions 1 and 2. This authority also exerts its influence on the region planning authority (*Ruhrsiedlungsverband*) to impede new industrial settlements in this area.

Owing to the strong restrictions on water quality (B.O.D.-load) and the relatively small amounts of dilution water, it is not surprising that the optimal solution of the model calls for total labour input in the manufacturing industries to be decreased in region 1.

In region 2 the manufacturing industries may increase by 4·2 per cent, owing to the larger amount of dilution water. In this region the tributary Lenne enters the Ruhr, and this tributary is regulated by the largest upland reservoir in the whole area, the Biggetal reservoir.

For these reasons, the greatest increase of production should take

TABLE 4.4

LABOUR (1970) AND OPTIMAL CHANGE (1970)

Sectors	Region 1		Region 2		Region 3		Region 4		Planning Area	
	Labour	Optimal change	Labour	Optimal change	Labour	Optimal change	Labour	Optimal change	Labour	Optimal change
	(Pers.)	(%)	(Pers.)	(%)	(Pers.)	(%)	(Pers.)	(%)	(Pers.)	(%)
1. Agriculture	8,459	20·0	5,070	20·0	1,253	20·0	2,541	20·0	17,323	20·0
2. Energy	780	20·0	1,464	20·0	1,223	20·0	3,635	20·0	7,102	20·0
3. Coal mining	0	0·0	0	0·0	0	0·0	6,397	−11·6	6,397	−11·6
4. Oil and gas extraction	0	0·0	0	0·0	0	0·0	0	0·0	0	0·0
5. Iron ore mining	0	0·0	0	0·0	0	0·0	0	0·0	0	0·0
6. Mining of other materials	52	20·0	1,018	20·0	0	0·0	72	0·0	1,142	18·7
7. Chemicals	769	20·0	1,335	20·0	132	20·0	3,336	20·0	5,572	20·0
8. Mineral oil processing	0	0·0	0	0·0	205	20·0	115	20·0	205	20·0
9. Artificial fibres	660	−20·0	4,373	−20·0	40	20·0	2,183	−12·3	7,256	−17·5
10. Rubber and asbestos	42	−20·0	486	−20·0	26	−7·7	381	−20·0	935	−19·6
11. Quarrying	1,718	−20·0	1,861	−12·5	1,381	20·0	4,256	20·0	9,216	6·0
12. Glass and pottery	1,184	20·0	440	20·0	161	20·0	1,024	20·0	2,809	20·0
13. Iron and steel production	0	0·0	10,220	−20·0	7,745	20·0	23,232	−5·7	41,197	−3·6
14. Iron and steel founding	1,040	−20·0	1,810	20·0	2,692	20·0	17,187	20·0	22,729	18·6
15. Non-iron metal	2,124	−20·0	14,212	−20·0	4,470	20·0	925	−5·0	21,731	−11·1
16. Sheet metal and cables	1,210	−20·0	10,771	20·0	1,774	−7·8	5,690	−20·0	19,445	−18·9
17. Steel manufacturing	3,206	20·0	25,320	3·4	6,446	20·0	12,413	20·0	47,385	11·0
18. Light engineering, ship building, motors and	6,126	20·0	18,226	20·0	4,506	20·0	50,472	20·0	79,330	20·0

20. Fine mechanical and Optical	526	14·1	898	20·0	880	20·0	1,408	20·0	3,712	19·0
21. Iron, plate and metal products	6,495	−20·0	21,185	4·5	174	20·0	8,769	20·0	37,223	4·2
22. Musical instruments, sport equipment, toys and jewellery	137	−20·0	176	−20·0	278	−20·0	497	−20·0	1,088	−20·0
23. Timber	1,438	−20·0	860	−20·0	109	−20·0	569	−20·0	2,976	−20·0
24. Timber products	3,433	20·0	1,174	20·0	128	20·0	817	20·0	5,552	20·0
25. Pulp and paper	1,326	−20·0	755	−20·0	1,480	−20·0	604	−20·0	4,165	−20·0
26. Publishing	2,113	20·0	2,925	20·0	620	20·0	2,702	20·0	8,360	20·0
27. Leather	42	−20·0	207	−20·0	65	7·7	54	−20·0	368	−14·4
28. Textiles	3,755	−7·4	2,441	20·0	322	20·0	1,678	20·0	8,196	7·4
29. Clothing	992	4·4	2,164	20·0	431	20·0	1,751	0·0	5,337	10·7
30. Milling	0	0·0	0	0·0	141	−20·0	197	−20·0	338	−20·0
31. Edible oils and margarine	0	0·0	0	0·0	247	−20·0	0	0·0	247	−20·0
32. Sugar	0	0·0	0	0·0	705	−20·0	0	0·0	705	−20·0
33. Brewing and malting	498	20·0	423	20·0	470	20·0	1,325	20·0	2,716	20·0
34. Food and tobacco	2,259	−20·0	4,809	−20·0	3,174	20·0	4·029	−0·9	14,271	−1·0
35. Craft and handwork										
36. Building	9,626	20·0	16,401	20·0	6,193	20·0	18,559	20·0	50,779	20·0
37. Commerce	12,648	20·0	30,087	20·0	17,607	20·0	41,982	20·0	102,324	20·0
38. Transport	4,488	20·0	8,132	20·0	9,356	20·0	10,882	20·0	32,858	20·0
39. Services	14,388	20·0	27,761	20·0	12,975	20·0	34,195	20·0	89,319	20·0
40. Government	8,287	20·0	15,251	20·0	7,704	20·0	24,573	20·0	55,815	20·0

TABLE 4.5

REALLOCATION OF PRODUCTION

Sectors	Leaving or prevented from entering regions	Entering regions	Increasing in all regions	Decreasing in all regions
1. Agriculture			+	
2. Energy			+	
3. Coal mining	1, 2, 3, 4			
4. Oil and gas extraction	1, 2, 3, 4			
5. Iron ore mining	1, 2, 3, 4			
6. Mining of other materials	3, 4	1, 2		
7. Chemicals			+	
8. Mineral oil processing	1, 2	3, 4		
9. Artificial fibres	1, 2, 4	3		
10. Rubber and asbestos				−
11. Quarrying	1, 2	3, 4		
12. Glass and pottery			+	
13. Iron and steel production	1, 2, 4	3		
14. Iron and steel founding	1	2, 3, 4		
15. Non-iron metal	1, 2, 4	3		
16. Sheet metal and cables	1, 3, 4	2		
17. Steel manufacturing			+	
18. Light engineering, ship building, motors and aerospace			+	
19. Electrical and electronics			+	
20. Fine mechanical and optical			+	
21. Iron, plate and metal products	1	2, 3, 4		
22. Musical instruments, sport equipment, toys and jewellery				−
23. Timber				−
24. Timber products			+	
25. Pulp and paper				−
26. Publishing			+	
27. Leather	1, 2, 4	3		
28. Textiles	1	2, 3, 4		
29. Clothing			+	
30. Milling				−
31. Edible oils and margarine				−
32. Sugar				−
33. Brewing and malting			+	
34. Food and tobacco	1, 2, 4	3		
35. Craft and handwork				
36. Building			+	
37. Commerce			+	
38. Transport			+	
39. Services			+	
40. Government			+	

place in regions 3 and 4. As mentioned above, there is more dilution water, standards for B.O.D. are lower, and massive measures for re-aeration of the river water take place.

The optimal changes for each sector and each region are given in Table 4.4. Having a closer look at this table, we see that we may divide the sectors into three groups:

(a) sectors increasing in all regions;
(b) sectors decreasing in all regions;
(c) sectors increasing in some and decreasing in other regions.

The optimal allocation of production for all sectors and regions is summarised in Table 4.5. The column 'Leaving or prevented from entering regions' shows in which regions the production of a given sector ought to be reduced or to be kept at the present level. The column 'Entering regions' shows to which regions the industries should be reallocated. The last two columns show in which regions the production of a sector should be increased or reduced, respectively.

As mentioned in the introductory description of the model, there are also constraints demanding a minimum income per head in each region, and the amounts of value-added differ from sector to sector.

In addition to allocating production in the total planning area in a manner which takes account of the relative amounts of waterborne residuals, the model determines the industrial mix which maximises regional income. This is the reason why a sector like musical instruments (22) should decline, whereas a rather heavy polluting industry like chemicals should increase in all regions. Further, it must be remembered that effluent charges are based on waste load and that the water authority in the Ruhr area has built central plants for the detoxification of industrial sludges. Many industries for these reasons pre-treat their waste water and deliver the sludge to these plants in order to reduce the charges they have to pay. This has led to a reduction of waterborne discharge by former heavy polluting industries, so that in the Ruhr area these industries often have discharge coefficients which lie substantially below the average coefficients at many other locations in the Federal Republic of Germany.

In Table 4.6 the sectors are listed for which a relocation within the area could be considered. The last column shows the optimal change for the total planning area. All sectors in this table belong to the heavy polluting category. It is clearly shown that a shift towards regions 3 and 4 should be considered.

What might be gained from structural changes in the above-mentioned directions can be learned from Table 4.7, which gives the

TABLE 4.6

REALLOCATION OF PRODUCTION

Sectors	Sectors decreasing (−) and increasing (+) in different regions								Total planning area optimal change
	Region 1		Region 2		Region 4		Region 3		
	+	−	+	−	+	−	+	−	(%)
6. Mining of other materials	+		+		const. (0)		const.		18·7
8. Mineral oil processing	const. (0)		const. (0)		+		+		20·0
9. Artificial fibres		−		−		−	+		−17·5
11. Quarrying		−		−	+		+		6·0
13. Iron and steel production		−		−		−	+		−3·6
14. Iron and steel founding		−	+		+		+		18·6
15. Non-iron metal		−		−		−	+		−11·1
16. Sheet metal and cables		−	+			−		−	−18·9
21. Iron, plate and metal products		−	+		+		+		4·2
27. Leather		−		−		−	+		−14·4
28. Textiles		−	+		+		+		7·4
34. Food and tobacco		−		−	const.		+		−1·0

marginal rates of productivity of labour at either the upper or the lower constraint level of labour activity. As described above, for each sector in each of the regions we introduced an upper and lower boundary on changes in employment, limiting it to ±20 per cent of the 1970 figures. In Table 4.7 sectoral labour at upper limit is denoted by 'U'; sectoral labour at lower limit is denoted by 'L'; and sectoral labour lying between the boundaries is denoted by 'B'. The coefficients in the R.C.-columns (shadow prices) denote the amounts (mill. DM) by which the flexible target, i.e. regional income, would change if labour restrictions in the corresponding sectors and regions were changed by one unit. Thus a sector with labour at lower limit and a corresponding negative shadow price indicates that a *reduction* would lead to an *increase* in the flexible target; whereas an *increase* in the labour of the sector would lead to a *decrease* in the flexible target. On the other hand, a sector with labour at upper limit and a corresponding positive shadow price indicates that an *increase* would lead to an *increase* in the flexible target, whereas a decrease in the labour of the sector would lead to a *decrease* in the flexible target.

By looking at Table 4.7 it is obvious that region 3 is the one in which further expansion of productive capacity should take place for most sectors. For instance, a further increase in labour input in sector 12 in region 3 by one unit would lead to an increase in the income for that region by about 700,000 DM, whereas an increase in sector 25 in the same region would lead to a decrease in income by about 90,000 DM. Looking at the manufacturing industries in region

TABLE 4.7

LABOUR ACTIVITY LEVELS AND REDUCED COSTS (R.C.)

Sectors	Region 1		Region 2		Region 3		Region 4	
	At	*R.C.*	*At*	*R.C.*	*At*	*R.C.*	*At*	*R.C.*
Agriculture	U	·02508	U	·02556	U	·02644	U	·02580
Energy	U	·15106	U	·15154	U	·15242	U	·15178
Coal mining	—	—	—	—	—	—	B	—
Oil and gas extraction	—	—	—	—	—	—	—	—
Iron ore mining	—	—	—	—	—	—	—	—
Mining of other materials	U	·02645	U	·02904	L	·03381	B	0
Chemicals	U	·39187	U	·42363	U	·42687	U	·41187
Mineral oil processing	L	− ·00012	B	0	U	·97634	U	·79241
Artificial fibres	L	− ·28728	L	− ·10126	U	·00082	B	0
Rubber and asbestos	L	− ·09862	L	− ·07050	B	0	L	− ·00332
Quarrying	L	− ·00439	B	0	U	·00807	U	·00221
Glass and pottery	U	·69605	U	·70169	U	·71204	U	·70452
Iron and steel production	L	− ·00195	L	− ·00065	U	·00173	B	0
Iron and steel founding	L	− ·00145	U	·00130	U	·00635	U	·00268
Non-iron metal	L	− ·00303	L	− ·00101	U	·00269	B	0
Sheet metal and cables	L	− ·00252	U	·00031	B	0	L	− ·00769
Steel manufacturing	U	·00961	B	0	U	·04093	U	·03301
Light engineering, ship building, motors and aero-space	U	·02846	U	·03349	U	·04273	U	·03602
Electrical and electronics	B	0	U	·00744	U	·02111	U	·01118
Fine mechanical and optical	B	0	U	·090701	U	1·00214	U	·99923
Iron, plate and metal products	L	− ·00760	B	0	U	·01395	U	·00382
Musical instruments, sport equipment, toys and jewellery	L	− ·04872	L	− ·03699	L	− ·01544	L	− ·03109
Timber	L	− ·41725	L	− ·46415	L	− ·54984	L	− ·56023
Timber products	U	1·95350	U	1·96593	U	1·98877	U	1·97218
Pulp and paper	L	− ·09946	L	− ·09738	L	− ·09157	L	− ·09634
Publishing	U	1·16448	U	1·17161	U	1·18472	U	1·17520
Leather	L	− ·04240	L	− ·02745	B	0	L	− ·00994
Textiles	B	0	U	·23251	U	·40470	U	·39628
Clothing	B	0	U	− ·00975	U	·02588	B	0
Milling	L	− 2·10017	L	− 2·09638	L	− 2·08942	L	− 2·09448
Edible oils and margarine	B	0	L	− ·50506	L	− ·49843	L	− ·50324
Sugar	B	0	B	0	L	− ·41444	L	− ·01920
Brewing and malting	U	·13722	U	·14212	U	·15144	U	·14459
Food and tobacco	L	− ·01318	L	− ·00441	U	·01170	B	0
Craft and handwork								
Building	U	2·60340	U	2·62253	U	2·65768	U	2·63216
Commerce	U	·26701	U	·28615	U	·32130	U	·29577
Transport	U	2·30330	U	2·32243	U	2·35759	U	2·33206
Services	U	2·31692	U	2·33606	U	2·37121	U	2·34568
Government	U	1·06954	U	1·08868	U	1·12383	U	1·09830

3 we can see that expansion of productive capacity would pay most in sectors 24 and 26, and least in sectors 9 and 13.

It should be mentioned, however, that the total gains and losses to be achieved by changing productive capacity can only be measured after testing the sensitivity of the model, i.e. to what extent a change may take place without changing the optimal solution. None the less, the shadow prices give good advice about directions in which we ought to move.

APPENDIX

NOTATIONS

$'$	denotes a row vector
$\wedge$	denotes a diagonal matrix
$-$	denotes an average value

Symbol	Definition
	Primal Variables
X	gross output vector in mill. DM per year, 40×1
C_{pr}	total private consumption in mill. DM per year, 1×1
C_{St}	total public consumption in mill. DM per year, 1×1
ΔK	private and public investment in mill. DM per year, 1×1
E_A	foreign exports in mill. DM per year, 1×1
E_I	vector of sectoral exports to the rest of the German Federal Republic in mill. DM per year, 40×1
M_A	foreign imports in mill. DM per year, 1×1
M_I	vector of sectoral imports from the rest of the German Federal Republic in mill. DM per year, 40×1
$E_A - M_A$	balance of trade surplus in mill. DM per year, 1×1
ΔV	vector of increase or decrease in sectoral stocks in mill. DM per year, 40×1
Y	regional income in mill. DM per year, 1×1
P	population in thousand, 1×1
$_0^2O$	household sewage in cubic metres per day (treated), 1×1
$_1^1O$	cooling water vector in cubic metres per day (untreated), 40×1
$_2^1O$	processing water vector in cubic metres per day (untreated), 40×1
$_2^2O$	processing water vector in cubic metres per day (treated), 40×1
$_3^2O$	personnel water vector in cubic metres per day (treated), 40×1
^2O	total treated industrial and household sewage in cubic metres per hour, 1×1
R	total amount of B.O.D. discharged into the river in grammes per day, 1×1

T	vector of auxiliary activities necessary for the computing of the construction costs of treatment plants, 4×1
G	total construction cost of a treatment plant in mill. DM, 1×1
F	land requirement for treatment plants in hectares, 1×1
S	streamflow immediately before waste water discharge in cubic metres per day, 1×1
Q	streamflow immediately after waste water in cubic metres per day, 1×1
W	water withdrawal in cubic metres per day, 1×1
Z	water inflow from natural sources in cubic metres per day, 1×1

Parameters

I	unit matrix
i	unit vector
A	direct coefficient matrix in mill. DM per year per mill. DM per year; an element a_{ij} in the matrix $(i=1,\ldots, 40)$; $(j=1,\ldots, 40)$ expresses the input value from sector i into sector j for the latter to produce one mill. DM of output, 40×40
d_{pr}	private consumption coefficient vector, dimensionless, 40×1
d_{St}	public consumption coefficient vector, dimensionless, 40×1
b	private and public capital coefficient vector, dimensionless, 40×1
e	foreign export coefficient vector, dimensionless, 40×1
w	value added coefficient vector, dimensionless; a coefficient $w_j(j=1,\ldots, 40)$ expresses payments to households (e.g. wages, salary, interest, profits and depreciation) per mill. DM of output from sector j, 40×1
m_x	foreign sectoral import coefficient vector, dimensionless, 40×1
m_{pr}	foreign private import coefficient, dimensionless, 1×1
m_k	foreign private and public capital import coefficient, dimensionless, 1×1
L	sectoral labour vector in thousand persons, 40×1
g	vectors of elements describing the relation between production and population in thousand persons per mill. DM, 40×1
l	labour coefficient vector in thousand persons per mill. DM, 40×1
τ	private average propensity to consume, dimensionless, 1×1
ν	public average propensity to consume, dimensionless, 1×1
φ	private and public average propensity to invest, dimensionless, 1×1
Ψ	ration between foreign trade surplus and gross regional product, dimensionless, 1×1
$\bar{\omega}$	minimum average income per head in mill. DM per year per thousand persons, 1×1
$_0 u$	household sewage per head in cubic metres per day, 1×1
$_1 u$	cooling water coefficient vector in cubic metres per mill. DM, 40×1

$_2u$ processing water coefficient vector in cubic metres per mill. DM, 40×1

$_3u$ personnel water coefficient vector in cubic metre per mill. DM, 40×1

B^h B.O.D. coefficient vector for all types of waste water, with and without treatment respectively in grammes per cubic metre per day, 122×1

π B.O.D. degradation coefficient in grammes B.O.D. per day, 1×1

γ dissolved oxygen deficit coefficient in grammes dissolved oxygen deficit per gramme B.O.D. per day, 1×1

ϕ coefficient denoting the k-th capacity stage of the treatment plant in cubic metres per hour, $k = 1, \ldots, 4$

Γ_k construction cost coefficient for the k-th capacity stage in mill. DM, 4×1

ϵ land requirement coefficient for treatment plants in hectares per cubic metre per day, 1×1

β B.O.D.-river standard in grammes per cubic metre water, 1×1

δ dissolved oxygen deficit river standard in grammes per cubic metre of water, 1×1

ζ B.O.D.-concentration of withdrawn water in grammes per cubic metre, 1×1

ξ dissolved oxygen deficit of withdrawn water in grammes per cubic metre, 1×1

Interpretation of Indices

i, j denote sectors ($i, j = 1, \ldots, 40$)

r denotes regions ($r = 1, \ldots, 4$)

$*$ denotes summation over all regions

k denotes the number of linear segments necessary for linearising the non-linear treatment plant construction cost function ($k = 1, \ldots, 4$)

s denotes a control point for the B.S.B.-load ($s = 1, \ldots, 4$)

p denotes a control point for the D.O.-deficit ($p = 1, \ldots, 6$)

REFERENCES

[1] Kneese, A. V., 'Environmental Pollution: Economics and Policy', *American Economic Review*, vol. 61 (1971) p. 156.

[2] Russell, C. S., and Spofford, W. O., 'A Quantitative Framework for Residuals Management Decisions', in Kneese, A. V., and Bower, B. T., *Environmental Quality Analysis: Theory and Method in the Social Sciences* (Baltimore: The Johns Hopkins University Press, and London, 1972) pp. 115–79.

[3] Kneese, A. V., and Bower, B. T., *Managing Water Quality: Economics, Technology, Institutions* (Baltimore: The Johns Hopkins University Press, 1968) pp. 237–53.

[4] Tinbergen, J., *Centralisation and Decentralisation in Economic Policy* (Amsterdam: North Holland Publishing Co., 1954) pp. 7–14.

[5] Theil, H., *Economic Forecasts and Policy* (Amsterdam, 1970) pp. 385–90.

[6] Stäglin, R., and Wessels, H., 'Input-Output-Tabelle für die Bundesrepublik Deutschland, 1966', *Vierteljahreshefte zur Wirtschaftsforschung* (1971) pp. 215–20.

[7] Statistisches Bundesamt Wiesbaden, Industrie und Handwerk, Reihe 5, *Energie und Wasserversorgung*, II: *Wasserversorgung der Industrie* (Wiesbaden, 1967).

[8] Bucksteeg, W., 'Problematik der Bewertung giftiger Inhaltsstoffe im Wasser und Möglichkeiten zur Schaffung gesicherter Bewertungsgrundlagen', *Münchner Beiträge zur Abwasser-, Fischerei- und Flußbiologie*, vol. 6 (1959) pp. 1–21.

[9] Meinck, F., Stooff, H., Kohlschütter, H., *Industrie-Abwasser*, 4th ed. (Stuttgart, 1968) pp. 694–9.

[10] Pürschel, 'Der Wasserverbrauch von Industrie, Handwerk und Kleingewerbe', *Zeitschrift des Deutschen Vereins von Gas- und Wasserfachmännern*.

[11] Streeter, H. W., and Phelps, E. B., 'A Study of the Pollution and Natural Purification of the Ohio River', *Public Health Bulletin*, no. 146 (Washington, D.C., 1925).

[12] Günther, W., 'Die Planung eines Systems von Kläranlagen in einem Flußgebiet mit Hilfe der linearen Optimierung', *Wasserwirtschaft–Wassertechnik*, vol. 18 (1968) pp. 217–22.

[13] Günther, W., 'Optimierungsmodelle zur Planung und Betreibung von Anlagensystemen zur Abwasserbehandlung in Flußgebieten', *Wasserwirtschaft–Wassertechnik*, vol. 20 (1970) pp. 81–4.

[14] Grava, S., *Urban Planning Aspects of Water Pollution Control* (New York and London, 1969) p. 63.

[15] Federal Water Pollution Control Administration. *Examination into the Effectiveness of the Construction Grant Program for Abating, Controlling, and Preventing Water Pollution* (Washington, D.C., 1969) Appendix 1, pp. 1–28.

[16] Schmidt, U., Über die Kosten der biologischen Abwasserreinigung', *Veröffentlichungen des Instituts für Siedlungswasserwirtschaft der TH Hannover* Hannover, 1964).

[17] Kehr, D., and Teichmann, H., 'Bau- und Betriebskosten öffentlicher Kläranlagen in der Bundesrepublik', *Veröffentlichungen des Instituts für Siedlungswasserwirtschaft der TH Hannover* (Hannover, 1961).

[18] Bucksteeg, K., 'Beitrag zum Thema: Baukosten von Kläranlagen', *Das Gas- und Wasserfach*, vol. 112 (1971) pp. 163–72.

[19] *IBM Application Program, Mathematical Programming System/360, Version 2, Linear and Separable Programming: User's Manual*, Program No. 360A–CO–14x, 3rd ed. (1971) pp. 165–88.

[20] Ruhrverband, *Wasserversorgung und Abwasserbeseitigung im Ruhrgebiet. Bericht über die Arbeiten von Ruhrtalsperrenverein und Ruhrverband.* Unpublished report by the Ruhr Water Resources Association.

Discussion of the Papers by Allen V. Kneese, and Rainer Thoss and Kjell Wiik

Formal Discussant: Beckerman (Chapter 3, A. V. Kneese). This paper is one that we shall all applaud, whatever the details and irrespective of how far we may approve of the contents, because the main conclusion that emerges is that environmental policy should not be left to administrators or to engineers. It needs to be determined in conjunction with economists. It discusses two detailed studies of water quality management, the first of which (the Delaware study) shows that some economic incentive to water quality control should be cheaper than the sort of direct regulation conventionally favoured by administrators; the second shows that, if the range of options is opened up beyond that normally considered by engineers operating within a conventional political and legal framework, superior arrangements will be indicated.

When there are several possible sources of waste, and hence of pollution abatement, a given amount of abatement will be achieved at minimum cost when the marginal costs of abatement are equal in all abaters (subject to the usual qualifications, such as the possibility of multiple equilibria and so on). Otherwise it would obviously be possible to reduce costs further by switching some abatement from high (marginal) cost abaters to low marginal cost abaters. Broadly speaking, this is why some price mechanism, such as pollution or effluent charges, is superior to the direct regulation often used by administrators, and which takes the form of directing all polluters to reduce their pollution by x per cent. While this is clear in principle, officials and politicians are suspicious of economic advice based on purely theoretical models of the way firms operate (some people just have very suspicious natures). Hence it is extremely useful to be able to point to some empirical evidence of the degree to which uniform direct regulation would be more costly than some charging scheme.

Of course, the Delaware study by no means provides *all* the evidence one would need for such a purpose, since it does not contain any evidence concerning the behavioural patterns of the various dischargers of effluent into the river, so that it is not possible to say that they would react to an effluent charge in the optimal manner. Hence, as the author points out, the estimates in the study only represent the cost of achieving target quality in the river by means of effluent charges on the assumption that firms are rational, and that they are rationally trying to maximise profits. But I doubt whether the rationality assumption is all that important in the long run, since (i) the hidden hand is likely to operate to some extent in weeding out those firms that are bad at making profits, and (ii) in so far as firms are not profit-maximisers, this would apply to all their operations, not just their failure to maximise profits with respect to the degree to which they continue to pollute the river in the presence of effluent charges. Hence, if irrationality of firms were to be an argument against the use of effluent charges and in favour of administrative regulation, it should equally well be an argument for, say, using direct regulation to tell firms how much investment they should carry out, or how much raw materials they should use, and so on.

Given all the data in the Delaware study, it is possible to obtain a minimum-cost allocation of pollution abatement among different effluent dischargers, and the study includes an estimate of what such an optimal allocation would cost. This is then compared with three alternative procedures for allocating the reduction in waste discharges among various dischargers. Firstly, the cost of the uniform reduction by direct regulation (e.g. everybody reduces their waste discharged by x per cent); secondly, the cost of achieving the same target for river quality if a uniform charge (i.e. a given charge per unit of waste) were imposed on all polluters so that all polluters would reduce effluent up to the point where the marginal costs of further abatement were equal to this uniform charge. Thirdly, an estimate is made of the cost of achieving the same quality target if the charge was varied from one zone to another to take account, not only of the differences in the cost curves between different dischargers, but also of the different damage functions corresponding to the fact that discharge from one zone will not have the same effect on the rest of the river as discharge from another.

The results show that the uniform direct regulation method would be over twice as costly as the minimum-cost allocation of abatement that could be specified in the light of all the data available (and the computer programme used to minimise costs subject to the target constraints for water quality and so on). But the calculations also show that the third system, namely zone effluent charges, is almost as cheap as this minimum-cost allocation, and that the second system, namely the uniform charge, is still less expensive than uniform direct regulation, though more expensive than zoned effluent charges.

The superiority of the latter method arises, of course, from (i) the point made above, namely that the damage done by any discharge depends on location and other physical parameters that enter into the determination of the self-purification capacity of the river, and (ii) the fact that the charges are specified in terms of physical units, not in units of damage done. As regards the former point, when there are several dischargers, each having a different function relating its physical flow of effluent to the damage flow, as well as different abatement cost functions, it is desirable, in principle, to vary the charge according to the different level of cost (and damage) at which each discharger's marginal abatement cost function equals the marginal social damage function. While, therefore, a uniform effluent charge in terms of physical units is better than uniform direct regulation, it is not as efficient as a method which discriminates between zones with respect to the different functional relation between the physical flow of their effluent and the damage done by it. For example, it would be absurd to apply a uniform charge per unit of biochemical oxygen demand to some discharger who was located at such a point in the river that the whole of the effects of his waste was eliminated by the self-purification capacity of the river by the time the water entered into the utility or production functions of any subsequent user. As the author puts it, 'In effect, this zone charge procedure "credits" waste dischargers at locations remote from the critical point with degradation of their wastes in the intervening reach of a stream before they arrive at the critical reach'. The converse of this is that there should be an incentive for pollution to be transferred to areas where the social damage incurred is least, and this is what tends to be achieved by a system of zoned effluent charges.

Another way of presenting the argument is to say that what is required *is* a uniform charge, provided the charge is in terms of units of damage done, not in terms of units of physical flow of effluent, or B.O.D., or some other physical parameter. Since the relation between the physical units and the economic damage done will vary from one point of the river to another, uniform charges per unit of damage imply variable charges per unit of physical characteristic of the effluent. Conversely, a uniform charge in terms of physical units will mean a variable charge per unit of damage, so that it will not be possible to obtain the equalisation of marginal costs with marginal social damage that corresponds to the usual economists' prescription for optimal allocation. I think that if the paper had made the point in terms of the difference between uniform charges per unit of damage and uniform charges per physical characteristic, it would not appear to be advocating conclusions which, at a superficial glance, seem to be intuitively unattractive, but which are, in fact, perfectly sound.

The author is careful to highlight some of the rougher edges in the Delaware study. These include the very simplified models that had to be used to estimate the amenity value of different standards of water quality. The author does not give the details of these estimates, but references are provided to the main sources of further information. The author also points out that the cost data underlying the exercise do not cover all possible technological methods for waste reduction, including process changes in industry, so that the results probably overestimate the total cost.

On the other hand, his paper makes very little reference to the costs of implementing and monitoring an effluent charge scheme. This is a subject which is given a lot of attention in another paper presented to this conference (Chapter 6), and which reaches the conclusion that the amount of information needed to operate a charging scheme is no greater than that required to achieve the same targets by means of regulation. While I am prepared to accept this proposition, since it is almost self-evident that the data required to identify the optimum degree of pollution do not depend on the particular methods adopted to attain this desired level of pollution, it is still possible that the data required to pursue optimisation policy by *any* means could be very costly. It has been increasingly recognised that the costs of obtaining information have tended to be neglected in the past in economics, and where we are dealing with what is essentially a non-marketed externality, namely pollution damage, the chances of the right data being easily available are remote. What it amounts to, I suppose, is that the costs of obtaining the requisite data should be incorporated into the abatement cost function, so that the optimum in the light of these functions may be inferior to sub-optimisation on the basis of lower-cost abatement functions.

The last part of the author's paper is about the Potomac study, where a plan for water quality control had originally been drawn up by the Corps of Engineers. The author shows that the so-called benefit–cost calculation accompanying this original plan was very defective, largely because it failed to take account of more than the most conventional techniques of controlling river conditions, and was also influenced by financial considerations that were irrelevant to any social accounting criteria.

There are a few other minor points that I should like to make. Firstly, the author follows the original Delaware study in talking about the effluent charge method as if it is the only method that leads to optimum resource allocation, whereas *any* method that creates the correct opportunity cost of pollution, whether by charging for it or by paying a subsidy to reduce pollution, should have the same allocative effects.

Secondly, the author discusses the use of the Delaware study results for purposes of obtaining an order of magnitude for other river basin policies, and he indicates that the charge could then be varied in a somewhat iterative manner if the initial charge proved to be inappropriate. This raises a point that is frequently made by those who prefer regulation, namely that with direct regulation the controllers can at least know what result they will achieve, whereas with charges they cannot know in advance how much firms will react. But the weakness of this argument is that although they may be certain to hit their target by direct regulation, they have no means of knowing – and no incentive to find out – if the target is the correct one. By contrast, with the charge, it will suffice to compare the charge with the marginal social damage at the point reached after firms have been allowed to adjust to the charge. If there is a divergence it will be clear that the charge could not have been correct, and according to whether the gap is positive or negative the direction in which the charge should be changed will be indicated. Pollution control by direct regulation may ensure that the administrators will always be able to hit whatever target they aim at, but pollution policy should not be a sort of play-therapy whereby the self-confidence of administrators can be boosted by allowing them to win at easy games.

Thirdly, the paper does not bring out at all clearly the role of public purification and restoration facilities, though there is a passing reference to the payments that firms should make to public facilities for their services. More analysis of the way that such facilities enter into the determination of the optimum allocation between private and public effluent control or treatment, as well as the analysis of the allocation between different firms in response to an effluent charge, would have been particularly useful for those concerned with policy.

Fourthly, the reader is left in suspense concerning the outcome of the study – i.e. how far did the Delaware River Basin Commission adopt the conclusions of the study?

In his concluding remarks, the author speculates on the reasons for some of the hostile reactions that are aroused by suggestions to use pollution charges, rather than direct regulation, as a method of attaining optimum levels of pollution abatement, and I should like to confirm that these reactions are not confined to the United States but are just as common in the United Kingdom. Our conservationists, too, fear that a charging scheme would not completely eliminate pollution and do not appreciate that we wicked economists don't want to eliminate pollution; we only want to maximise social welfare by reducing pollution to the point where it would cost more to reduce it by a further unit than the damage inflicted by it.

Formal Discussant: Russell (Chapter 4, R. Thoss and K. Wiik). There is no doubt that this paper, along with that of Førsund and Strøm (Chapter 2), represents a valuable contribution to the rapidly growing field of environmental modelling. In order to make clear the nature of that contribution, I intend to begin by commenting briefly on the paths which previous work has followed. Then I shall bring up my reservations about this approach.

I believe it is desirable to begin by establishing a schematic view of the system underlying what may be called residuals management decisions. This may be thought of as consisting of three boxes. The first is the economy, in which residuals are generated as an inevitable result of man's production–consumption activities. These residuals may be changed in form or location or stored over time, but ultimately all material not re-cycled in production, and all energy, is discharged to the natural environment which is represented by the second box in our system.

In the environment, the discharges are again transformed, transported, accumulated and so forth, and at any particular moment the result is a spatial distribution of environmental quality, defined in terms of such elements as SO_2 concentrations in the air, dissolved oxygen levels and fish populations in watercourses, rate of deposition of soot, and so forth. This ambient quality level (and, of course, its time path) gives rise to damages and benefits to the receptors who live in the environment; and it may also be true that governments have imposed, or will impose, con-straints on the 'badness' of the environment. The third box, then, we can visualise as a comparison and evaluation box for the ambient environ-mental quality.

The problem for public policy, if we confine ourselves for the moment to a region (or problem shed), is to find the optimal levels of production, consumption and residuals discharge, taking into account the damages imposed on receptors through the common property resources of the air and water (and the visual aspect of land). Models may be distinguished on the basis of how they treat each of these boxes and, consequently, which features of the broadest problem they reflect and which they ignore or slight.

Models of the economy used in previous work in this field have fallen into two groups: general equilibrium models (usually of the input–output type) which *simulate* the effect on discharges of an assumed set of final demands and inter-industry structure; and partial equilibrium models (usually in the linear programming form) designed to find the optimal way of meeting given constraints on ambient environmental quality, where the optimum is most frequently defined in terms of costs for given production or consumption targets. (Some models have been designed to minimise costs of meeting discharge constraints, the impact of which on the natural environment is then simulated. However, there is usually no internal evaluation of the resulting ambient concentrations [1].)

Quite simply, models of the first type provide us with no efficient way of exploring the set of feasible ambient qualities and production levels, while models of the second type cannot tell us the pervasive impacts on the economy of a set of ambient standards which will presumably change relative prices, affect factor markets, etc.

This distinction is not meant to cover every model, but only to bring out the (quantitatively) most important lines of inquiry. In particular, there have been a number of non-linear general equilibrium models, but only at the abstract levels, I believe. Another caveat is that I confine myself to static models, though, as Førsund and Strøm point out (Chapter 2), the dynamic aspects of the problem are potentially very important.

Considerable effort is going into building models of the environment itself. On the aquatic side, the operational models, such as the one described by Kneese (Chapter 3), have generally relied on the work of Streeter and Phelps [2], who constructed a simple model of the relation between discharges of biochemically oxidisable organic material and dissolved oxygen levels in streams. This work has been advanced to allow, under the assumption of steady-state conditions of flow and temperature, the derivation of constant coefficients relating discharges to dissolved oxygen levels downstream (and upstream in tidal estuaries).

On the atmospheric side, work has generally started from the assumption of a Gaussian distribution of the residuals in a plume (or puff) of 'smoke' [3]. By following individual plumes using an average wind rose (directions and speeds averaged over a season or other period) plus other meteorological information, it has been possible to estimate constant coefficients relating discharges at a point to concentration at some larger or smaller number of points, usually arranged in a grid pattern, throughout the region.

Clearly, the fact that constant coefficients have been the product of most environmental modelling fits in very neatly with the stress on linear models of the economic box. It is a relatively simple matter to use these coefficients to include within a single large linear programme constraints on ambient quality. If the input–output table of a general equilibrium regional model can be disaggregated to the plant level, the simulation of regional quality resulting from a given bill of final demand is also easy to arrange [4].

Recent work, particularly in aquatic systems, has begun to go in the direction of models which are more complex and, hopefully, more accurate and informative. In particular, the aim has been to expand from the two-compartment Streeter–Phelps models (B.O.D. and D.O.) to models including as exogenous variables more important parts of the system – for example, algal and fish populations [5]. The aim is to make models which are more robust in the face of wide variations in discharge and stream flow conditions and which provide more meaningful inputs to the decision process. These models take the form of non-linear differential equation sets which, in general, do not yield constants, or even analytic functions, connecting discharge and concentrations at other locations. Analogous effort is now going into constructing models of the formation of photochemical smog in the urban atmosphere. Clearly, such models will force us to develop and use optimisation techniques designed to deal with non-linear constraint sets even if we maintain our linear economic models. For example, one such technique has been developed by Fiacco and McCormack [6] in which penalty functions are introduced to translate constraints into objective function components.

The third box (comparison or evaluation) may be essentially empty, if we are simply simulating with an input–output system linked to discharge coefficients and environmental models. In the optimisation models we have the choice between comparison against standards (using constraints or penalty functions) or evaluation via damage functions. While there has been considerable research in this latter area, at the present time there are no results in which confidence can be placed for most of the damage effects we suspect exist, nor is there even a highly developed methodology. Over at least the near term we shall almost certainly have to content ourselves with using environmental constraints and reporting to decision-makers the minimum-cost (or maximum net benefit) ways of meeting alternative sets of such constraints.

From this very brief survey of a rapidly growing field, it seems clear enough that the contribution of the authors' paper lies in its wedding of the optimisation and input–output approaches in the economic box. (The Førsund–Strøm paper (Chapter 2) also combines input–output and optimisation, but it does not deal with ambient quality, being a national aggregate, rather than a regional, model.) This wedding allows the tracing of indirect impacts through the general equilibtium framework, while at the same time permitting the efficient search for the 'best' ways to meet alternative sets of ambient quality constraints. The rest of the authors' model – that is, their use of the Streeter–Phelps model and of a combination of ambient and effluent standards – is very much in the tradition of other models in this area.

Now let me spell out my reservations about the use of input–output tables in an optimisation setting for solving regional pollution control problems:

(i) One of the arguments for using input–output tables is that they frequently already exist, thus drastically reducing research cost and time. But existing input–output tables (with only one exception that I know of [4]) are based on data aggregated for sectors, and sectors in general contain more than one plant. Consequently, the spatial distribution of discharges, a key feature of real-world pollution problems, gets washed out.

(ii) Input–output tables, by their nature, do not allow for substitutions of processes, inputs, etc., as part of the response to ambient quality standards. The standard table simply gives us a particular technology for each sector; if we want to add substitution possibilities, we run into the following problems:

(a) Data availability – one of the advantages of input–output analysis disappears. (Indeed, we might as well go to a completely disaggregated model.)
(b) Costing of alternative processes to the extent that the opportunity costs of the substitution are not explicitly reflected in the matrix structure.

(iii) Once residuals are generated in the input–output model, only treatment is a feasible alternative for discharge reduction, and that treatment must generally be assumed to be carried on in regional plants because of the aggregation mentioned in (i). Under special circumstances, a treatment plant for a sector might make sense, but generally something similar to the authors' approach will be necessary, that is, the model will have one, or only a very few, regional plants. Regional plants, however, may or may not be desirable in a real problem, for one is here trading economies of scale against less efficient use of the river's ability to 'purify' itself. (Note that treatment cost does *not* enter the objective function in this paper but is only calculated and reported at the end of the optimisation. This peculiarity arises because the treatment cost function is concave, and the model would always run out to the quality constraints; it would never be stopped by rising marginal costs of treatment. The inclusion of the treatment cost piecewise approximation would only raise the possibility of multiple local optima.)

(iv) In the process of solving the optimisation programme, it is entirely possible to push the regional product mix (and relative price structure) rather far from that which obtained where the input–output table was estimated. This will stretch the 'equilibrium' concept, perhaps beyond the breaking-point.

None the less, I think the authors have made a valuable start down a road that definitely requires exploration. I hope they will experiment with ways to refine the approach to reduce the force of some of my objections and to increase their model's utility in actual problems.

Rothenberg felt that charges and subsidies were not really equivalent. Any equivalence rested on the assumption that there was a unique combination of functions which was independent of any administrative arrangements. However, the functions must vary – depending on institutional arrangements – because charges represented a change in *ex ante* incentives to polluters; subsidies represented a change in *ex post* incentives. Potential polluters had to commit themselves in advance before anyone was willing to subsidise them or else everyone would want to pollute. The damage functions would therefore differ because they would be based on different initial conditions.

Mills added that if you use subsidies instead of fees, you lower the relative cost of the polluting commodity in the production function. This would produce a higher level of consumption of this commodity and hence lead to more pollution.

Kolm noted that, since France was a laboratory for Kneese's ideas on regulating water pollution through taxation, a brief comment on the situation there could be of interest. The rules of the *Agences de Bassin* corresponded to Kneese's scheme. The tolerable levels of pollution were set by a Basin Committee on which all the interested parties were supposed to be represented. These levels were then implemented by taxing polluters. However, clues from various sources suggest that the tax rates were lower, and the resulting pollution levels higher, than the social optimum. This implied that the quasi-political process by which these levels were determined was less than perfect: the polluters exercised too much power (the Administration often sided with them, and many victims were not really represented). One of these indications was derived by applying the theorems of 'qualitative returns to scale' (Chapter 5) to the fact that the proceeds of the tax covered only a small part of the cost of public purification (one-fifth on the lower Seine). It thus seemed that tax rates should be raised by a substantial amount. The use of the theorems of 'qualitative returns to scale', together with estimates of the relevant elasticities, thus led to the conclusion that the pollution tax rates on the lower Seine should be three times higher than at present. This change was presently being implemented.

Beckerman disagreed with Rothenberg's statement that there was a difference, from the point of view of resource allocation, between pollution charges and subsidies, provided both were based on units of pollution (or pollution abated). It was essential to distinguish between short-run and long-run effects. As Abba Lerner had pointed out [7], one normally only wanted to use subsidies rather than charges on equity grounds – i.e. to avoid penalising firms established before pollution was regarded as socially undesirable. One could therefore take it for granted that pollution abatement subsidies would only apply to existing levels of pollution, so that, as Lerner pointed out, one could ignore the problem of providing an incentive for new potential polluters to come into the pollution business as a sort of protection racket in order to qualify for the pollution abatement subsidy. This meant that one could ignore the long-run effects on the number of polluters and concentrate on the short run.

Beckerman also disagreed with what Mills seemed to be saying, namely that the use of subsidies added to pollution (compared with charges, though not compared with no restriction at all) on the grounds that it lowered the prices of the products which gave rise to pollution. In fact, the prices of these products would be raised by a subsidy to pollution abatement as much as by pollution charges. The reason was that the opportunity cost of pollution was exactly the same under either method, and it was the opportunity cost which determined the supply curves of the products concerned.

For example, consider a production function of the form $Q = f(K, L, S)$, where K was capital, L was labour and S was smoke (this could be looked at either as a joint product or as the counterpart of an input into the productive process, namely 'clean air' used up in the course of polluting it with smoke). Now, if S was subject to a tax, it would no longer be used up to the point where its marginal product was zero. Instead the firm would reduce S and to do this it would have to substitute K and L to amounts depending on the precise shape of the production function. If there was a subsidy, based on reductions in S, the opportunity cost of S was precisely the same; in the one case the firm pays more tax as it uses more S, and in the other case it forgoes the extra subsidy. The opportunity cost of S was the same in both cases, and the degree to which firms would substitute K and L would therefore be the same in both cases. As long as K and L had positive prices, increased use would raise costs, and the opportunity cost of S, which now also had to be added to the cost curve, was also the same in both cases. So the supply curve of the polluting products rose by the same amount in both cases. Hence, since there was no reason to assume any different behaviour of demand curves, the price would also rise by the same amount in both cases.

Lave said that adjustment costs were great – at least for air pollution. Iteration was very expensive, so that one had to hit the target quite closely the first time.

The cost of direct regulation for air pollution in Pittsburgh had been very great. The regulators had to know detailed information about production functions and abatement techniques. They had to have about as much knowledge as was required for central regulation. Presumably, effluent charges would be more efficient.

While uniform effluent (emission) charges or standards were presumably inefficient, they were more acceptable politically and perhaps more efficient economically. Regions were afraid of losing industry; they demanded uniform emission charges so that nothing was gained by moving. This would also tend to preserve some really clean (pure) areas or streams. He preferred some clean streams and some sewers rather than uniformly mucky ones. There was currently a development in the United States requiring areas not to allow their air to get worse than it was today; thus some areas must remain quite clean.

He would advocate uniform national emission charges which could be increased by area authorities.

Kolm thought that Lave had not made enough use of the purely economic arguments at his disposition. Environment was a localised resource. Uniformity came from mobility either of other factors, or of products, through a 'factor-price equalisation' mechanism. But the shortcomings of factor-price equalisation were well known from international trade theory. And in the short run other factors were not mobile. For these reasons, the optimal allocation of resources called for geographical discrimination, in spite of the existence of tendencies for long-run equalisation.

Prud'homme said that Kneese's ideas were the foundation stone of a system of water quality management that had been in existence in France since 1964. France had been divided, for that purpose, into six water *bassins*; in each *bassin*, an agency levied effluent charges, based on the amount of B.O.D. and of suspended matter discharged into the rivers. The money raised was then lent, or given, to potential polluters for water treatment plants. There had not yet been a detailed assessment of the scheme. It was probably too early to undertake one because (i) the scheme had been slow starting and had only really been operating for the last two or three years, and (ii) the level of the effluent charges was still very low – and scheduled to be greatly increased by 1975.

He nevertheless asked some students to write term papers on the efficiency of effluent charges. They had interviewed about 50 potential polluters, 20 of whom were local communities. About half reported that they were already operating a water treatment plant before the tax was enacted, and that the tax had therefore not had any influence on their behaviour; the other half reported that they were engaged in, or contemplating, the establishment of water treatment plants, but mostly for 'civic consciousness', since the cost incurred for such plants was about five times greater than the tax they would save. The 30 others interviewed were industrialists. None reported any change of production processes induced by the tax. Only a few suggested that they had introduced some sort of water treatment because of the tax. Practically none of the 50 polluters interviewed viewed the tax as an inducement; all of them viewed the tax as a tax, i.e. as a means of raising revenue earmarked for the financing of water treatment equipment. Their tentative conclusion was that the efficiency of effluent charges in France did not seem to be very great.

Russell shared Lave's preference for having some areas of pristine beauty or natural wilderness and some areas of intense industrial and commercial development (as opposed to uniform mediocrity), but did not share his gloomy view that the establishment of emission charges based on damages would lead to the latter. He believed this view was based on a restricted concept of damage; he was perhaps only thinking of air pollution damages (to health and the like) which would necessarily be low in a lightly settled area. If, however, any significant number of people shared our preferences, then damage functions – or, more realistically, the political process – would reflect this, and industries would face very high emission charges in planning to move their operations into a scenic or wild area.

Thoss asked why there was any conflict between standards and charges. In the case of water quality standards it did not really matter whether you tried to reach this standard by means of charges or regulations. Once a standard had been set you could easily set up a market in 'rights to pollute' and sell them until the set standard had been reached.

Beckerman replied to Thoss's question by observing that one was aiming at the same optimum amount of pollution with both methods, but with a charge (or subsidy) the optimum amount would be obtained at a lower total cost. Secondly, he was correct in raising the issue of 'pollution rights' which was developed by Dales [8]. It was true that, given a desired amount of pollution, this amount could be allocated by auctioning pollution rights in the way that Dales had suggested. This would then also lead to the least-cost allocation, since the abatement cost function of firms was, in effect, their demand curve for pollution rights. If the optimum supply was therefore placed on the market, it would intersect the demand curve for pollution rights at the optimum point – i.e. where the marginal social costs of pollution intersect the marginal costs of abatement. But the pollution rights solution was not discussed in the Kneese paper, which was concerned with demonstrating that a given target level of pollution could be obtained more cheaply with pollution charges than with direct regulation.

Kolm replied to earlier points made by Lave and Rothenberg, particularly those by Lave, by saying that if there were two identical rivers, and a given amount of pollutant which had to be discharged into the water, he would prefer to put it all into one river (thus keeping the other one clean) rather than spread it evenly between them. He wished to point out that this revealed preference was non-convex. The variables associated with this preference were the pollution levels in each river (variables which were preferred when lower), and these two variables entered symmetrically into the preferences (since these rivers were identical in all other respects). With convex preferences, the choice was thus an equal sharing of pollution between the two rivers. To prefer to have one river very polluted and the other completely clean implied non-convex preferences (and then, because of symmetry, it did not matter which river was clean and which was polluted).

Kavanagh felt that Kneese's paper provided us with an excellent assessment of the progress of the application of economic analysis to water quality management using the Delaware and Potomac rivers as case studies. His six conclusions merited individual discussion, but he wished to pick out conclusion 4, which stated, 'Even the few efforts at "comprehensive" planning for water quality improvement have not sufficiently examined the "production function" to permit a least-cost solution to be found', for attention, because it illustrated the point he had made when discussing the paper by Førsund and Strøm (Chapter 2).

If we were going to make cost–benefit analysis an operational tool for decision-making in water resources development, we had to integrate the economic and engineering data of the systems under examination. We had to get engineers and economists to work together to produce operational models and, as a longer-term goal, to push for the development of courses in economics for engineers.

In discussing some models of the natural system, the Streeter–Phelps model had been discussed. This model seemed to be favoured in the U.S. studies, but some engineers in Britain believed that it was not a good predictor of water quality in U.K. river systems (partly because of differences in relative sizes).

In Britain, a model of the Trent river system has been developed [9]. One outcome of this model was the availability of empirical estimates of angling benefits for part of the river system. These estimates, shown in Table 4D.1, were unlike those reported by Kneese for the Delaware river in that they were estimated by measuring the demand for outdoor recreation using the method suggested by Marion Clawson. The figures in Table 4D.1 were monetary benefits in terms of consumers' surplus. These suggested an annual value of angling benefits of over £300,000 for the

TABLE 4D.1

PROVISIONAL ESTIMATES OF ANGLING BENEFITS, RIVER TRENT MODEL STRETCHES, MARCH 1968–MARCH 1969

River	Limits of stretch of river	Aggregate benefits £	Benefits per km (mile) of river £	
Trent (T1)	Hanley to Hanford	nil	nil	
Trent (T2)	Hanford to Great Haywood	nil	nil	
Trent (T3)	Great Haywood to Yoxall	250*	13·1	(21)*
Trent (T4)	Yoxall to Burton	400*	17·5	(28)
Trent (T5)	Burton to Shardlow	4,965	158·2	(253)
Trent (T6)	Shardlow to Trent Bridge	61,650	2,920	(4,671)
Trent (T7)	Trent Bridge to Kelham	75,770	2,067	(3,308)
Trent (T8)	Kelham to Dunham	62,880	2,245	(3,593)
Trent (T9A)	Dunham to Littleborough	17,000	1,328	(2,125)*
Trent (T9B)	Littleborough to Wildsworth	630*	33·1	(53)*
Trent (T9C)	Wildsworth to Keadby	160*	8·8	(14)
Soar (S1)	Birstall to Stanford	6,020	294	(470)
Soar (S2)	Stanford to Keyworth	205	20	(32)
Soar (S3)	Keyworth to confluence with Trent	660	121·2	(194)
Anker (Ank.B)	Weddington to St. Helena	150*	8·1	(13)*
Anker (Ank.A)	St. Helena to Tamworth	5,050	420	(673)
Tame (UTA)	Tame Bridge to Perry Barr	nil	nil	
Tame (UTB)	Perry Barr to Water Orton	nil	nil	
Tame (MT)	Water Orton to Lea Marston	nil	nil	
Tame (LT)	Lea Marston to Chetwynd Bridge	nil	nil	
Churnet (CH)	Bridgend Leek to Rocester	860*	22·5	(36)
Dove (D1B)	Rocester to Aston Bridge	4,950	289	(463)
Dove (D1A)	Aston Bridge to Martin's Bridge	6,015	278	(446)
Amber (AM1)	Shirland to Bull-Bridge	280	29·4	(47)
Amber (AM2)	Bull-Bridge to Ambergate	120	53·7	(86)
Derwent (DT1B)	Matlock Bath to Milford	10,325	576	(922)
Derwent (DT1A)	Milford to St. Mary's Bridge	4,040	316	(505)
Derwent (DT2)	St. Mary's Bridge to confluence with Trent	6,560	286	(458)

Items marked * indicate that no statistical testing was possible due to lack of sufficient degrees of freedom.

Source: N. J. Kavanagh and J. G. Gibson, 'Measurement of Fishing Benefits on the River Trent', *Proceedings of Symposium on the Trent Research Programme* (Nottingham, Apr 1971).

period under present water quality states and socio-economic conditions. Of course, these figures were subject to the reservations usually associated with the Clawson method, but they did provide some idea of the value of angling benefits in the river stretches considered. Where nil values were stated, the water was 'heavily' polluted.

Russell did not believe that either the Clawson–Knetsch or the 'participation probability' approaches had yet been applied in such a way as to get at the fundamental question of how benefits change with *small* changes in quality. Both methods presently required us to assume that large changes would be made, e.g. that the Delaware estuary would be 'cleaned up', and then we treated the body of water as an addition to regional recreational water area, just as a new Corps of Engineers reservoir would be treated. This was clearly inadequate. Firstly, it was most often the case that water-based recreation was taking place before any change in quality was accomplished (witness the fishermen and boaters who use the Potomac estuary within the city of Washington where the river is subject to large raw sewage loads). Changes in quality generally changed the types of recreation available, their location and the number and identity of participants. The effects were very seldom captured by the assumption that the water body was now 'available' whereas before it was 'unavailable'. Secondly, the method would produce, at most, one or two point estimates of benefits along the entire axis of stream quality. The marginal benefit would be undefined by virtue of the assumptions.

REFERENCES

[1] T.R.W. Inc., *Air Quality Implementation Program* (Environmental Protection Agency, 1970) vols. I, II.

[2] Streeter, H. W., and Phelps, E. B., 'A Study of the Pollution and Natural Purification of the Ohio River', *Public Health Service Bulletin*, no. 146 (Washington, D.C., 1923).

[3] Martin, D. O., and Tikvart, J. A., 'A General Atmospheric Diffusion Model for Estimating the Effects on Air Quality of One or More Sources', *Air Pollution Control Ass. J.* (1968) pp. 64–148.

[4] Isard, W., *et al.*, *Ecologic-Economic Analysis for Regional Development* (Free Press, New York, 1972).

[5] Di Toro, D. M., O'Connor, D. J., and Thomann, R. V., 'A Dynamic Model of the Phytoplankton Population in the Sacramento–San Joaquin Delta', *Advances in Chemistry Series*, no. 106 (American Chemical Society, 1971).

[6] Fiacco, F. A., and McCormack, G. P., *Nonlinear Programming Sequential Unconstrained Minimisation Techniques* (Wiley, New York, 1968).

[7] Lerner, A. P., 'Priorities and Efficiency', *American Economic Review* (1971) pp. 527–30.

[8] Dales, J. H., *Pollution, Property and Prices* (Univ. of Toronto Press, 1968).

[9] *Proceedings of Symposium on the Trent Research Programme* (Institute of Water Pollution Control, Nottingham, 1971).

Part II

Problems of and Approaches to Public Policy

5 Qualitative Returns to Scale and the Optimum Financing of Environmental Policies

Serge-Christophe Kolm

5.1 *THE PROBLEM*

This paper deals with the financial aspects of an environmental policy which utilises two tools: a tax on polluters; and the implementation, or financing, of the restoration, maintenance or improvement of environmental qualities.

This seems to be the situation confronting numerous agencies set up to deal with environmental problems. For instance, in France, the *Agences Financières de Bassins* established for each of the nation's river basins tax water polluters and finance water purification plants; the proceeds of this tax are earmarked for this expenditure, and it seems that the authors of this law thought that this revenue should be sufficient to provide all the financial resources required. The famous river agencies of the Ruhr region function along similar lines. Recreation parks of various kinds charge visitors a fee and carry out some cleaning, restoration or extension expenditures. The tax on aeroplane noise around airports, just instituted in France, is explicitly earmarked for expenditures to abate or compensate this nuisance [39]. And in the spring of 1972 the Social and Economic Council of France decided to adopt the taxation of polluters, and the use of the product of this tax for public improvements of the environment, as the general rule for all national environmental policies.

We are interested in optimising. But the optimisation of environmental policies requires the consideration of several interdependent criteria, whose synthesis is anything but straightforward. Among them are efficiency (i.e. 'Pareto optimality'), distribution (of welfare), effects on public finances, and the kind of social organisation required for informational reasons. Usually several criteria compete in the same decision, which is therefore commonly a compromise between their various requirements. However, in the field under study, we shall see that there is a broad convergence, rather than opposition, of these objectives.

The main social preference which has been revealed about the distributional aspects of environmental policies is the principle 'the polluters must be the payers'. It seems that all Western governments have officially enunciated or adopted this rule (which does not mean that their actions always conform to it). There is no doubt that the aim of this principle is distributional. More precisely, its basic meaning is to deny environmental property rights to the polluters. The obvious application of this principle to the environmental policies under consideration is that the proceeds of the pollution tax, or user's fee, must cover (exactly or at least) the expenditure on restoration or protection.

In spite of this, in all countries the present rapid development of environmental policies is seen as being necessarily a new and important drain on public funds. But taxing polluters, on the other hand, provides a new source of public funds. If both were equal, the overall environmental policy would have no direct effect on the overall government budget and hence on other taxes and public expenditures (environmental fees might even yield more than the corresponding expenditures, so that it would represent a net addition to public funds). (For an analysis of the effects of environmental policies on public budgets, see [38].)

The net financial result of the treatment of an environmental problem is also important for the organisation of the implementation. To entrust this implementation to an autonomous, decentralised, *ad hoc* agency offers many advantages: it will probably minimise the cost and delay of control and of bureaucratic procedures, and decisions will be taken by those in possession of the relevant local information, etc. This is clearly easier to do if the agency, which both raises the fees and finances the expenditure, has a balanced budget, or at least has no deficit which calls for control by the upper-level administration which has to pay for it (see [4], [10], [17], [18]).

The requirements for the above three objectives – distribution, the public budget, and organisation – focus on only one variable: the net financial result of the operation. The first question to raise is, thus: what is the level of this variable when the system is optimised from the viewpoint of traditional, narrow, economic efficiency? More precisely, these first problems make it of special interest to know if there is any reason to suppose that this financial result might be zero or, if not, would have a definite sign. In other words, we have to compare the yield of an efficient pollution deterrent tax with the optimum level of the environmental improvement expenditures; in particular, we wish to know whether they are equal and, if not, which is the larger.

5.2 *GENERAL THEORY*

5.2.1 THE MODEL

(a) *Presentation*. We shall first present a general theory of the problem, then use it to classify the situations, and finally apply it to several relatively specific cases.

The theory considers indexes of environmental qualities, E^i, which are defined up to a monotonic transformation (increasing or decreasing).

These indexes depend upon two types of variable, which respectively reduce and improve the quality of the environment.

The first variables, x_i, are the levels of polluting or deteriorating activities and the bases of the pollution taxes or environmental fees. An x_i can be a quantity of an output, of an input, of a pollutant, of a certain kind of deterioration, or a number of users who have similar deteriorating effects on the quality of the environment, etc. (see [28] for a theory of the choice of the tax base and the way in which the results of this paper are altered when the costs of implementation influence this best choice).

The environmental improvement variables, X_i, can, for instance, be levels of cleaning, repairing, protecting, improving, etc., activities, or surfaces, flows or volumes which can dilute a pollutant or disperse a deterioration, etc. A typology of their nature and its consequences will be presented below.

The x_i and X_i's are quantities. If they are numbers of indivisible units, these numbers are assumed to be large enough to be treated as continuous variables.

The dependence of the E^i's on the x_i's and X_i's is the *environment function*. Its structure will turn out to be the heart of the problem.

The tax on x_i will be a flat-rate tax of rate t_i, and the price of X_i will be a given p_i. In some cases the structures of these tax and cost functions can be different. But it is very easy to see how the differences would affect the results (see [12], [19], [10], [17], [18]). In other cases these tax and cost structures have to hold (see [22], [12], [19]).

v_i is the marginal social monetary value of E^i. If E^i is a collective concern ('public' good or bad), v_i is the sum of the agents' marginal willingnesses to pay for it, possibly weighted so that distributional objectives can be taken into account in this way too. v_i is a function of the variables of the system, and in particular of E^i.

There are various constraints on the variables x_i and X_i. First of all, these variables are non-negative since they are quantities. But there may also be other constraints on this set of variables; some of them will be specified below.

The general results will be derived from a model which does not look so general. This is for the sake of simplicity of writing. It is very easy to guess, or to prove, that these results hold for much more explicitly general representations. For example, we could explicitly consider all agents, households and firms, and allow all of them to experience and influence environmental qualities (this is done in [8], [9], [11] and [19]). Also, x_i's cost, defined in terms of a given unit price, could have a different structure, etc.

(b) *Notation.* We use the following notation:

x_i = the quantity of a deteriorating activity
X_i = the quantity of an improving activity
E^i = the index of an environmental quality
f^i = a constraint function
p_i = the price of X_i
q_i = the price of x_i
t_i = the environmental fee on x_i.

A letter without an index represents the vector, which has the same letter with indexes as components. x, t and q on the one hand, and X and p on the other, have the same dimension. Other vectors have different dimensions. One has

$$E = E\,(x, X)$$
$$f = f\,(x, X)$$

Examples of constraints involving f^i functions will be given in paragraph 5.2.2 below.

We use the traditional notations for first derivatives for, and of, vectors. Vector and vector–matrix multiplications will be obvious without writing the lines and columns differently.

The superscripts 1 and 2 represent polluters and pollutees, r^1 and r^2 represent their incomes (other commodities), $u^1\,(x, r^1)$ and $u^2\,(E, r^2)$ their utility functions, with $u_x{}^1$, $u_r{}^1$, $u_E{}^2$, $u_r{}^2$ as their first derivatives. We assume r^1 and r^2 are unconstrained and in demand so that $u_r{}^1 > 0$ and $u_r{}^2 > 0$. We call $v_x = u_x{}^1/u_r{}^1$ and $v = u_E{}^2/u_r{}^2$.

$F\,(x, X, r^1 + r^2)$ is the production possibility function. All its first derivatives are positive; F_r denotes its last one. One thus has $F_X/F_r = p$, and $F_x/F_r = q$.

λ_i', μ_i', v and γ are non-negative Lagrange multipliers; $\lambda_i = \lambda_i'/u_r{}^1$ and $\mu_i = \mu_i'/u_r{}^1$.

(c) *Optimisation.* The social optimum occurs when the following expression is maximised:

$$u^1(x, r^1) + u^2[E(x, X), r^2]$$

subject to

$$f(x, X) \geq 0 \quad : \lambda'$$
$$F(x, X, r^1 + r^2) \leq 0 : v$$

This gives

$$u_x{}^1 + u_E{}^2 E_x + \lambda' f_x - v F_x = 0$$
$$u_E{}^2 E_x + \lambda' f_x - v F_X = 0$$
$$u_r{}^1 = u_r{}^2 = v F_r$$
$$\lambda' f = v F = 0$$

Since $u_r{}^1$, $u_r{}^2$ and F_r are positive, we shall have $F = 0$ and $v > 0$. Therefore, the conditions are:

$$v_x + v E_x + \lambda f_x - q = 0 \tag{1}$$
$$v E_X + \lambda f_X - p = 0 \tag{2}$$
$$\lambda f = 0 \tag{3}$$

Equation (3) is equivalent to $\lambda_i f^i = 0$ for all i's.

(d) *Polluters' behaviour.* R being the initial income of polluters, these agents maximise $u^1(x, r^1)$ subject to

$$(q + t)\, x + r^1 \leq R : \gamma$$
$$x \geq 0 : \mu'$$

This gives

$$u_x{}^1 - \gamma(q + t) + \mu' = 0$$
$$u_r{}^1 - \gamma \qquad\qquad = 0$$
$$\gamma \cdot [(q + t)x + r^1 - R] = 0$$
$$\mu' x = 0$$

Since $u_r{}^1 > 0$, and therefore $\gamma > 0$, the first and last conditions are:

$$v_x - q - t + \mu = 0 \tag{4}$$
$$\mu x = 0 \tag{5}$$

(e) *The optimum tax.* Comparing (1) and (4) gives the optimum tax rate t:

$$t = - v E_x - \lambda f_x + \mu$$

(f) *Optimum financial result.* Equation (2) gives $p = v E_X + \lambda f_X$. Therefore, $B = tx - pX = - v \cdot (E_x x + E_X X) - \lambda \cdot (f_x x + f_X X) + \mu x$, that is, from equation (5):

$$\boxed{B = - v \cdot (E_x x + E_X X) - \lambda \cdot (f_x x + f_X X)} \tag{6}$$

We shall now study the effects on B of these two terms, i.e. the effect of the structure of f and of E.

5.2.2 FINANCIAL CONSEQUENCES OF THE CONSTRAINTS

The above expression for B leads to the following theorem: *A homogeneous or unbinding constraint is financially neutral.*

By financial neutrality we mean the absence of any direct effect on B, i.e. we say that constraint $f^i \geq 0$ is financially neutral when $\lambda_i(f_x^i x + f_X^i X) = 0$. However, such a constraint, when binding, can still influence B indirectly because it influences the optimum x and X.

The property is obvious for unbindingness since that makes $\lambda_i = 0$.

By the homogeneity of a constraint $f^i \geq 0$, we mean homogeneity of the function $f^i(x, X)$ in the set of all x_i's and X_i's. If f^i is homogeneous of degree zero, from Euler's Theorem, $f_x^i x + f_X^i X = 0$, and this constraint is financially neutral. If f^i is homogeneous of any degree δ, $f_x^i x + f_X^i X = \delta f^i$ and $\lambda_i . (f_x^i x + f_X^i X) = \delta \lambda_i f^i = 0$ from equation (3), which proves the property.

In particular, the following constraints are homogeneous:

(i) Non-negativity constraints $x_i \geq 0$ and $X_i \geq 0$.
(ii) A pollutant only exists because of activity x_i, which produces a quantity ax_i of it, and it can be removed by activity X_i which reduces this quantity by bX_i (a or b are 1 if x_i or X_i are themselves these quantities). Then $bX_i \leq ax_i$.
(iii) x_i per unit of a space dimension X_i, or improving activity X_i per unit of x_i, are limited: $x_i/X_i \leq A$ or $X_i/x_i \leq B$.
(iv) X_i cleans a surface or volume βX_i (β is a constant) per period, whereas the total variable dimension is $X_j : \beta \leq X_i X_j$.

This property has the following consequences:

If all constraints are homogeneous

$$\boxed{B = -v . (E_x x + E_X X)} \tag{7}$$

From the first property, this holds if all the constraints reached at the optimum are homogeneous. But we also know *a priori* that it holds, without any need to know which constraints are reached, which are not reached, and where the optimum is, if all the problem's constraints are homogeneous.

Let us now suppose that a constraint has an f^i of the form $f^i = \phi^i + k^i$, where ϕ^i is a homogeneous function of (x, X) and k_i is a constant. If ϕ^i's degree of homogeneity is zero, f^i is also homogeneous of degree zero, and this constraint is then financially neutral. Let δ be ϕ^i's degree of homogeneity. Then $\lambda_i . (f_x^i x + f_X^i X) = \lambda_i . (\phi_x^i x + \phi_X^i X) = \lambda_i \delta \phi^i = \delta \lambda_i . (f^i - k_i) = -\delta \lambda_i k_i$. For $\delta = 0$, this is the above remark. For $\delta = 1$, this expression is $-\lambda_i k_i$.

This latter case is found, for instance, when X_i is a cleaning or improving activity removing bX_i of pollutant or deterioration, and when, in addition to the quantity ax_i of pollutant or deterioration created by x_i, there is an initial or untaxed such quantity or amount

k_i, or when there can be a net improvement of this amount (for instance, plantations or reafforestation). Then, $bX_i \le ax_i + k_i$.

Now, all the constraints which will be found in the examples presented below will either be homogeneous of degree zero or one, or will have $f^i = \phi^i + k_i$ with k_i constant and ϕ^i linear homogeneous. The f^i's can thus all be written as $\phi^i + k_i$ where ϕ^i is homogeneous of degree zero or one and k_i is a constant, possibly zero. Then

$$B = -v \cdot (E_x x + E_X X) + \lambda k \qquad (8)$$

with $\lambda_i k_i = 0$ if constraint i is unbinding or homogeneous.

Furthermore, we may have constraints of the form $E \le \bar{E}$ (a constant). If all others are homogeneous or known not to have been reached, (6) becomes

$$B = -(v - \lambda^e)(E_x x + E_X X) \qquad (9)$$

where λ^e is the vector of corresponding λ_i's. If, in addition, all other constraints are unbinding or do not contain X, equation (2) shows that $(v - \lambda^e)E_X = p$. And if, finally, E is unidimensional, $p > 0$ shows that $(v - \lambda^e)$ has the sign of E_X if $E_X \ne 0$. Since, in this case, v also has this sign, we see that v and $(v - \lambda^e)$ have the same sign, and that the B of equation (9) has the same sign as that in equation (7).

5.2.3 QUALITATIVE RETURNS TO SCALE

Consider simultaneous proportional variations of all the x_i's and X_i's. If this does not change the quality of the environment, we say that there are constant qualitative returns to scale. If the quality of the environment is reduced by an increase, or improved by a decrease, we say that there are decreasing qualitative returns to scale. And if the quality of the environment is improved by an increase, and reduced by a decrease, we say that there are increasing qualitative returns to scale. (This idea was introduced in [11] and [19]; it is applied to many problems in these books, to the economics of waiting lines in [16] and [19], to the relation between growth and the quality of environment in [14] and [19], and to the static macroeconomics of the environment in [25]. It may be noticed that the x_i's and X_i's being *quantities*, the nature of the returns to scale at a point is well defined; by definition of the word 'quantity' it is a magnitude which is non-negative and for which zero, unity, addition, and multiplication by a scalar, have a meaning; we are thus not free to replace an x_i or X_i by a function of it which is not a mere multiplication by a positive scalar, i.e. a change of unit: if this function is the base of the tax or pricing, this means that there is quantity discrimination, which is an other problem.)

We apply these definitions, at a given point (x, X) for small variations around it. Now, with $dx = \epsilon x$ and $dX = \epsilon X$, where ϵ is scalar,

$$dE = E_x \cdot dx + E_X \cdot dX = \epsilon \cdot (E_x x + E_X X)$$

Then

$$\frac{du^2}{u_r^2} = v \cdot dE = v \cdot (E_x x + E_X X)\epsilon$$

The term u^2 therefore varies in the opposite direction to this variation, is unaffected by it or varies in the same direction, depending on whether $v \cdot (E_x x + E_X X) \gtrless 0$, i.e. qualitative returns to scale at (x, X) are decreasing, constant or increasing according to this condition.

It follows that in the case of equation (7), or in the case mentioned at the end of the last section, $B \lesseqgtr 0$ depending on whether the qualitative returns to scale are decreasing, constant or increasing at the optimum. This leads to the theorem: *If all constraints are unbinding or homogeneous, the environmental fees must yield more, as much as, or less than the environmental expenditures, depending on whether the qualitative returns to scale are decreasing, constant or increasing at the optimum.*

In addition, equation (8) shows that: – if $k_i \geq 0$ for all i, $B \geq 0$ if qualitative returns to scale are not increasing at the optimum; – if $k_i \geq 0$ for all i and $k_i > 0$ for at least one binding constraint, $B > 0$ if qualitative returns to scale are decreasing at the optimum.

With these results, the financial consequences of an environmental problem can be determined very easily and speedily, by a cursory examination of the environment function and constraints, or simply by glancing at the structure of the practical problem without even having to write these functions and constraints. Several examples of these possibilities are shown below. The point is to identify the nature of the qualitative returns to scale and of the constraints, eventually depending upon the value of the problem's parameters.

5.3 THE TWO BASIC TYPES OF ENVIRONMENTAL IMPROVEMENTS

We must now be more specific about the nature of the variables X_i. There are basically two types of such variables, i.e. two types of environmental improvements, which can be broadly described as 'cleaning' and 'dilution' (this very important distinction was suggested to me by a remark from Jean-Michel Grandmont).

From now on, all the variables will be unidimensional.

5.3.1 CLEANING OR REPAIRING

(a) *No external pollution.* The first case is that of a cleaning or repairing activity. We call y such an X_i. y is, for instance, the capacity of a water purification plant, or the number of workers in charge of a public park, or even the funds allocated to this activity. In the standard situation, each unit of x leaves a quantity a of pollutant, or trash, or deterioration, and each unit of y removes an amount of b of it. The quantity $Q = ax - by$ remains.

E is any monotonic function of Q. Let us take $E = -Q$, so that v is positive and is the marginal social (external) cost of Q. Since E is homogeneous of degree one in (x, y), $xE_x + yE_y = E$, so that equation (7) results in:

$$\boxed{B = vQ}$$

But what about the constraints? The constraints $x \geq 0$ and $y \geq 0$ are both homogeneous and, in addition, the nature of Q, as described, imposes $Q \geq 0$, i.e., $ax - by \geq 0$. This constraint is also homogeneous, so that the given form of B is correct. But this constraint also shows that $B \geq 0$, and $B > 0$ if $Q > 0$ and $B = 0$ when $Q = 0$.

That is to say: *The expenditure on improving environmental quality must not exceed the proceeds of the pollution tax; it must be lower if the cleaning or restoration must be less than complete; both must be equal if the cleaning or restoration must be complete.*

This latter condition, which is an especially interesting case of a balanced budget, may seem fortuitous. But it is not, since it holds when the constraint $Q = ax - by \geq 0$ is reached in the optimisation problem.

Let us find conditions under which this happens. We write λ_x, λ_y and λ_Q as the λ_i's for the constraints $x \geq 0$, $y \geq 0$ and $Q \geq 0$. Conditions (1) and (2) are

$$v_x - va + \lambda_x + \lambda_Q a - q = 0 \tag{10}$$
$$vb + \lambda_y - \lambda_Q b - p = 0 \tag{11}$$

In the usual case where v is a function of E alone, we can write it as $v(Q)$. The usual decreasing returns in this social cost means that v is an increasing function of Q. This is likely to hold for small Q's, and we assume it for $Q = 0$, although the opposite certainly holds for large Q's [24, 27]. We suppose that the constraint $Q \geq 0$ is reached at the optimum: $Q = 0$, and $\lambda_Q \geq 0$, with $\lambda_Q > 0$ if the constraint is binding. That is, $ax = by$, which shows that x and y can only be simultaneously zero, in which case $x = y = Q = 0$. Let us put this case aside. Then $\lambda_x = \lambda_y = 0$, and equations (10) and (11) give the conditions

$$v(0) \geq \frac{p}{b}, \quad v(0) \geq \frac{v_x - q}{a}$$

with strict inequalities if the constraint is binding.

(b) *External pollution.* However, one can imagine situations in which $by > ax$ might be possible: for instance, if there is originally some quantity of pollutant, or if some pollution or deterioration escapes the tax and is thus not included in x, or if the restoration consists of plantations or reafforestation. But in these cases there is generally an upper limit on $(by - ax)$. Call it k $(k > 0)$. The quantity of pollutant, or the shortfall from the most perfect environment, is then $(k + ax - by)$, and the constraint becomes $ax - by + k \geq 0$. Keeping the same definition for E and v, equation (8) then gives $B = vQ + \lambda k$, where $\lambda \geq 0$ and $\lambda = 0$ if $ax - by + k > 0$. But when $ax - by + k = 0$, $Q = -k$. The net result is thus

$$B = vQ \qquad \qquad \text{if } ax - by + k > 0$$
$$B = -(v - \lambda)k \qquad \text{if } ax - by + k = 0$$

Furthermore, the constraint can be written as $E \leq k$, and a previous result then shows that $(v - \lambda) > 0$.

The result is that B has the sign of the optimum Q. The environment fee must thus yield more, as much as, or less than the improvement expenditure depending on whether the overall operation must deteriorate, leave untouched or improve the environment. But now, $B = 0$ would be purely fortuitous, as is $Q = 0$ at the optimum. It is $Q = -k$ at the optimum which is now not fortuitous, since it holds when the constraint $(by - ax) \leq k$ is reached in the optimisation problem. However, $B = vk$ would nevertheless be fortuitous, in contrast with the previous case which is yet a special case of this one in which $k = 0$.

5.3.2 DILUTION

The second type of variable X and function E is what may be called dilution. A relevant index of the quality of the environment is often a density which is a quantity per unit of space (volume or area or length), e.g. quantity of pollutant per volume of water or air; quantity of trash or environmental degradation per unit area of a public park, forest or beach, etc. When the total volume or surface is given, and is not a variable, it makes no difference whether we consider this density, or the total quantity of pollutant, trash, deterioration, etc. The preceding section must be interpreted in this manner. But the spatial dimension is often a variable which can be increased (at a cost), e.g. the volume of a public hall can be increased to dilute the polluted air; the lower flow of a polluted river can be increased by means of dams and reservoirs; the surface of a recreation area can be increased by renting or buying or reclaiming more land, etc. We call z the X variable which measures this quantity of space (volume, surface, flow, etc.). An increase in z diminishes the pollution because it dilutes the pollutant.

x is still the level of the polluting or deteriorating activity, and each of its units produces the quantity a of pollutant or depredation so that the total quantity produced is ax ($a=1$ if x measures this quantity directly). The density of pollutant or deterioration is $P = ax/z$.

E is any monotonic function of P. Since P is homogeneous of degree zero in the two variables x and z, so is E. There are therefore constant qualitative returns to scale everywhere, and $xE_x + zE_z = 0$. It follows that $B=0$, i.e. *the proceeds of the environment tax and the expenditure on environmental expansion must be equal*. An agency in charge of both must therefore balance its budget.

The non-negativity constraints $x \geq 0$ and $z \geq 0$ are of course consistent with this result since they are homogeneous. But a constraint such as $x/z \leq k$ is also homogeneous (of degree zero) and is also consistent with this result. Such a constraint would, for instance, imply that the number of users of a public place per unit area has an upper limit. Another example of dilution where such a constraint holds is that of a road with a traffic flow x, of width (number of lanes) z, and traffic speed E: the standard flow–speed diagram shows an upper limit of traffic flow per lane. (This financial aspect of road management, omitting the discussion of the effect of this constraint, is mentioned in [8], [9], [1], [11], [19] and [3].)

If there were an initial, or untaxed, quantity of pollutant or deterioration k ($k>0$), one would have $P = (ax+k)/z$. Writing $P = ax/z + k/z$ shows that when x and z incur a simultaneous proportional variation, P varies in the opposite direction. That is to say, there are *increasing qualitative returns to scale* everywhere. Therefore, if the constraints are of the type above mentioned, $B<0$: *the expenditures on the expansion of space must exceed the proceeds of the pollution tax*. More precisely, to compute $xE_x + zE_z$ we can replace P by its non-homogeneous of degree zero part, k/z; it is homogeneous of degree -1, and therefore, taking $E = -P$, this expression is k/z. Hence $B = -vk/z$, which shows a loss equal to the value of the initial or untaxed pollution or deterioration.

5.3.3 BOTH CLEANING AND DILUTION

Of course, cleaning or restoring and dilution can exist simultaneously. With the same notations, there are three variables, x, y and z.

The simplest case is

$$Q = ax - by$$

$$P = \frac{Q}{z} = \frac{ax - by}{z}$$

and E is any monotonic function of P. Since P is homogeneous of degree zero in (x, y, z), so is E. There are thus constant qualitative returns to scale everywhere, and $xE_x + yE_y + zE_z = 0$. All constraints such as non-negativity in the variables, $ax - by > 0$, $x/z \leq k'$ (a constant), $y/z \leq k''$ (a limitation on cleaning or restoring activity per unit of surface, k'' being a constant), are homogeneous and therefore financially neutral. When there is no other constraint, $B = 0$, i.e. *the proceeds of the tax must equal the expenditures for both cleaning or restoring, and dilution through the expansion of volume or space.*

If, in addition, there is an initial or untaxed pollution or deterioration, or a possibility of net improvement, in quantity $k > 0$, the quantity of pollutant or deterioration is $k + ax - by$, and $P = (ax - by + k)/z$. E is any monotonic function of P. It is no longer homogeneous of degree zero. Writing $P = (ax - by)/z + k/z$ shows that when x, y and z simultaneously vary in the same proportion, P varies in the opposite direction. There are therefore *increasing returns to scale* everywhere. More precisely, to compute $xE_x + yE_y + zE_z$, one can replace P by its non-homogeneous of degree zero part, k/z; it is homogeneous of degree -1, and therefore, taking $E = -P$, this expression is k/z. Constraints such as non-negativity in the variables, $x/z \leq k'$, $y/z \leq k''$, are financially neutral and any of them may exist. But we now have in addition the non-homogeneous constraint $ax - by + k \geq 0$. If it is not binding, $B = -vk/z$, which is negative, and corresponds to a loss equal to the value of the initial or untaxed pollution or deterioration. If it is binding, $B = -vk/z + \lambda k = [\lambda - (v/z)]k$. Now, the constraint can also be written as $P \geq 0$. i.e. $E \leq 0$. A previous result then shows that this B still has the *sign* of $-vk/z$, i.e. it is negative. Hence, in all cases, *the environmental improvement expenditures must exceed the proceeds of the pollution tax.*

5.4 EXAMPLES

From this point, we must proceed towards more specific and practical cases. Let us take one step in this direction by considering two groups of examples drawn from water quality management and the management of the quality of public areas.

5.4.1 CASES IN WATER QUALITY MANAGEMENT

The problem is illustrated in Fig. 5.1. (More detailed studies of this problem can be found in [26] and [32], the former being more partial and the latter more complete than the model considered here.) Pollution is measured by density of pollutant per unit volume of water. The polluting source (city, plants, etc.) produces a flow x of waste water with pollution a. A water purification installation can treat a flow of y and remove pollution b from each unit volume of it; as a special case, this does not exist, i.e. $by = 0$. All water is then discharged into a river, the initial flow of which is z; it is the final pollution of this river, P, which causes the social (external) cost; as a special case, there is no such river, i.e. $z = 0$, and P is the 'effluent standard' after treatment of flow y. The initial pollution of the river is c; an important special case is when this incoming water is pure, i.e. $c = 0$. Part of the water flow x used by the polluting source ($1 - \alpha$ of it) originates from an upstream point in the river; the remaining fraction α is drawn from other sources; $0 \leq \alpha \leq 1$ (the two extreme cases $\alpha = 0$ and $\alpha = 1$ are of special importance). When $\alpha < 1$ and $c > 0$, the water coming from the river may be purified of the quantity d of pollutant per unit volume, before being used; the cost of this purification is charged directly to the users; the need for this purification depends upon the water use; complete purification $d = c$ and no purification $d = 0$ are two special cases. One has

$$0 \leq y \leq x \leq \frac{z}{1 - \alpha}$$
$$0 \leq d \leq c$$

and

$$0 \leq b \leq a + (1 - \alpha)(c - d)$$

(if the water from the outside sources is pure or completely purified). All variables except x, y and z are parameters. (Cases where α is variable, and where the water from external sources can be polluted are studied in [32].)

E is any monotonic function of P. The mere structure of the problem shows the nature of the qualitative returns to scale, and therefore the main financial results, in the various cases.

When x, y, z are all variable, there are constant qualitative returns to scale and homogeneous constraints, and therefore $B = 0$.

When $by = 0$, and z and x are the only two effective variables, there are constant qualitative returns to scale and homogeneous constraints, and therefore $B = 0$.

When the river flow z is a constant, an equiproportional increase in both x and y increases, leaves unchanged, or decreases P depending on whether the pollution of the effluent reaching the river is higher than, the same as, or lower than the river's own pollution, c, i.e. depending on whether c is lower than, the same as, or higher than

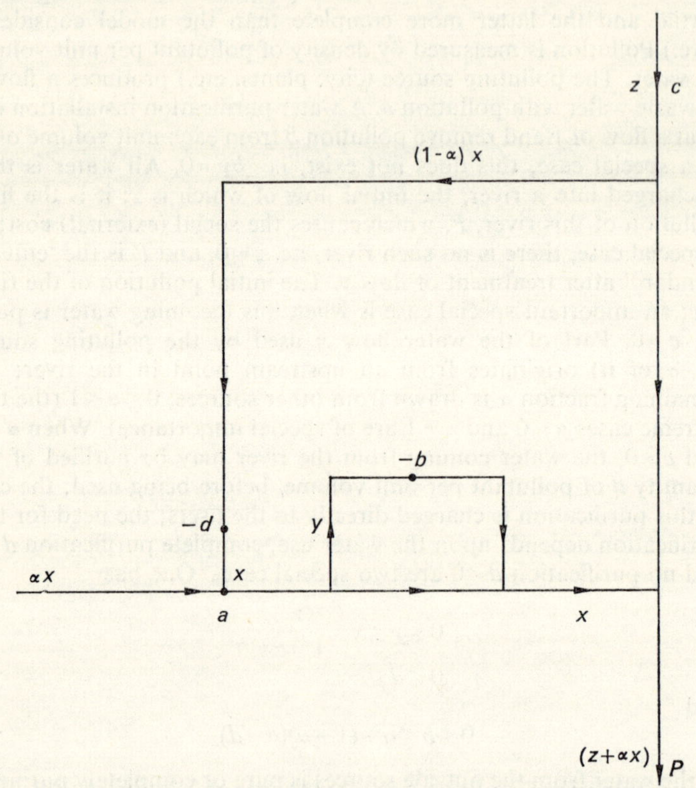

Fig. 5.1   An example of water quality management

P. That is to say, qualitative returns to scale in x and y are decreasing, constant, or increasing, depending on whether $c \lessgtr P$. Therefore:

$$B \lessgtr 0 \Leftrightarrow \text{optimum } P \lessgtr c$$

If, in addition, the river initially contains none of this pollution, $c=0$, and since $P \geq 0$ by nature, $B \geq 0$. More precisely, $B>0$ when optimum $P>0$; and $B=0$ when optimum $P=0$.

In particular, this is valid when $z=0$, and hence when the relevant pollution is the 'effluent standard' rather than the 'stream standard'.

However, an equiproportional increase in both x and y is possible only if the non-homogeneous constraint $(1 - \alpha)\, x \le z$ is not reached. For $\alpha = 1$, this is always the case, and in particular $\alpha = 1$ in 'effluent standard'. The attainment of this constraint means that all the river's water will be diverted for the polluters' use. Then, from equation (8), a term λz must be added to B. It is non-negative and we thus still have

$$P \ge c \Rightarrow B \ge 0$$
$$B < 0 \Rightarrow P < c$$

And, when $c = 0$, $B \ge 0$.

These results yield relationships between the parameters which are sufficient conditions for B to have a definite sign. Let us mention them for $(1 - \alpha)x < z$, knowing that one must add $\lambda k \ge 0$ to the B found when this constraint is reached, which keeps B positive or non-negative if it already has this property. The condition that the effluent's pollution is higher than, equal to, or lower than c is

$$\frac{[a + (c - d)(1 - \alpha)]x - by}{x} \gtrless c$$

or $a \gtrless by/x + \alpha c + (1 - \alpha)d$. Therefore, $B \gtrless 0$ according to this condition at the optimum. But $0 \le y \le x$, $0 \le \alpha \le 1$, $d \le c$. Therefore:

$a > b + \alpha c + (1 - \alpha)d \Rightarrow B \ge 0$, and $B > 0$ if $y < x$ at the optimum

$a > b + c \Rightarrow B \ge 0$, and $B > 0$ if either $d < c$ and $\alpha < 1$ or $y < x$ at the optimum

$a < \alpha c + (1 - \alpha)d \Rightarrow B \le 0$, and $B < 0$ if $y > 0$ at the optimum

$a < d \Rightarrow B \le 0$, and $B < 0$ if either $d < c$ and $\alpha > 0$ or $y > 0$ at the optimum.

More precisely,

$$P = \frac{cz + [a - (1 - \alpha)d]x - by}{z + \alpha x}$$

When z is constant, and is not paid for, and $(1 - \alpha)x < z$, taking $E = -P$, $B = v(xP_x + yP_y)$. But since P is homogeneous of degree zero in the three variables x, y, z,

$$xP_x + yP_y + zP_z = 0, \text{ and } B = -vzP_z$$

But, calling N and D the numerator and the denominator of P, and taking logarithmic derivatives,

$$\frac{P_z}{P} = \frac{c}{N} - \frac{1}{D} \text{ i.e. with } P = \frac{N}{D}, \; P_z = \frac{1}{D}(c - P),$$

and finally,

$$B = v\,\frac{z}{z + \alpha x}\,(P - c)$$

In this expression, v is the marginal social (external) cost of pollution, $z/(z + \alpha x)$ is the 'dilution factor', which shows improvement in water quality due to dilution by the extra water αx, and $(P-c)$ is the 'pollution added' to the river.

If all the water comes from the river, $\alpha = 0$ and $B = v(P-c)$.

If the river is initially clean, $c = 0$ and $B = \dfrac{vx}{z+\alpha x}P$.

If both hold, $B = vP$.

5.4.2 ANOTHER CASE OF CLEANING TECHNOLOGY

Until now, we have assumed that the effect of one unit of the cleaning activity was to remove a given amount of pollutant. However, the cleaning technology may have a different effect. For instance, a frequent case is one in which the effect of one unit of cleaning activity cleans a given surface, or volume, or length (path, façade, etc.). This tends to apply to the ventilation of a public hall; to the use of a vacuum cleaner when there is not too much dust; and to water purification within narrow limits (the example given in the last section is not of this type because a technologically given amount of pollutant b is removed from each volume of y).

We use the same notation as in section 5.3. The only difference is that a unit of y now cleans the surface, or volume, or length β, which is a technological constant; that is to say, it removes all the pollutant or trash from β whatever its quantity within given limits. We make obvious assumptions about the time dimension and the homogeneity of pollution.

If π is the pollution before, or without, the cleaning (quantity of pollutant per unit volume or surface area), writing the total quantity of pollutant in two different ways, one has $ax = \pi z$, so that $\pi = ax/z$. The quantity of pollutant per volume or surface β being $\beta\pi$,

$$Q = ax - \beta\pi y = ax - \frac{a\beta xy}{z} = ax\left(1 - \frac{\beta y}{z}\right), \text{ and}$$

$$\boxed{P = a\frac{x}{z}\left(1 - \beta\frac{y}{z}\right)}$$

The constraints are the non-negativity of the variables and $\beta y \leq z$. E is any monotonic function of P.

When the three variables x, y and z can vary, P, and therefore E, being homogeneous of degree zero in this set of variables, there are constant qualitative returns to scale everywhere. And the constraints are homogeneous and therefore financially neutral. Hence, $B = 0$.

When $y = 0$, we have the classical dilution situation studied above.

The case where z is given and is not paid for is a frequent one. We take $E = -P$ and note that P's homogeneity of degree zero in (x, y, z)

shows that $xP_x + yP_y = -zP_z$. It thus follows that $B = -vzP_z$ if $\beta y \le z$ is not binding.

If dilution of the pollution or deterioration is the only use for z, we cannot have $P_z > 0$ at the optimum, since then it would be fruitful to abandon, or to abstain from using, a part of the given z: we would make the space used shrink so as to be able to clean it better. Therefore, $P_z \le 0$ at optimum. But $P_z = ax/z^2[2(\beta y/z) - 1]$, so that $P_z \le 0$ implies, if $x > 0$, $\beta y \le z/2$, and therefore, for $z > 0$, $\beta y < z$, so that the only non-homogeneous constraint is not binding. Finally, $B \ge 0$, with $B = 0$ or $B > 0$ *depending on whether half, or less, of the total surface area is cleaned.*

Note that in the case where x, y, z are all three variables, and y and z are both paid, the same argument shows that at the optimum $\beta y < z/2$, i.e., the *optimal cleaning is less than half the total surface.*

Now, z very often has another usefulness than just diluting pollution or deterioration. In particular, this space may have a pure recreational value, and it could, in this respect, be either a private, or a public, or a congested good. It has no such value at the margin when there is satiation in it.

When another value is attached to z in this way, it influences both the optimum level of the variables – in particular, of z – and the formula giving the global financial result. But the B given above still represents the net part of this result due to the existence of pollution. However, it is now possible for $P_z > 0$ at the optimum, and this net financial contribution from pollution may be negative.

5.5 CONCLUDING REMARKS

In the cases of cleaning or repairing, the costs of which are proportional to the 'quantity' of improvement, we found that the optimum public budget (for this environmental problem) is balanced if the quality of the environment is exactly restored; it is in surplus if the quality incurs a net deterioration; and in deficit if it shows a net improvement. Furthermore, the surplus and the deficit in some sense represent the 'value' of the deterioration or improvement. This suggests that we might give the surplus to the victims of the deterioration as a compensation, or that we ask the beneficiaries of the improvement to finance the deficit. One then wonders why these transfers, and the corresponding actions, are not achieved by an ordinary market process between the polluters, the pollutees and the cleaning or repairing firms, i.e. why the government has to intervene at all. The most important answer is clearly the collectivenesses of the concern for the quality of the environment, which makes both the deteriorating and the restoring variables collective concerns. A second reason is that the environmental rights, which would be the objects of such exchanges, frequently lack initial explicit allocation by society (because the underlying scarcity has only recently been recognised). A third reason could lie in the non-convexities which arise in spaces encompassing both the environmental variables and the instrumental variables of the victims of environmental deterioration. (For an exhaustive analysis of this question, see [27] and [43] and the literature quoted there.) However, this reason is of very limited practical importance, since either there is only one or, more generally, a small number of pollutees, and there is then no reason for the exchange to take place with unitary prices imposed upon them (it is this type of exchange that the non-convexity impeaches); or these agents are numerous and any exchange with parametric prices is then generally impaired by a collective concern among pollutees (for an analysis of the causes of this collective concern, see [24]).

This explains the importance of the type of policy being stuided – if we add the advantage of using a tax rather than a constraint to control the nuisance. Apart from considerations of income distribution, of the flow of public funds, and of the organisation of the public sector, the advantage of a tax is that it decentralises to the polluters the necessity of knowing information about themselves. It avoids the necessity of having information about pollutees (v in the previous sections), but there are many ways of dealing with this question (see [22], [37], [43], [44]). However, to this end, taxing pollution is not different from subsidising abstention from polluting or, more generally, from any measure which defines a given level of pollution and taxes any excess or subsidises any abstention.* The value of the given level must be subtracted from the net public financial results shown above. The difference is fundamentally a question of allocation of environmental right to the polluters. It is distributional. However, it may also affect efficiency when the polluter has a non-convex set of production possibilities. This question is of considerable practical importance. When one wishes to tax polluting firms, they often argue that this would force them to close down, thereby leaving local unemployment. And when their production shows in-

---

* There are also other possibilities. For instance, if x is the level of the polluting activity (base of the intervention), one may require from the polluter the amount of money $t_1 . (x - x_1) - t_2 . (x_2 - x)$, or give him the opposite if this expression is negative, with $t_1 + t_2 = t$ (the efficient rate), i.e. if $x_1 < x < x_2$, altogether tax $x - x_1$ at rate t_1 and subsidise the abstention $x_2 - x$ at rate t_2. In particular, x_1 may be zero and x_2 may be the level of x which the polluter chooses in the absence of any public anti-pollution measure bearing upon him. The text's cases are also particular cases of this one with t_1 or t_2 being zero and with specific values of x_1 or x_2. Finally, these functions of x may be non-linear, this sum being $s(x)$ with $s'(x) = t$ at the optimum.

creasing returns to scale at some level of output (fixed capital, for instance), it may be true that the taxation would force them into deficit and that the continuation of their operation might be socially optimal (see [27], [38], [34]). In these cases, the discussions between the firm and the pollution control authority usually relate to taxing pollution at a lower rate t. This is a very bad practice. The discussions should be about an untaxed amount of pollution, the correct tax rate being applied to the extra pollution (this being a subsidy if the firm pollutes less than the chosen amount). This is equivalent to defining the firm's rights to the environment, and providing it with the possibility of buying more, or of selling a part, of these established rights.

In addition, other 'imperfections' of the economic reality make the optimum tax different from that considered above. They thus impose a second-best policy. It is in particular the case of costs or obstacles met by the operations of taxation – information, collection, discrimination (see [29], [36], [28], [20], [13], [21], [18]) – of market imperfections such as the case when the polluter is a monopoly (see [29], [5], [20], [21], [18]), of imperfections of the environment policy in fields related to the one under consideration such as in other locations or for alternative pollutions (see [28], [20], [21], [18], [13]), of general imperfections of economic policy such as the absence of lump-sum transfers which creates the 'cost of public funds' (see, for instance, references [18], [19], [21]) and forces to consider the whole incidence of the tax (or other measure) in order to analyse the distribution effect (see [42]). The financial consequences of these differences may be computed (see [28], [29], [13], [21], [18]). But some of these phenomena may encourage the choice of other policy tools (an exhaustive analysis of the instruments available for the environment policy, and of the criteria according to which they must be selected, is provided in [31], [30], [35], [33]).

Finally, strict economic efficiency may require an action on the victims of the environmental deterioration (for an opposing view see [6]), and when this action consists in financial incentives (taxes or subsidies), it has a direct effect on the global financial result of the management of this environmental quality. But this action on pollutees is not yet completely studied (see, however, [3], [40], [7], [41]).

REFERENCES

[1] Strotz, R. H., 'Urban Transportation Parables', in J. Margolis (ed.), *The Public Economy of Urban Communities* (Johns Hopkins Press, Baltimore, 1965).

[2] Lévy-Lambert, H., 'Tarification des services à qualité variable', *Econometrica*, vol. 36 (1968) no. 3–4.

[3] Mohring, H., and Boyd, J. H., 'Analysing Externalities: "Direct Interaction" *vs.* "Asset Utilisation" Frameworks', *Economica* (Nov 1971).

[4] Nora, S., *et al.*, *Rapport sur la gestion des entreprises publiques* (1968).

[5] Buchanan, J., 'External Diseconomies, Corrective Taxes, and Market Structure', *American Economic Review* (Mar 1969).

[6] Baumol, W., 'On Taxation and the Control of Externalities', *American Economic Review* (June 1972).

[7] Mohring, H., and Kraus, M., 'The Role of Pollutee Taxes in Externalities Problems' (*Notes for Dalhousie and Toronto University Seminars*, 1972).

[8] Kolm, S.-Ch., 'L'économie des services publics', *Économie Appliquée*, vol. 28 (Dec 1965).

[9] ——, *Introduction à la théorie du rôle économique de l'État: les fondements de l'Économie Publique* (I.F.P., Rueil-Malmaison, 1965).

[10] ——, 'Le rôle social ambigu des prix publics', *Économie Appliquée*, vol. 21, no. 2 (1968).

[11] ——, *La théorie économique générale de l'encombrement* (SEDEIS, Paris, 1968).

[12] ——, *Prix publics optimaux*, Monographie d'Économétrie (C.N.R.S., Paris, 1968).

[13] ——, 'Le déficit optimal des transports communs', *Revue d'Économie Politique* (July–Aug 1968).

[14] ——, 'La croissance et la qualité de l'environnement', *Analyse et Prévision* (Aug 1969).

[15] ——, 'L'encombrement pluridimensionnel', *Revue Économique* (Nov 1969).

[16] ——, 'L'économie normative des services de masse', *Revue d'Économie Politique* (1970).

[17] ——, *La valeur publique* (Dunod-C.N.R.S., Paris, 1970).

[18] ——, *L'État et le système des prix* (Dunod-C.N.R.S., Paris, 1970).

[19] ——, *Le service des masses* (Dunod-C.N.R.S., Paris, 1970).

[20] ——, 'La vérité des prix dans un monde imparfait', *Revue Économique* (July 1969).

[21] ——, *La théorie des contraintes de valeur et ses applications* (C.N.R.S. Paris, 1970).

[22] ——, 'Politiques anti-pollution optimales en présence d'incertitude', *Kyklos* (1971).

[23] ——, 'La pollution des lieux publics', *Recherches Économiques de Louvain*, no. 1 (Mar 1971).

[24] ——, 'Possibilités et limites de la régulation de l'environnement par ententes directes entre les intéressés', *Consommation* (July–Sep 1971).

[25] ——, 'La macro-économie de l'environnement', *Revue d'Économie Politique* (July 1971).

[26] ——, 'Les pollueurs doivent-ils être les payeurs?', *Revue Canadienne d'Économique* (Nov 1971).

[27] ——, 'La non-convexité d'externalité' (CEPREMAP, 1971).

[28] ——, 'Le moindre mal en politique d'environnement' (CEPREMAP, 1971); *Consommation* (1973).

[29] ——, 'Effets des coûts de réalisation de la politique d'environnement sur le choix de celle-ci' (CEPREMAP, 1971).

[30] ——, 'L'économie de l'environnement', *L'Expansion* (Oct 1971).

[31] ——, 'Instruments et critères d'une politique d'environnement', *Analyse et Prévision* (Dec 1971).

[32] ——, 'La pollution des eaux' (mimeo, CEPREMAP, Jan 1972).

[33] ——, 'Pour une planification de l'environnement', *Le Monde* Apr (1972).

[34] ——, comments on Karl-Göran Mäler's contribution (pp. 217–222 below).

[35] ——, 'Introduction à l'économie de l'environnement', *Revue Suisse d'Économie Politique* (Sep 1972); and *Revue des Ingénieurs des Ponts et Chaussées et des Mines* (Sept 1972).

[36] ——, 'Taxation sans discrimination', *Kyklos*, vol. xxv, fasc. 1 (1972).

[37] ——, 'Connaissance des coûts et valeurs d'environnement', in *Évaluation des risques et dommages d'environnement* (O.E.C.D., Paris, 1973).

[38] ——, 'Une économie écologique', *Revue Politique et Parlementaire* (Jan 1973); and *Nuisances et Environnement* (1973).

[39] ——, 'Une occasion manquée', *Le Monde* (27 Dec 1972).

[40] ——, 'Les pollués doivent-ils payer?', *Kyklos*, vol. xxvi, fasc. 1 (1973).

[41] ——, 'Comments on the Mohring–Kraus Models' (CEPREMAP, Feb 1973).

[42] ——, 'Un impôt sur les pollueurs est-il finalement payé par les consommateurs?', *Le Monde* (Feb 1973).

[43] ——, 'Problèmes et méthodes de l'estimation de la demande de biens publics avec référence particulière aux éléments de l'environnement public' (CORDES, Paris, Aug 1973).

Discussion of the Paper by Serge-Christophe Kolm

Formal Discussant: Hicks. I should like to start off by asking what can be meant by 'qualitative returns to scale'? This is not, I think, an issue confined to public-sector economics; and since I find it easiest to trace one difficulty at a time, I should like to begin by looking at it in private-sector terms. What are the units in which we are to measure commodities? We always start with the firm which produces commodities measurable in physical units, so that its output is in units of commodity X and so on. But this is not the general case. Take the case of the firm which delivers milk to consumers. A litre of milk delivered at one place is not the same (to producers or to consumers) as a litre delivered in another place: if we are interested in delivery costs we cannot treat them as equal. Nor can we say that there is a joint supply of two products – milk and delivery for the units in which delivery is to be measured are quite obscure. Now, I would maintain that as long as we are solely concerned with optimum conditions, it does not matter in what units delivery is measured. What matters for the determination of the optimum is that there should be no way of varying the milk run which would give a net advantage to both producer and consumer concerned, and which is not taken. But this idealised condition can be stated in various ways according to the index of 'delivery' which is chosen. According to one index there may be constant; according to another, increasing; according to another, decreasing returns to scale; but the same optimum conditions will result. Of course, this is not (even potentially) a perfectly competitive structure; it is quite probable that there will be no index of delivery which makes it possible to reach an optimum by charging identical prices per unit of 'delivery' – as measured by some index. I am therefore led to the possibly nihilistic conclusion, when we are concerned with quality variations, that there is usually no possibility of perfect competition – and also no possibility of unambiguous 'returns to scale'.

I turn now to the substance of the author's paper, putting aside this possibly pedantic point. He will excuse me if I restate what I think is his problem in my own terms. There is a river, which (for convenience) I shall call the Danube, which passes through a territory, which we shall call Hungary. It is already a dirty river when it enters Hungary. The activities of the Hungarians may either clean it or dirty it further. What the Austrians have been doing is no concern of ours. But it is arranged (perhaps in Moscow) that the Hungarians have to pay attention to the welfare of the Romanians. We are looking for a Pareto optimum for Hungary and Romania.

The only way in which the Hungarians can affect the welfare of the Romanians is by affecting the state of the river where it enters into Romania (say at the Iron Gates). There are both positive and negative ways in which the Hungarians can affect the state of the river. They may pollute it, or they may depollute it. It is just the net effect on the state of the river at the Iron Gates which matters.

We may take the state of the river as it would be if the Hungarians paid no attention to the Romanians as a 'zero point' from which to measure. If the Hungarians are to adopt some other policy – voluntarily – they must be given an incentive to do so. Is it conceivable that the Romanians might be such dirty fellows that they would pay the Hungarians to pollute beyond the zero point? This would only happen if the pollution had positive utility to the Romanians, and that (I think) may be ruled out. Otherwise the optimum must require a transfer payment from the Hungarians to the Romanians. Such a payment must diminish the incentive to Hungarians to pollute, whatever the formula for payment.

Suppose, first of all, that there are just three alternative plans by which Hungarians could reduce this pollution, if they were given the incentive to do so. If the only form of pollution was biochemical oxygen demand (B.O.D.), we could order them according to their effect on dissolved oxygen (D.O.), but in general there will be no such objective ranking available. Let us simply rank them in terms of the maximum that the Romanians would be prepared to pay in order to have the advantage of each form of depollution. Let OA, OB, OC, on the horizontal axis in Fig. 5D.1, measure these amounts. Then let Aa, Bb, Cc measure total costs, to Hungarians, of installing these kinds of pollution measures. If we now draw a 45° line through the origin, then the optimum solution is simply that in which the gap $a\alpha$, $b\beta$, $c\gamma$ is the largest – provided it is positive

Now suppose that there are a large number of possible plans, so that the whole quadrant would be covered with dots, in declining cost of providing depollutant on the vertical scale, and maximum Romanian payment on the horizontal scale. Corresponding to every such abscissa there will be a minimum cost of providing depollutant which would justify that payment, and a further curve could be drawn representing these minimum costs. There would be a maximum total payment that would be worth while for the Romanians to pay in any instance – the maximum, perhaps (but not necessarily) what they would be prepared to pay for a perfectly clean river – so that there will be some maximum L which will set a limit to where we may want to draw the curve. The relevant abscissa would be between O and L. But between O and L there will be a cost curve that might have all sorts of shapes.

It might be wholly above the 45° line, so that the zero solution would be optimal. It might be wholly below, or it might intersect at various points. Obviously the important distinction is whether it lies above or below the origin. If it lies below or near the origin, there must be some plan for positive depollution which is optimal, but there is also the possibility that it will be above the origin, but will come down below somewhere to the right. The marginal conditions for optimality are then sufficient.

It would obviously be possible to redraw the diagram classifying by cost instead of 'utility'. Basically, the same would show up. Or we may classify in some other way. If there is only one physical index in question (B.O.D.), we could classify plans according to this depollution effect

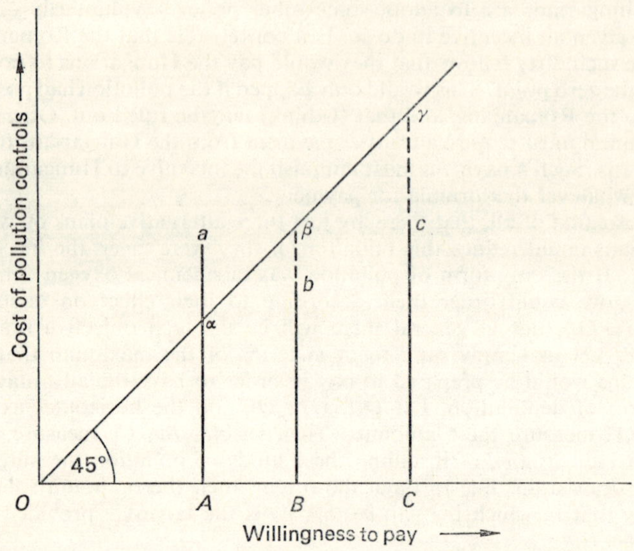

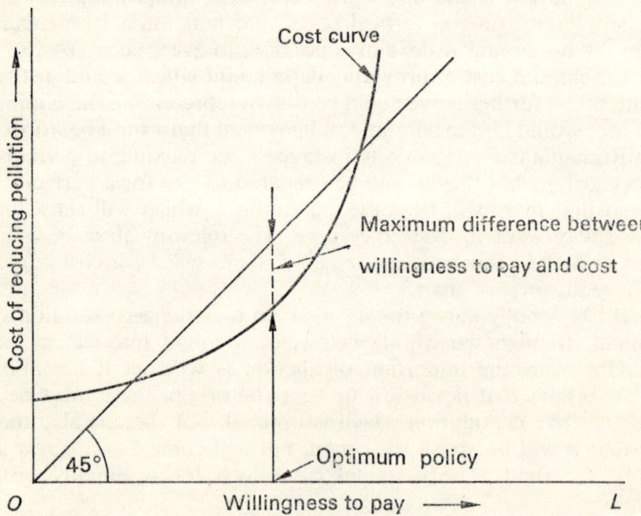

FIG. 5D.1   Evaluation of alternative pollution schemes

relative to that index, and then construct a cost curve and a utility curve (in money terms) maximising the difference between them. If we use two such indexes, one will still be doing the same in three dimensions, maximising money difference between utility and cost surfaces. And presumably what much of the author's analysis amounts to is doing the same in a number of different dimensions.

I am sure that there is more in his analysis than that – but I am not sure how much more. I should like him to tell us.

One final point in conclusion. It would obviously be possible for the Romanians to do the depolluting themselves. Their costs of doing this would set a maximum, at each stage, to what they were prepared to pay.

Tulkens pointed out that, in the author's model, the polluter's utility functions did not include the quality of the environment as an argument. He then asked to what extent the analysis would be modified by introducing this additional variable.

Hicks concluded by observing that the author's paper addressed itself to showing that there were many cases within the field in which a particular difficulty did not arise. That was a useful thing to do. Unfortunately the other side of the problem had not received very much attention, although it was equally worthy of discussion: what did we do when the author's conditions were not satisfied – when, according to his own analysis, the budget would not balance? In such cases a straightforward shadow price would not be sufficient. Some kind of formula would have to be discovered and it was difficult to imagine that the construction of that formula was going to be an easy matter. The constraints which would have to be imposed in the combined interests of equity and efficiency are difficult to reconcile. It was useful to show how the problem could be dealt with without involving these concepts, but we should at least consider the problems that were likely to arise when we did introduce them.

Kolm replied to Hicks' comments by observing that it was possible to provide a diagrammatic exposition of qualitative returns to scale, and of their financial implications, in the case where the variables x, X and E were all three unidimensional. Figs. 5D.2 to 5D.7 all had x and X as co-ordinates and showed a set of *isoquals*, i.e. curves $E(x, X) = $ constant (Fig. 5D.2). Qualitative returns to scale depended upon the effect on E of an equiproportional variation in x and X. More precisely, the relevant 'local' qualitative returns to scale depended upon whether a small equiproportional increase (decrease) in both x and X improved, reduced or had no effect on environmental quality. The qualitative returns to scale at a point A thus depended upon the relative position, in the neighbourhood of this point, of the isoqual which passed through A and of the straight line OA. In Fig. 5D.3 these two curves were smoothly tangent and there were constant qualitative returns to scale in A, when the quality index E was chosen so that society preferred it to be larger (i.e. $v > 0$). Then, if A represented the optimum, $B = 0$ for Fig. 5D.3, $B > 0$ for Fig. 5D4 and $B < 0$ for Fig. 5D.5. It may seem fortuitous that the optimum is in the case of Fig. 5D.3, but this is not so, as is shown below.

In many practical cases the qualitative returns to scale were the same

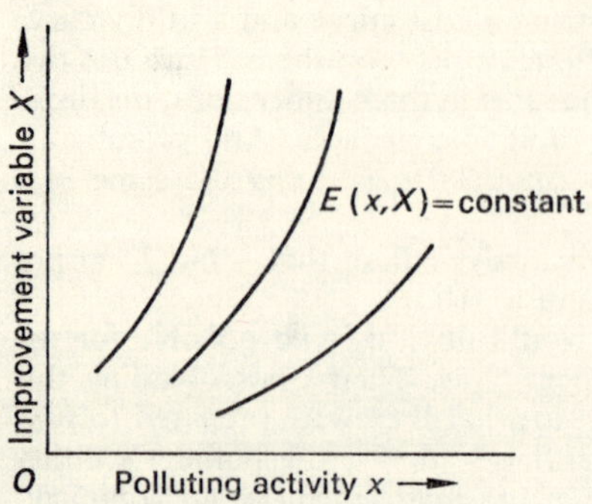

FIG. 5D.2   Isoquals representing constant environmental quality

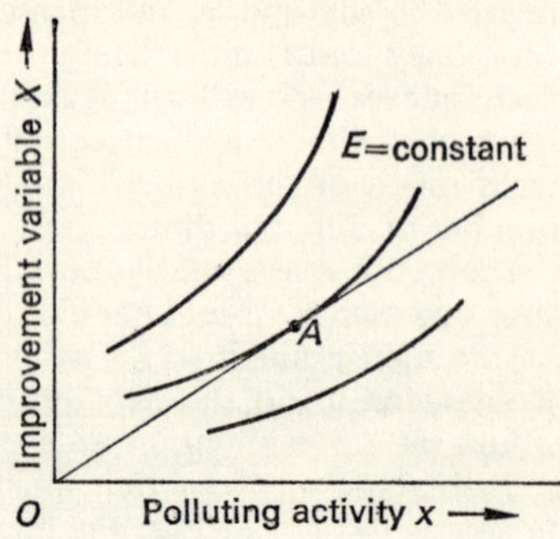

FIG. 5D.3   Constant qualitative returns to scale in A, i.e. $B=0$

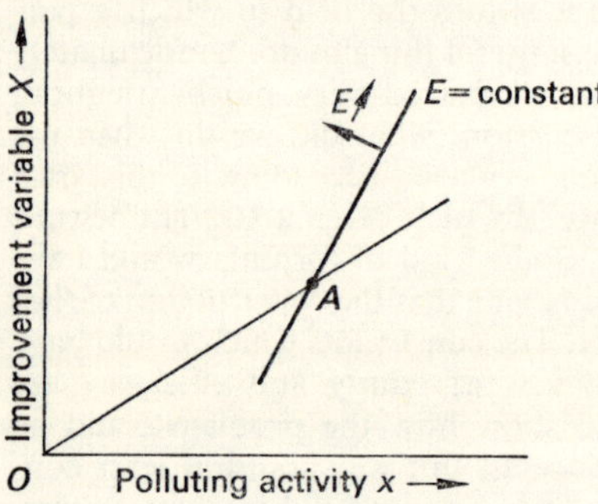

FIG. 5D.4   Decreasing qualitative returns to scale in A, i.e. $B>0$

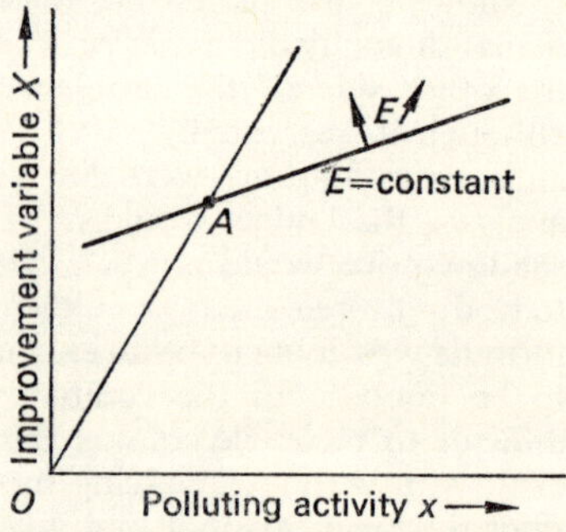

FIG. 5D.5   Increasing qualitative returns to scale in A, i.e. $B<0$

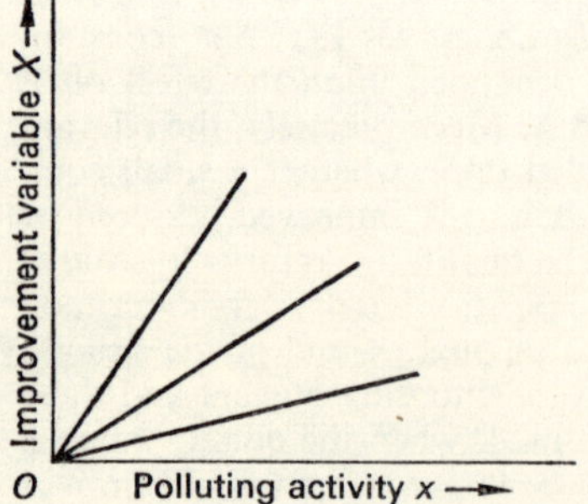

FIG. 5D.6   Constant qualitative returns to scale, i.e. $B=0$

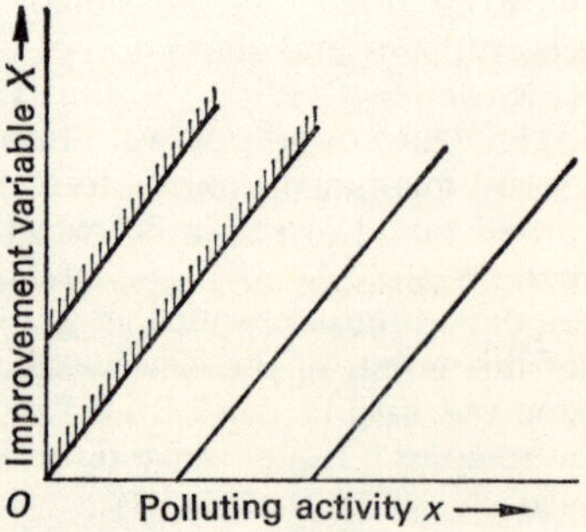

FIG. 5D.7   Qualitative returns to scale decreasing, constant or increasing depending on whether $(ax - bx) \gtrless 0$

at every possible point of the plan, so that we knew *a priori* what they were at the optimum. The simplest situations of cleaning or dilution were cases in point.

For dilution, E depended upon x and X only through their ratio X/x. The isoquals were straight lines from the origin (Fig. 5D.6). Qualitative returns to scale were constant everywhere and consequently $B=0$.

For cleaning or repairing, the isoquals were parallel straight lines of the form $(ax-bX)=$ constant. The desired quality E decreased when the residual pollution $(ax-bX)$ increased. Fig. 5D.7 then showed that the qualitative returns to scale were decreasing, constant or increasing depending upon whether $(ax-bX) \gtrless 0$. In the absence of initial or untaxed pollution, $(ax-bX) \geq 0$, and thus $B \geq 0$, with $B=0$ if the optimum A lies on the line $(ax-bX)=0$ (the constraint is reached); otherwise, $B>0$.

Other cases could be similarly studied and discussed; in particular, the examples of paragraphs 5.4.1 and 5.4.2. For the water pollution discussed in paragraph 5.4.1, if the variables were x and y we were dealing with 'cleaning' with parallel isoquals; while if they were x and z the isoquals were straight lines passing through a point which was generally not the origin. For the kind of cleaning technology presented in paragraph 5.4.2, the isoquals in the variables x and y were rectangular hyperbolas with the same asymptotes, and a classical property of such curves gave the results found when more or less than half the total surface was cleaned. Such graphical analyses were performed in references [11], [19], [23], [26] and [32].

Finally, Tulkens' remark that the same agents may be both polluters and pollutees was examined in references [8], [9], [11], [15] and [19]. The result was not changed provided the pollution created by each polluter was small relative to the total pollution (a situation also noted in reference [1] for the case of road congestion). When this condition did not hold, the tax rate had to discriminate among different polluter–pollutee agents.

Førsund, in a written note, offered the following comments.

The economic literature on externalities is gradually changing towards more policy-oriented questions of environmental pollution, reflecting, with the usual time-lag, the needs of society [10].

One such question is the optimum financing of environmental improvement to which the author addresses himself in the present paper.

The list of references demonstrates the author's extensive and impressive contributions on various aspects of the economics of environmental pollution. In references [8] and [9] it is shown that a wide variety of topics can be fruitfully analysed in a similar way to the author's treatment of environmental pollution: topics such as the economics of queueing or waiting lines, transportation (road, rail, air), telecommunications, random demand and risk of shortage, peak-load pricing, etc.

I shall here consider the present paper on its own merits and advise the reader to consult the author's own paper for references to all relevant previous papers.

Firstly, it seems worth while to gather and review some other models scattered in the literature containing the central relationship of the author's paper.

Models of Generic Congestion

The key relationship in the author's model is the environmental function relating an environment quality index to the amounts of those activities generating pollution and environmental improvement goods. Analogous relationships can be found in other models representing either congestion or pollution. The term 'generic congestion' is introduced in Rothenberg [13] to cover both phenomena. His model also, in fact, contains a generic congestion function analogous to the environmental function (see pp. 116–17, equations (3)–(4)), but the model is not suitable for analysing optimum financing, so that it will not be reviewed below.

To bring out the similarities, two models from each of the fields of transportation, water pollution and environmental pollution are presented.

Transport congestion:

(a) Mohring ([11] pp. 80–7)

(1) $D = f(C)$

(2) $rK = rK(S)$

(3) $g = g(D, S)$

(4) $C = T + g$

D = total number of trips per unit time

C = individual cost of a trip

rK = annual capital expenditure (r = rate of interest)

S = index of size or capacity of the highway

g = time cost of the trip

$g(\cdot)$ = time cost function, $\partial g / \partial D > 0$, $\partial g / \partial S < 0$

T = toll on the trip.

(b) Strotz [14]

(1) $U^i = U^i(t^i, d, x^i) \ (i = 1, \ldots, n)$

(2) $\Sigma x^i = X(R - E)$

(3) $d = D(\Sigma t^i, E)$

u^i = utility of individual i

t^i = number of trips per unit time

d = driving time per trip

x^i = amount of composite commodity ('bread')

R = given amount of resources

E = resources spent on roads per unit time

$D(\cdot)$ = congestion function, $\partial D / \partial(\Sigma t^i) > 0$, $\partial D / \partial E < 0$.

Water pollution:

(c) Upton [15]

(1) $C = C(F)$

(2) $P_i = P_i(\bar{L}_i - L_i) \ (i = 1, \ldots, n)$

(3) $D = D(L_1, \ldots, L_n, F)$

C = cost of low-flow augmentation

F = rate of stream flow

P_i = treatment cost of firm i

$\bar{L}_i$ = amount of pollutant created by firm i

L_i = amount of discharged pollution

D = measure of water pollution at a point downstream,
$\partial D/\partial L_i > 0$, $\partial D/\partial F < 0$.

(d) Hayden Boyd [5]

(1) $Y_i = Y_i = Y_i(X_i, Z_i, Q_i)$ $(i = 1, \ldots, n)$

(2) $C = C(I)$

(3) $Q_i = Q_i(Z_i, \ldots, Z_{i-1}, I)$

Y_i = output of firm i

X_i = input

Z_i = quantity of waste removal services used by firm i

Q_i = water quality enjoyed by firm i,
$\partial Q_i/\partial Z_j < 0$, $\partial Q_i/\partial I > 0 (j = 1, \ldots, i - 1,$ the upstream firms)

I = level of investment in river capacity

C = cost of I.

Environmental pollution:

(e) Førsund [3]

(1) $U^i = U^i(Y^i; Z)$ $(i = 1, \ldots, n)$

(2) $F(X, R) \leqq 0$

(3) $Z = Z(X, Y^1, \ldots, Y^n, R)$

(4) $\sum_{i=1}^{n} y_j{}^i, \leqq x_j(j - i, \ldots, m)$

U^i = utility level of individual i

Y^i = consumption vector of i

Z = measure of level of pollution and congestion or measure of environmental services, $\partial Z/\partial X$, $\partial Z/\partial Y > 0$, $\partial Z/\partial R < 0$ when Z is interpreted as the level of pollution, and opposite signs when Z is interpreted as environmental services

X = vector of outputs

R = vector of environmental improvement goods.

(f) Kolm (Chapter 5)

(1) $u^1 = u^1(x, r^1)$, $u^2 = u^2(E, r^2)$

(2) $F(x, X, r^1 + r^2) \leqq 0$

(3) $E = E(x, X)$

(4) $f(x, X) \geqq 0$

u^1, u^2 = utility levels of polluters and pollutees

x = level of pollution

X = level of improvement activity

r^1, r^2 = consumption of ordinary commodities

E = index of environmental quality, $\partial E/\partial x < 0$, $\partial E/\partial X > 0$.

The key relationship in common is the one numbered (3) in the models. In the transport congestion models the cost of a trip increases with the number of trips and decreases with the capacity of the transportation network; in water pollution models, pollution increases with increasing individual discharges and decreases with investment in riverflow augmentation; and in environmental pollution models, environmental services decrease with increased levels of economic polluting activities, and increase with inputs of environmental improvement activities.

I shall return to a further comparison of the models after a closer examination of the special features of the author's model.

2 *Comments on the Author's Model*

In addition to the usual production possibility function incorporating three types of goods – polluting, non-polluting and environmental improvement – the author also introduces special constraints on the relationship between the bundle of goods which generates external costs, and the environmental improvement goods. To be of any interest, and to justify their mathematical and interpretative complications, these constraints must be binding in some realistic cases. Fig. 5D.8 illustrates the examples (in the unidimensional case). The first constraint is simply a non-negativity condition. Constraint 2 represents the obvious, that you cannot remove more pollutants than were originally introduced. As it stands, example 3 represents a constraint on the level of improvement activity, given the level of polluting activity, or a constraint on the polluting activity, given the level of improving activity. To give a better meaning, an improving activity is in this case interpreted as increasing a spatial dimension of the medium

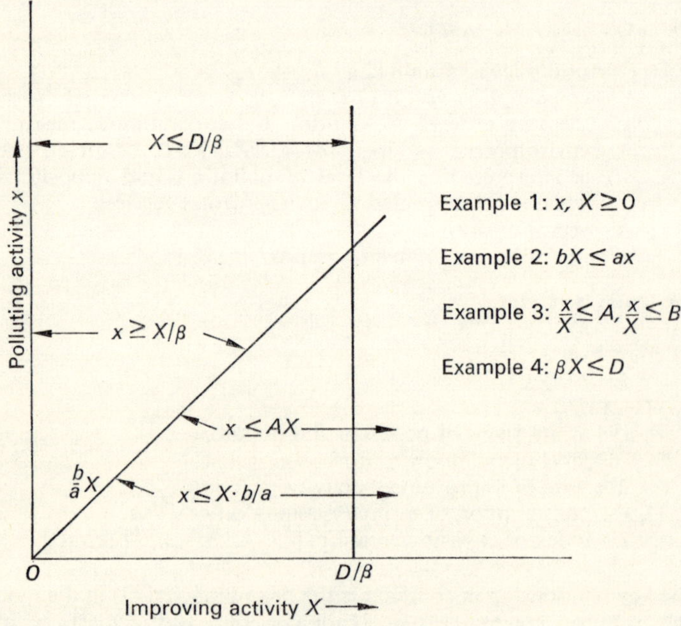

FIG. 5D.8   Kolm's examples illustrated

in question, i.e. the case of dilution. Example 4 may be interpreted as a special case of 2 or 3. The point is (see the author's last example) that one unit of the improving activity removes *all* pollutants, independently of the amount, from a given volume, or surface, or length β of the medium in question with the total volume, etc., of D.

Now, I think the significance of such constraints must be considered in the light of the level of aggregation used in the analysis. Consider, for example, a high level of aggregation and situations in which the first two constraints are binding. With suitable specifications, the environmental function $E(x, X)$ and the utility functions, i.e. specifying the values of $\lim u_x^1$, $x \to 0$, $\lim u_E^2$, $X \to 0$, identical solutions can be obtained. For instance, zero pollution and improvement would be the case if the damage evaluated through the utility function of example 2 was greater than the utility function of example 1 for all levels of polluting activity, and if improvement activity had a higher social cost than reducing the level of polluting activity at all levels of this activity. Complete purification would be the case if the damage was always greater than the improvement costs, viz. the case of highly poisonous substances, and the polluting activity will be positive when the total social cost of improvement activity is less than the cost of cutting back the polluting activity.

The explanations of constraints 3 and 4 show that they are more appropriate on disaggregated levels and for more specific problems.

3 *The Optimal Solution and the Market Solution*

It is a usual classroom principle to distinguish between the optimal solution in the 'real' model and the systems capable of implementing the rea solution. Because the author's presentation is so compact, the division is made here to facilitate the understanding of the analysis.

The real model is formulated as a non-linear model with the (un-weighted) sum of utilities as objective function. To bring out the impact of the special constraint, the necessary optimality condition is presented both with and without the binding constraint (equations (2) and (1) below). (The author's assumption that the index good is always desired means that the production possibility frontier will be realised, i.e. the F-constraint is binding.)

$$\frac{u_x^1 + u_E^2 . E_x}{u_E^2 . E_X} = \frac{F_x}{F_X} \tag{1}$$

or

$$\frac{u_x^1 + u_E^2 . E_x}{u_r^1} = \frac{F_x}{F_r}, \quad \frac{u_E^2 . E_X}{u_r^2} = \frac{F_X}{F_r} \tag{1'}$$

$$\frac{u_x^1}{u_r^1} + \frac{u_E^2}{u_r^2}\left(E_x - E_X . \frac{f_x}{F_r}\right) = \frac{F_x}{F_r} - \frac{F_X}{F_r} . \frac{f_x}{f_X} \tag{2}$$

$$(u_r^1 = u_r^2).$$

For the sake of simplicity, regard the variables as unidimensional. For the questions considered in this comment, the case of unidimensional variables suffices to provide insights into the problems.

The conditions express the general rule that the social rate of substitution is equal to the rate of transformation: the numerator on the left-hand side of equation (1) is the social marginal benefit of environmental improvement. The same interpretation applies to the necessary conditions of the form shown in equation (1′). Equation (2) shows the complications of assuming that the special constraint is binding: the marginal rate of substitution between environmental quality and the non-polluting good for the pollutee is multiplied by the net marginal change in environmental quality, corrected for the relationship between the polluting and the environmental improvement good, via the rate of transformation; on the right-hand side we have the marginal rates of transformation between the polluting and improvement goods and the non-polluting good, adjusted, again, by the rate of transformation between the polluting and the improvement goods.

Now let me turn to the method of implementation. As usual, the private decentralised decision-making units are guided towards the optimal solution by means of a price system. The author is, in the present paper, only explicit regarding the behaviour of the polluter, and only implicitly treats the behaviour of the pollutee and the producer of the three types of man-made goods.

The prices are defined *before* deriving the optimal conditions, and the definitions just inserted in the optimal condition and in the adjustment conditions of the polluter when deriving the optimal tax, without showing how the prices are *derived* in the market model. The general question is how to determine a set of prices that implements the optimal solution, given the specified type of decentralised decision-making. In the market model of the polluter's behaviour an initial income is introduced. As the distribution of goods and services is *endogenously* determined in the social optimisation, an initial income distribution introduced in the market model must also be adjusted or specified in such a way that the optimal solution is realised.

Since the author is not very explicit on certain aspects of the market system in the present paper, it may be of interest to offer some speculation about possible difficulties.

For instance, who considers the special constraint in the adjustment process: the pollutee, the producer, or neither of them? And what about the pollutee: does he only adjust the amount of non-polluting good, or does he also vary the amount of improvement good, or, perhaps, does he pay directly for environmental quality? The answers to these questions have certain implications for the optimal tax on the polluting good.

Let us, for the sake of the argument, assume that the pollutee maximises utility with non-polluting and improvement goods as decision variables and with given prices and income, and that the producers maximise profit with given prices and resources, with the transformation function and the special relationship between polluting ʂund improvement goods as constraints.

The market solution then yields:

$$\frac{u_x^1}{u_r^1} = q + t, \ \frac{u_E^2\,Ex}{u_r^2}p, = q = \frac{F_x}{F_r} - \chi f_v, \ p = \frac{F_X}{F_r} - \chi f_x \tag{3}$$

where q is the price of x and p the price of X.

Eliminating the prices and the Lagrangean parameter χ gives:

$$\frac{u_x{}^1}{u_r{}^1} - \frac{u_E{}^2}{u_r{}^2} . E_X . \frac{f_x}{f_X} - t = \frac{F_x}{F_r} - \frac{F_X}{F_r} . \frac{f_x}{f_X} \tag{4}$$

Comparing the real solution (2) with the market solution (4) gives the optimal tax:

$$t = -\frac{u_E{}^2 . E_x}{u_r{}^2} \tag{5}$$

The corresponding optimal financial budget is:

$$B = tx - pX = -\frac{u_E{}^2}{u_r{}^2} E_x . x - \frac{u_E{}^2}{u_r{}^2} E_X . X \tag{6}$$

$$= -\frac{u_E{}^2}{u_r{}^2} (E_x . x + E_X . X)$$

i.e. the special constraint does not enter into the optimal budget when the producer takes it into consideration.

But it must be admitted that this is a strange assumption when the pollutee pays for environmental improvement. For instance, letting the pollutee (or a public authority) take the special constraint into consideration, as considered by the author, yields the result of equation (6) in his paper.

4 *An Environmental Agency*

In the introduction to his paper, the author considers a public authority concerned with, and paying for, improvement activity and collecting pollution taxes. References are made to the 'Agences Financières de Bassins' and the river agencies of the Ruhr region.

The question of optimal institutions is so important that it seems worth while to consider it somewhat more extensively and explicitly than in the author's present paper, and also from a somewhat different point of view.

In the tradition of regarding environmental pollution as a problem of managing a common property resource [6, 7], let us introduce an environmental agency, EA. The EA maximises the 'social' profit by 'selling' environmental improvement goods:

Maximise $v . E(x, X) + tx - pX$
subject to $f(x, X) \geq 0$

where v is the price per unit of environmental quality. It is natural to regard environmental quality as a public good [2] so that no real payments need to be considered as long as the authority acts *as if* it is selling environmental quality.

Let us first assume that EA has the polluting and improvement goods as decision variables, and regard the prices as given. The first-order conditions are then:

$$\left.\begin{array}{r} v . E_X + t + \lambda f_x \leqq 0 \\ v . E_X - p + \lambda f_X \leqq 0 \\ \lambda . f(x, X) = 0 \end{array}\right\} \tag{7}$$

A sufficient second-order condition is that the environmental function (and the constraint function) should be concave. This is an interesting question to pursue empirically.

Assuming an interior solution, and that the constraints are binding, yields the maximised profit function:

$$\hat{\Pi}_{EA} = q.E - q(E_x.x + E_X.X) - \lambda(f_x.x + f_X.X) \tag{8}$$

The financial budget $B = (tx - pX)$ considered by the author corresponds to the last two terms in (8) and is identical with his result (6).

In the above formulation the condition for a balanced budget is *independent* of the price of environmental services, provided EA maximises profit. But for this market solution to simulate the optimal solution, the price of environmental quality must equal the pollutee's rate of substitution between environmental quality and the non-polluting good, i.e. $v = u_{E^2}/u_r{}^2$.

Let me now assume that the constraint is homogeneous (see the author's paper) or unbinding. The maximised profit function can be written [4]:

$$\hat{\Pi}_{EA} = v.E(1 - \epsilon) \tag{9}$$

where ϵ is the elasticity of scale of the environmental function. (To derive this, the *passus equation* is employed. For readers unfamiliar with this Frisch relationship, it can be mentioned that the passus equation is a generalisation of Euler's equation from homogeneous functions to inhomogeneous functions, i.e. the constant degree of homogeneity is replaced with the scale elasticity [1].) In the author's terminology there are decreasing, constant and increasing qualitative returns to scale according to $\epsilon \lesseqgtr 0$ (for $E > 0$).

As regards the first four examples of models, only homogeneous environmental functions are considered when analysing the budgetary question. In reference [8] inhomogeneous functions are discussed, as is the case in the present paper. However, inhomogeneity is sometimes obtained here in another special way by introducing an initial, or untaxed, quantity of pollutant.

When the scale elasticity is zero, the financial budget is then balanced, as the author states. From (8) we see that this situation can alternatively be characterised by the elasticity of substitution being identically equal to -1 (see [8], pp. 103–7):

$$B = 0 \Leftrightarrow S_{Xx} = Ex/E_X = -X/x \text{ for all admissible values of } x, X \tag{10}$$

where S_{Xx} is the rate of substitution.

Considering an inhomogeneous environmental function with variable scale elasticity, the question arises as to which scale is optimal. The general rule is that the scale is determined by the equalisation of the product price, i.e. the price on environmental services, and the marginal cost of expanding output. This has the very important implication that the sign of the financial budget depends on the evaluation of environmental services reflected through the corresponding 'product' price. Taken as it stands, the expression (9) for the optimal profit function shows that the sign of the financial budget $B = -vE.\epsilon$ is independent of the value of the positive environmental price, v. But the point is that the optimal scale, i.e. the value of ϵ, depends on the environmental price. Restricting the budget to be balanced implies a special 'revealed' price on the environment.

It should be noted that the value of a variable elasticity to scale depends on the index function chosen for measuring environmental services, except for the case of the key value where it is zero. The property of a balanced budget is not changed by a monotonic transformation of the environmental function, as stated by the author.* This can, for instance, be shown by employing the passus equation:

$$\epsilon E = \frac{\partial T}{\partial E} \cdot \epsilon' E'$$ (11)

where $T(E)$ is the transformation function, and $\epsilon' E'$ the scale elasticity and level of environmental quality after transformation. When E, $E' \neq 0$ and $\partial T/\partial E > 0$, $\epsilon = 0 \Rightarrow \epsilon = 0$.

Considering the introduction of 'environmental agencies' into the models reviewed in section 1, relationships equivalent to (9) are obtained. In fact, the connection between budget surpluses and deficits, and the elasticity of scale, may be termed the 'Mohring paradox' (p. 80): a technically favourable situation – with a decrease in congestion when the number of trips and the capacity of a highway are increased proportionately – implies, at the optimum, a budget deficit. A surplus results in the unfavourable situation of increased congestion following a proportionate increase of the factors.

Let a management agency maximise the following profit function in model (a):

$$\Pi = T.D - rk(S) - p.g(D, S)$$

where the new magnitude, p, is the given unit price the agency must consider for the time cost. The number of trips is determined by:

$$\frac{\partial \Pi}{\partial D} = T - P\frac{\partial g}{\partial D} = 0, \quad T = P\frac{\partial g}{\partial D}$$ (12)

and the capacity of the highway by:

$$\frac{\partial \Pi}{\partial S} = r\frac{\partial k}{\partial S} - p\frac{\partial g}{\partial S} = 0, \quad r = -p\frac{\partial g}{\partial S}\bigg/\frac{\partial K}{\partial S}$$ (13)

The financial budget now becomes:

$$B = T.D - r.K(S) = p\frac{\partial g}{\partial D}D + p\frac{\partial g}{\partial S}S.\frac{1}{\epsilon_k}$$ (14)

where $\epsilon_K = \partial K/\partial S$, $S/K > 0$.

Quite apart from the shadow price p on congestion, the budget is balanced when the scale elasticity $\epsilon_g = [(\partial g/\partial D)D + (\partial g/\partial S)S]/g$ is zero (assuming $g \neq 0$), and deficits and surpluses occur according to whether $\epsilon_g \lessgtr 0$. Remember that g measures congestion, i.e. the public 'bad'; $\epsilon_g < 0$ is the technically favourable case, and $\epsilon_g > 0$ the favourable one.

* Special communication to Karl-Göran Mäler.

The same result is obtained by considering model (b). Maximising

$$\Pi = \tau \sum_{i=1}^{n} t^i - E - pD(\Sigma t^i, E)$$

where τ is the tax per trip and p the shadow price on congestion (the price on improvement resources is normalised to 1), gives the following relationship determining the *total* number of trips and the level of improvement activity:

$$\frac{\partial \Pi}{\partial(\Sigma t^i)} = \tau - p\frac{\partial D}{\partial(\Sigma t^i)} = 0 \tag{15}$$

$$\frac{\partial \Pi}{\partial E} = -1 - p\frac{\partial D}{\partial E} = 0 \tag{16}$$

The financial budget becomes:

$$B = \tau \sum_{i=1}^{n} t^i - E = p\frac{\partial D}{\partial(\Sigma t^i)}Zt^i + p\frac{\partial D}{\partial E}E = p\,\epsilon_D \tag{17}$$

A technically unfavourable situation again results in a surplus being optimal.

Finally, let us consider the water pollution model (c). The model (d) yields equivalent results, but the locational aspects specified in the model by identifying upstream and downstream firms make the expressions more complicated. The locational aspect is disregarded in reference [12], making the model similar to Upton's [15]. Maximising the profit of a river agency,

$$\Pi = \Sigma t_i L_i - C(F) - pD(L_1, \ldots, L_n, F)$$

yields the following condition for determining the discharge from each of the polluters and the resources needed for flow augmentation:

$$\frac{\partial \Pi}{\partial L_i} = t_i - p\frac{\partial D}{\partial L_i} = 0 \ (i = 1, \ldots, n) \tag{18}$$

$$\frac{\partial \Pi}{\partial F} = \frac{\partial C}{\partial F} - p\frac{\partial D}{\partial F} = 0 \tag{19}$$

The optimal budget is:

$$B = \Sigma t_i L_i - C(F) = p\Sigma\frac{\partial D}{\partial L_i}L_i + p\frac{\partial D}{\partial F}F \cdot \frac{1}{\epsilon_c} \tag{20}$$

where $\epsilon_c = \partial C/\partial F . F/C > 0$. The favourable case when $\epsilon_D = [\Sigma(\partial D/\partial L_i)L_i + (\partial D/\partial F)F]/d$ is negative yields a deficit; $\epsilon_D = 0$ yields a balance; and $\epsilon_D > 0$ yields a surplus.

Considering the environmental agency introduced in the environmental pollution model (e) yields the same result as shown by equation (9), when Z is treated as environmental services, and the income of the agency results from 'selling' these services at positive prices.

Summing up, the optimal value, when it is variable, of the scale elasticity of the congestion or environmental service functions is dependent on the evaluation of environmental deterioration. Balancing the financial budget cannot always be the optimal policy. Management decisions should be based on explicit considerations of 'selling' environmental services or 'paying' for environmental deterioration, i.e. treating environmental services analogously to man-made goods, with the price reflecting society's evaluation of the environmental services.

Considering specific situations, the nature of the environmental function can be such that, for instance, positive values of the scale elasticity are excluded. (This seems reasonable in many cases, especially if 'untouched' nature implies the highest level of environmental services, and the variation of the scale elasticity is monotonic.) One can then, *a priori*, conclude that the budget deficit is not optimal, independently of the price of environmental services.

Kolm replied to Førsund's highly interesting and valuable 'comments', by mentioning his own remark about the 'aggregation level'. His point was that the 'aggregation level' of the model's variables was not a choice open to the model-maker since it had to correspond to what had in reality to be the base of the tax or of the pricing. The 'aggregation level' thus had to be very low (i.e. detailed). This was one of the reasons which prompted him to use a 'multidimensional' analysis (several x_i's and X_i's), the other being the existence of 'special constraints' linking these variables. Of course, taxing and pricing costs, particularly information costs, may make non-optimal, or may even prevent, the most detailed (otherwise optimal) discrimination among units of pollutants. The subsequent second-best problem was treated in references [28] and [43] of his paper. But the resulting optimal disaggregation model was still very low.

In answer to a question by Førsund he also argued that, in the model, each pollutee was perfectly free to undertake any cleaning or protection for himself. This behaviour of his was only implicit since it was not the public sector's problem.

He finally pointed out that Førsund's review of the literature in English interestingly revealed Mohring and Harwitz's anteriority about the central idea of 'qualitative returns to scale' as applied to road transportation. However, there also existed more remote forerunners in other languages and about other applications.

REFERENCES

[1] Førsund, F. R., 'A Note on the Technically Optimal Scale in Inhomogeneous Production Functions', *Swedish J. Econ.*, vol. 73, no. 2 (1971).

[2] ——, 'Allocation in Space and Environmental Pollution', *Swedish J. Econ.*, vol. 74, no. 1 (1972).

[3] ——, 'Generelle økonomiske prinsipper for løsning av forurensingsproblemene', *Statsøkonomisk Tidsskrift*, vol. 86, no. 2 (1972).

[4] ——, 'Externalities, Environmental Pollution and Allocation in Space: A General Equilibrium Approach, *Regional and Urban Studies*, no. 1 (1973).

[5] Hayden Boyd, J., 'Collective Facilities in Water Quality Management', appendix to chap. 10 in Kneese, A. V., and Bower, B. T., *Managing Water Quality: Economics, Technology, Institutions* (Johns Hopkins Press, Baltimore, 1968).

[6] Kneese, A. V., and Bower, B. T., *Managing Water Quality: Economics, Technology, Institutions* (Johns Hopkins Press, Baltimore, 1968).
[7] ——, 'Environmental Pollution: Economics and Policy', *Amer. Econ. Assoc., Papers and Proceedings* (1971).
[8] Kolm, S.-Ch., *Le service des masses* (Dunod, Paris, 1971).
[9] ——, *La théorie économique générale de l'encombrement* (SEDEIS, Paris, 1968).
[10] Mishan, E. J., 'The Postwar Literature on Externalities: An Interpretative Essay', *J. Econ. Literature* (Mar 1971).
[11] Mohring, H. D., and Harwitz, M., *Highway Benefits: An Analytical Framework* (Northwestern Univ. Press, Evanston, Ill., 1962).
[12] ——, and Hayden Boyd, J., 'Analysing "Externalities": "Direct Interaction" as "Asset Utilisation" Frameworks', *Economica* (Nov 1971).
[13] Rothenberg, J., 'The Economics of Congestion and Pollution: An Integrated View', *Amer. Econ. Assoc., Papers and Proceedings* (1970).
[14] Strotz, R. H., 'Urban Transportation Parables', in J. Margolis (ed.), *The Public Economy of Urban Communities* (Johns Hopkins Press, 1965).
[15] Upton, C., 'Optimal Taxing of Water Pollution', *Water Resources Res.*, vol. 4, no. 5 (1968).

6 Effluent Charges versus Effluent Standards

Karl-Göran Mäler

6.1 *INTRODUCTION**

In their book, *Managing Water Quality* [6], Kneese and Bower suggested that the use of effluent charges may be more efficient than the use of regulations in environmental policy. The reason for this is that charges induce those firms, for whom it is cheap to reduce activities harmful to the environment, to undertake large-scale reductions, while higher-cost firms are induced to undertake only lesser reductions. This argument was formulated mathematically by Baumol and Oates [2], and they proved rigorously that under certain assumptions effluent charges are efficient while effluent standards may be inefficient.

This paper is an attempt to discuss systematically this argument in favour of effluent charges. In order to simplify the discussion, only one kind of regulation is considered, namely effluent standards (i.e. upper limits of the amounts of wastes that firms are allowed to discharge into the environment).

The difference between effluent charges and effluent standards is first considered in terms of information requirements, and it is shown that in general the determination of optimal effluent charges does not require more information than the determination of optimal effluent standards. In some important cases effluent charges require strictly less information than effluent standards.

But it is also shown that in some cases (cases in which the waste load is a random variable) effluent charges and effluent standards are not comparable from an efficiency point of view because the former are based on a maximisation of the expected economic value of the environment, while the latter are based on a set of probabilistic environmental or ambient standards.

* This paper is a report of some results I obtained when I was at Resources for the Future, Inc., Washington, D.C., in the spring of 1971 as a visiting scholar. I wish to thank Blair Bower, Allen Kneese and Clifford Russell for many discussions and comments on earlier drafts. I accept of course the responsibility for any criticism raised against this paper, but I believe that they should share the responsibility for any credits.

6.2 *THE BASIC MODEL*

In this section we present the formal structure of the model that will be used in the subsequent discussions.

We assume first an economic region that is closed from an environmental point of view. This means that all environmental damages in the region are created by economic activities within the region alone and that activities within the region do not create any environmental damages outside the region. This is a rather restrictive assumption, but it can only be avoided at the cost of a much more complicated analysis. We assume also that the region can trade all goods at constant prices.*

All environmental damages to be discussed are assumed to be caused by the discharge of residuals or waste products into the environment. We shall not discuss other kinds of environmental distortion. Moreover we assume that all residuals are generated in firms and no residuals are generated in consumption.†

A residual is usually defined as a commodity that has no economic value at the current relative prices. A change in relative prices could give positive value to a commodity that previously had been classified as a residual, thereby converting it into an 'input' or 'output'. In spite of this possibility, let us assume that there is a fixed list of residuals, independent of any relative price configuration.‡ Assume that there are s different residuals. These residuals may differ in several respects. They may have different chemical and physical characteristics, and they may also differ in spatial distribution. A residual discharge at one place will in general have a different effect on the environment than a physically identical residual discharged at a different place. It is therefore important to keep the spatial distribution of residuals discharge in mind.

There are K firms in the region, each with a production function

$$x_k = f^k(S^k) \quad k = 1, ..., K \tag{1}$$

For simplicity, we have here assumed that each firm produces only one commodity, x_k. S^k is the vector of inputs in firm k. The amounts of the residuals generated by firm k are denoted by the vector z^k. z^k will in general be a function of the amount produced, the technology used and the input vector. This means that z^k is a function of S^k. But z^k will also be a function of waste treatment in firm k. Let the input in waste treatment in firm k be T^k (a vector). We can then write the residuals generating function as

$$z^k = g^k(S^k, T^k) + e^k \quad k = 1, ..., K \tag{2}$$

* The present paper is a brief summary of chap. 6 in a forthcoming monograph [7]. In this book more general models are studied.

† Consumption residuals are discussed in [7].

‡ For a more general model, see [7].

where g^k is a vector-valued function. e^k is a random variable, representing the stochastic nature of residuals generation. For the rest of this section and in the next section, e^k is assumed to be identically zero.

We assume that f^k and g^k are twice continuously differentiable, and that f^k is concave and g^k convex.

Let r be the vector of input prices and let p_k be the price on the output of firm k. We assume that all firms are profit-maximisers and that they are price-takers.

The profit π_k for firm k can then be written (superscript T denotes transposition) as

$$\pi_k = p_k f^k(S^k) - r^T(S^k + T^k) \tag{3}$$

if the firm is allowed to discharge its residuals at zero cost into the environment. If there are upper limits, i.e. effluent standards on the amounts of residuals they are allowed to discharge into the environment, firm k will maximise π_k subject to the restriction that $z^k \leq \bar{z}^k$, where $\bar{z}^k$ is the standard. If the firms are allowed to discharge any amount of residuals, but have to pay a constant price for the discharge (i.e. effluent charges) of each residual, the profit becomes

$$\pi_k' = p_k f^k(S^k) - r^T(S^k + T^k) - q^T z^k \tag{4}$$

where q is the vector of prices or effluent charges the firms have to pay.

Profit maximisation implies that each firm will generate a certain vector of residuals which depends on the prices. Together the K firms in the region generate the vector

$$z = \sum_{k=1}^{K} z^k \tag{5}$$

which is discharged into the environment.

Let us now assume that there is a list of m environmental qualities, represented by the vector $Y \in R^m$. The first component in Y may be the dissolved oxygen concentration at some point in a water body, the second the dissolved oxygen concentration at some other point, the third the concentration of sulphur dioxide at some point, etc. The vector Y will in general be a function of the vector z, but also of other human activities (flow regulation, for example). We shall not discuss these other activities, however. Let us write the environmental interaction function as

$$Y = F(z) \tag{6}$$

F is a vector-valued function, $F : R^s \rightarrow R^m$. We assume that F is twice continually differentiable, that each component in F is monotonically decreasing, and that F is concave. As each z^k depends on the prices, Y will also be a function of prices.

6.3 *THE OBJECTIVES IN ENVIRONMENTAL POLICY*

We assume that there are two objectives in environmental policy. The first is that a set of environmental standards must be met:

$$Y \geq \overline{Y} \tag{7}$$

$\overline{Y}$ is a vector of environmental qualities giving a lower limit below which the actual quality of the environment must never fall. The vector $\overline{Y}$ is assumed to be determined by a political process, based, perhaps, on a cost–benefit analysis of the use of the environment. We shall not discuss how $\overline{Y}$ is or should be determined, however. $\overline{Y}$ is a fixed vector given from outside.

The second objective is that these environmental standards should be met at lowest economic cost. We have assumed that all firms are price-takers, and that the prices are given from outside the region; this implies that there is no difference between the private costs incurred by the firms and the social costs incurred by the region for a reduction in the discharge of residuals. One of the main factors in the cost of meeting the environmental standards is thus the private costs incurred by firms for controlling the flow of their own residuals. Other factors are administration costs, costs for monitoring the wastes flows, etc. These factors will be disregarded, however, because it seems probable that the two policy measures to be discussed, effluent charges and effluent standards, will have much the same costs in this respect. Finally, there is one more significant type of cost, the cost of information gathering and analysis. These costs are important in deciding between the two policy measures, and the main purpose of the present paper is to compare the information requirements for the two policy measures.

In order to make this purpose more precise, let us for the moment neglect all costs except the private costs of controlling residuals flows. The costs incurred by firm k for discharging only the flow $\bar{z}^k$ is then determined indirectly from the firm's constrained maximisation problem:

$$\max \pi_k = p_k f^k(S^k) - r^T(S^k + T^k) \tag{8}$$
$$\text{subject to } g^k(S^k, T^k) \leq \bar{z}^k$$

Solving this optimisation problem gives $\max \pi_k$ as a function, B^k, of prices and the flow of residuals permitted:

$$B^k = B^k(p_k, r, \bar{z}^k) \tag{9}$$

Since we regard all prices as constant, we can as well write this benefit function as

$$B^k = B^k(\bar{z}^k) \tag{10}$$

This function gives the maximum profit (or benefit) from the production processes in firm k given the effluent standards $\bar{z}^k$. Disregarding income distribution problems (by assuming that lump-sum transfers are feasible, for example), this maximum profit can be identified as the social benefit from the production activities in firm k. If we add these benefit functions over all firms we shall obtain the total regional benefit from the production activities in the region:

$$B = B(\bar{z}^1, \ldots, \bar{z}^K) \equiv \sum_{k=1}^{K} B^k(\bar{z}^k) \tag{11}$$

Still neglecting all costs of information gathering and analysis, the objective for the environmental policy can be formulated as follows:

Find
$$\max B(\bar{z}^1, \ldots, \bar{z}^K)$$

$$\text{subject to } F(\sum_{k=1}^{K} \bar{z}^k) \geq \bar{Y} \tag{12}$$

If the responsible regional authority has all the information necessary to obtain a solution to this problem, it can obviously determine the optimal effluent standards and, by enforcing them, realise the optimal solution. But if there are costs associated with information gathering, it may be prohibitively expensive to gather all the information necessary to solve the problem, and the question arises: are there policy measures that require less information than effluent standards that can be used to obtain the optimal solution?

6.4 *EFFLUENT STANDARDS AND EFFLUENT CHARGES*

There are obviously many policy measures that can be used in environmental policy. In Sweden, for example, the Environmental Protection Board can subsidise investments in production and/or waste treatment processes that decrease the waste load on the environment. When a firm plans a major change of plant it must have a licence from the Board in order to make the change, and in that way it is possible for the Board to control to some extent the environmental damages the firm will create. But in general the discharges of residuals are not monitored, which means that the firm has no direct incentive to keep its waste load low after a new plant has been constructed, or after a new waste treatment process has been installed. A manager of a Swedish pulp mill has said that the relation between the amounts of actual discharge of waterborne residuals and the amount the plant was designed for is 4 : 1. This means that if the environmental policy is to be effective, the actual discharges of wastes must be monitored.* But if this is the case, it seems that the only real alternatives are effluent charges and effluent standards. We shall therefore discuss only these two approaches.

The first question that we shall analyse is whether it is possible to achieve the optimal solution by the use of effluent charges. This question is equivalent to the following: Is there a choice of a vector of effluent charges, q, such that the resulting residuals vector, $z = \sum_{k=1}^{K} z^k$, is the same as in the optimal solution?

Given a vector q, the corresponding discharge of residuals by firm k is called forth by the new maximisation situation, given by:

$$\max \pi_k' = \max p_k f^k(S^k) - r^T(S^k + T^k) - q^T g^k(S^k, T^k) \tag{13}$$

This maximisation problem will be called problem I. It can equivalently be written as

$$\max \left[B^k(z^k) - q^T z^k, \right] \quad z^k \geqq 0 \tag{14}$$

The first-order conditions for a maximum are:†

$$\text{(a)} \quad \frac{\partial B^k(z^k)}{\partial z_i^{\,k}} - q_i \leqq 0 \tag{15}$$

$$\text{(b)} \quad z_i^{\,k} \left[\frac{\partial B^k(z^k)}{\partial z_i^{\,k}} - q_i \right] = 0 \quad i = 1, \ldots, s$$

If B^k is strictly decreasing in each component z_i^k, which is strictly positive, the first-order conditions will have a unique solution. If the second-order conditions are satisfied, this unique solution will be a solution to problem I.

On the other hand, given optimal effluent standards $\bar{z}^k$, the behaviour of the firm is described by problem II, i.e. by

$$\max B^k(z^k)$$
$$\text{subject to } z^k \leqq \bar{z}^k, \ z^k \geqq 0 \tag{16}$$

Define the Lagrangean L by

$$L = B^k(z^k) - \bar{q}^T(z^k - \bar{z}^k) \tag{17}$$

---

* For some residuals, and especially consumption residuals, such monitoring may be too expensive. There exist in most such cases, however, possibilities to use taxes on other commodities as substitutes for effluent charges.

† See, for example [4].

where $\bar{q}$ is a vector of multipliers. The first-order conditions for a maximum are:

$$\text{(a)} \quad \frac{\partial B^k(z^k)}{\partial z_i{}^k} - \bar{q}_i \leqq 0 \tag{18}$$

$$\text{(b)} \quad z_i{}^k \left[\frac{\partial B^k(z^k)}{\partial z_i{}^k} - \bar{q}_i \right] = 0 \quad \bar{q} \geqq 0, \quad i = 1, \ldots, s$$

Formally, these conditions are exactly the same as the first-order conditions for problem I if we set the effluent charges equal to the multipliers in problem II.

This shows that to each effluent standard there is a vector of effluent charges which supports the standard, in the sense that discharges of residuals equal to or less than the standard are a solution to problem I. But there may exist solutions to problem I that violate the standard. If there is no unique solution to problem I, effluent charges may lead to a violation of the effluent standards and hence of the environmental standards. This situation can appear only if the benefit functions are piecewise linear, which can be seen in the following way.

As f^k is concave and g^k convex, it follows that B^k is concave in z^k. Let $z^{k\prime}$ and $z^{k\prime\prime}$ be two solutions to problem I. Then

$$B^k(z^{k\prime}) - q^T z^{k\prime} = B^k(z^{k\prime\prime}) - q^T z^{k\prime\prime}$$

The concavity of B^k implies that for $0 \leq t \leq 1$,

$$\begin{aligned} B^k(tz^{k\prime} + (1-t)z^{k\prime\prime}) - q^T(tz^{k\prime} + (1-t)z^{k\prime\prime}) &\geqq \\ t(B^k(z^{k\prime}) - q^T z^{k\prime}) + (1-t)(B^k(z^{k\prime\prime}) - q^T z^{k\prime\prime}) &= \\ B^k(z^{k\prime}) - q^T z^{k\prime} \end{aligned}$$

so that $B^k(z^k) - q^T z^k$ is constant on the segment connecting $z^{k\prime}$ and $z^{k\prime\prime}$.

The situation is illustrated in Fig. 6.1 (which is based on the assumption that there is only one residual under consideration). If the effluent charge happens to be $\bar{q}$, the firm will be indifferent to amounts of discharges that lie in the interval $|z^{k\prime}, z^{k\prime\prime}|$. The optimal effluent standard may, however, be $\bar{z}^k$, and it is clear that the use of effluent charges may lead to a violation of this standard. Moreover it may also happen that the firm discharges less than $\bar{z}^k$, and the region will incur unnecessary costs.*

* This objection against the use of effluent charges was raised by the Swedish Petrol Institute in discussions with the Swedish Environmental Protection Board about effluent charges on sulphur dioxide emissions. The main method to reduce these emissions is, according to the Institute, to desulphurise crude oil in the refineries, and as this can be done at constant costs, the aggregate benefit function will be linear over a large interval.

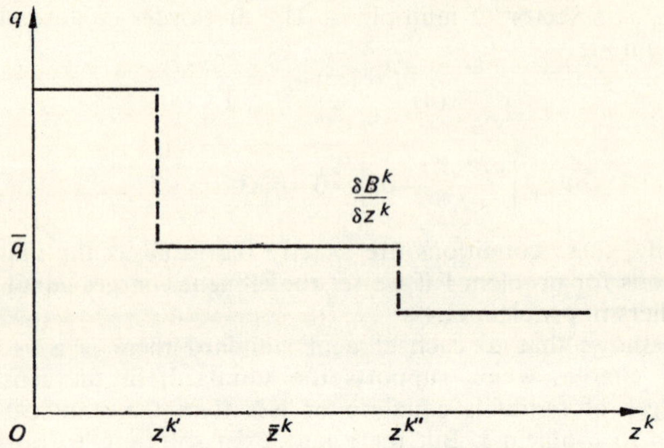

FIG. 6.1 Piecewise linearity of Benefit Function

Our first conclusion is thus: *If the individual benefit functions for waste discharges are piecewise linear, effluent charges may not lead to an optimal solution. In particular, they may cause a violation of the environmental standards.*

The relevance of this objection against the use of effluent charges depends on the answer to the empirical question: are the benefit functions piecewise linear and, if so, are the sets on which they are linear large in some practical sense? It is impossible to give a complete answer, but some empirical studies indicate that the firms have in general a large number of opportunities to reduce the flow of residuals. In some cases it seems that the number is almost infinite. If the number of opportunities is large, it is improbable that the sets on which the benefit functions are linear are large.

For the rest of this paper it will be assumed that the benefit functions are non-linear (or strictly concave).

Let us now assume that we have all the information necessary to set optimal effluent standards. These standards are determined from problem III:

$$\max \sum_{k=1}^{K} B^k(z^k) \tag{19}$$

$$\text{subject to } F\left[\sum_{k=1}^{K} z^k\right] \geqq \bar{Y},\ z^k \geqq 0 \quad k=1, ..., K$$

The first-order conditions for a maximum are

$$\frac{\partial B^k}{\partial z_i^{\,k}} + \sum_{j=1}^{m} \delta_j \frac{\partial F_j}{\partial z_i^{\,k}} \leqq 0 \tag{20}$$

$$z_i^k \left[\frac{\partial B^k}{\partial z_i^k} + \sum_{j=1}^{m} \delta_j \frac{\partial F_j}{\partial z_i^k} \right] = 0 \quad k = 1, ..., K; \ i = 1, ..., s$$

$$F \left[\sum_{k=1}^{K} z^k \right] \geq \overline{Y}$$

where δ is a vector of Lagrange multipliers. δ_j can obviously be interpreted as the imputed price on the environmental standard $\overline{Y}_j$.

The solution to these conditions determines the effluent standards $\overline{z}^k$. In order to set these standards so that the optimal solution is achieved, it is thus necessary to know the individual benefit functions $B^k(z^k)$ and have knowledge about the environmental interaction function F.

These same conditions also determine, however, optimal effluent charges q:

$$q_i = \sum_{j=}^{m} \delta_j \frac{\partial F_j}{\partial z_i^k} \tag{21}$$

because with the vector q defined in this way, it is immediately seen that the first-order conditions for problem III are identical to the first-order conditions for problem I.

Our second conclusion is thus: *The information necessary to set optimal effluent standards is sufficient for determining optimal effluent charges.*

Consider now problem IV:

$$\max \sum_{k=1}^{K} B^k(z^k) \tag{22}$$

$$\text{subject to } \sum_{k=1}^{K} z^k \leq z, \ z^k \geq 0 \quad k = 1, ..., K$$

The solution to this problem determines $\max \sum_{k=1}^{K} B^k(z^k)$ as a function of z:

$$B = B(z)$$

Given information on the individual benefit functions, it is thus possible to determine the aggregate benefit function $B(z)$ which tells us the maximum benefit that can be obtained in the region by discharging at most the vector z of residuals into the environment.

The optimal effluent charges are determined from problem V:

$$\max B(z) \tag{23}$$
$$\text{subject to } F(z) \geq \overline{Y}, \ z \geq 0$$

The first-order conditions for a maximum are

$$\frac{\partial B}{\delta z_i} + \sum_{j=1}^{m} \delta_j \frac{\partial F_j}{\partial z_i} \leq 0$$

$$z_i\left[\frac{\partial B}{\partial z_i}+\sum_{j=1}^{m}\delta_j\frac{\partial F_j}{\partial z_i}\right]=0 \quad i=1,\ldots,s \tag{24}$$

$$F(z)\gtreqless \overline{Y}$$

where the vector δ can again be interpreted as the imputed price vector on the environmental standards. The optimal effluent charges are as before determined by

$$q_i=\sum_{j=1}^{m}\delta_j\frac{\partial F_j}{\partial z_i} \tag{25}$$

In order to set optimal effluent charges it is therefore necessary to have information on the aggregate benefit function. Note, however, that this information is not sufficient for the determination of optimal effluent standards. In order to do that, one must have information on the individual benefit functions.

We can therefore state our third conclusion in the following way: *The information necessary to set optimal effluent charges is not sufficient for determining optimal effluent standards.*

Another, somewhat weaker way of formulating the last two conclusions is: *The information necessary to set optimal effluent charges does not exceed the information necessary to set optimal effluent standards.*

If the only way to obtain information on the aggregate benefit function $B(z)$ is from information on individual benefit functions, there is obviously no difference in information requirements between the two approaches. It is therefore of importance to investigate whether it is possible to obtain information about the aggregate benefit function without studying the individual functions.

It is clear that in some simple cases the information requirement will be the same. If, for example, only one firm discharges a certain residual, the aggregate benefit as a function of this residual only will be identical to the individual benefit as a function of this residual, and the information requirement is the same for the two approaches. This situation can be expected to prevail in many cases, because the location of the discharge will in general be different for different firms. Only when many firms discharge their wastes at the same location or when the spatial distribution of the discharge does not matter can a real difference between the information requirements for the two policy measures appear. There are many examples for which this is true. One example is furnished by exhausts from an internal combustion engine in a city. The exhausts can be regarded as the same kind of residual, independent of where the automobile is driving in the city. Another example is furnished by the effects of sulphur deposition resulting from emissions of sulphur dioxide. Because of the long time sulphur remains in the air, the effects will be diffused over a large area, and the spatial distribution of the discharges can be neglected (in contrast to the effects of sulphur dioxide concentrations which are mainly local).*

Even if there are many dischargers at the same location or if the spatial distribution does not matter, it is not clear that effluent charges will require less information. But if there are methods which can be used to estimate the aggregate benefit function (or its partial derivatives) without having to estimate the individual benefit functions, the effluent charge scheme will require less information. It seems probable that such methods must exist. In the case of consumption functions we may note that they are never built up from an individual basis but are directly estimated as aggregate functions, simply because it is much easier to get information on the aggregate level than on the individual. Even if there is no complete analogy between consumption functions and benefit functions, this example suggests that it may well be simpler in practice to obtain aggregate data.

One method that may be used in this context is presented in one of the papers delivered at this conference, 'The Management of the Quality of the Environment', by Russell, Spofford and Haefele. The regional model they propose may be applied to the problem calculating effluent charges and effluent standards. As such a regional model can never capture all the options open to individual firms, the resulting effluent charges or effluent standards will in general not be optimal. Therefore the question arises whether, in the event of imperfect information, effluent standards or effluent charges are more sensitive to changes in available information. It is impossible in general to answer this question, but some partial answers may be given.

Assume first that the lack of information can be represented by the random variable e^k introduced in the previous section. This means that the discharge of residuals is a random variable for each firm, and consequently so is the environmental quality vector Y. The environmental standards must therefore be reformulated. One way of doing this is to require that†

$$\text{prob}\ (Y_j < \overline{Y}_j) \le \rho_j \quad j = 1, ..., m \tag{26}$$

where 'prob' stands for probability and ρ_j is an arbitrary small number. This formulation of the environmental standard means that

* For a discussion of the effects of sulphur dioxide emissions, see Sweden's case study for the United Nations conference on the human environment [1].

† Such standards are used in the Swedish air pollution abatement policy as recommendations (see [8]).

a fall in environmental qualities may be tolerated for a small time given by ρ, but not for any longer time.

The effluent standards must also be reformulated. One way of doing this is to require that

$$\text{prob } (z_i^k > \bar{z}_i^k) \leq \rho_i^k \quad i=1, \ldots, s \tag{27}$$

for each firm k.

In order to determine optimal effluent standards, i.e. standards that maximise the expected regional benefit subject to the restriction that the environmental standards are met, it is necessary to have information on the distribution of e^k for each k. In order to determine optimal effluent charges, however, it is only necessary to have information about the distribution of $e = \sum_{k=1}^{K} e^k$. If $e^1, \ldots, e^K$ are independently distributed and if K is large, the central limit theorem in probability theory can be applied, and e will approximately be normally distributed.* This means that the effluent charge approach will require considerably less information than the effluent standards approach.

Our fourth conclusion is thus: *If our ignorance about individual benefit functions can be represented by a stochastic part of each firm's total discharge of residuals, and if these parts are independently distributed, and if the number of firms is large, effluent charges will require less information than effluent standards.*

As we have already noted, it will in general be impossible to get complete and accurate information about either the individual or the aggregate benefit functions. Some kind of approximation must be made, and the resulting effluent charges or effluent standards will in general be different from the optimal ones. Assume that a set of effluent standards has been determined and implemented, and that these environmental standards are met. After the effluent standards have been enforced, it is not possible to calculate the loss in regional benefit due to the suboptimal nature of the standards, without gathering more information about the individual benefit functions. Moreover it is not possible either to decide whether it is beneficial to the region to increase or decrease any of the standards.

---

* The relevant form of the central limit theorem is given by Feller ([3] Theorem 3, chap. VIII: 4). In loose terms, this theorem states that if e^k, $k=1, 2, \ldots$ are statistically independent with zero expected values and variances σ_k^2, and if these variances are bounded (bounded in a special sense), the random variable

$$\sum_{k=1}^{K} e^k \Big/ \sum_{k=1}^{K} \sigma_k^2$$

tends to be normally distributed with zero expected value and unit variances as K increases to infinity.

Our fifth conclusion is thus: *If effluent standards are determined from imperfect information, there is no way to improve these standards without improving the available information.*

On the other hand, if effluent charges are used, private profit maximisation will lead to an efficient reduction in the sense that the behaviour of the firms will satisfy the first-order conditions to problem IV. But the resulting aggregate discharge of residuals and environmental quality may not be consistent with the environmental standards. Any violation of the standards can, however, be immediately observed and corrective actions taken.

These corrective actions may take the following iterative form: Given the current vector of effluent charges q, we can calculate the imputed prices δ_j on the environmental standards as Lagrange multipliers in problem VI:

$$\max q^T z$$
$$\text{subject to } F(z) \geqq \overline{Y}, z \geqq 0 \tag{28}$$

The first-order conditions are

$$q_i + \sum_{j=1}^{m} \delta_j \frac{\partial F_j}{\partial z_i} \leqq 0 \tag{29}$$

$$z_i \left[q_i + \sum_{j=1}^{m} \delta_j \frac{\partial F_j}{\partial z_i} \right] = 0 \quad i = 1, ..., s$$
$$F(z) \geqq \overline{Y}$$

Next, the imputed prices on standards that are violated are increased and the prices on standards that are met as strict inequalities are decreased. These changes in the imputed prices can then be used for calculating a new vector of effluent charges, and the process can be repeated until all environmental standards are met.

Our sixth conclusion is thus: *If effluent charges, determined from imperfect information, are used, the realised environmental quality will be achieved at minimum cost. Differences with the optimal solution will appear as violations of the environmental standards, and corrective actions may be taken.*

There is one aspect of the information requirement that has been lost owing to our assumption of a fixed number of firms in the region. In reality, the number of firms in the region depends on the profitability of the different industries, which to some extent depends in turn on the environmental policy followed. The determination of optimal standards and charges must therefore be based not only on the benefit functions for existing firms, but also for those of firms that might move to the region. *This means in particular that if effluent standards are used, the regional authority must control entries to each of the industries in the region.* This is a very restrictive information requirement, because it means that the authority must not only have information about firms already in the region, but also about firms outside the region. In spite of this objection, the Swedish Government is now preparing a law that would make such control on the choice of localisation possible.* If, on the other hand, effluent charges are used, these may be determined from a knowledge of the aggregate benefit function only for firms currently in the region. If a firm plans to move to the region, the calculated profitability of this movement will depend on the charge, and it is up to the firm to decide about the movement. If entry results in a violation of the environmental standards, some of the effluent charges will be raised, exactly as when the price on a scarce resource increases when new firms start to demand the resource.

Finally, one aspect that goes beyond the objectives for the environmental policy as we have described them deserves to be mentioned; that is, the effect on income distribution. If there is no environmental policy, the waste disposal capacity of the environment is considered as a free resource, and no rent due to this capacity will appear. But as soon as environmental standards are introduced, the waste disposal capacity of the environment will become a scarce factor and environmental rents are created. The effects on income distribution of different environmental policies are exactly the same as the effects of different distributions of these environmental rents, since we are assuming that the environmental standards set are independent of the different policies. When effluent standards are used, the rents accrue solely to the owners of the firms, while if effluent charges are used, these rents are socialised. If the proceeds of the charges are distributed differently than the profits of the firms, there will arise differences in income distribution between the two policy measures. This implies in particular that the structure of demand for different commodities will be different with the two approaches, and consequently the ultimate overall allocation of resources will also be different.

We can summarise these findings in the final conclusion of this section: *Given perfect information on all relevant benefit functions, it is possible to construct a system of optimal effluent charges and effluent standards that will have the same short-run effects on resource allocation. The long-run effects will in general be different owing to different distributions of the environmental rents.*

This discussion also shows that it is possible to interpret the vector of effluent charges as a rent gradient. Problem VI can thus be interpreted as a problem of maximising environmental rents.

* In Sweden the national physical planning has as an objective to determine which industries may be permitted to be established in different regions (see [5]).

The discussion has been based so far on the assumption that the environmental interaction function F is completely known. This is a very dubious assumption, however, and it seems improbable that environmental policy in the near future can be based on anything like a perfect knowledge of the effects of residuals discharge on the environment. If we go to the other extreme and assume that no knowledge whatsoever is available concerning the function F, it is natural to define the objectives of environmental policy in terms of aggregate flows of residuals themselves, i.e. the environmental standards should be formulated as

$$\sum_{k=1}^{K} z^k \leqq \bar{z} \tag{30}$$

Our conclusions are still valid, however, with this formulation of the environmental objectives, because it is identical to the former if the function F is assumed to be the identity function and if $\bar{Y}$ is written $\bar{z}$ Therefore we shall continue to maintain the more general formulation.

6.5 RANDOM WASTE LOAD

In the previous section some remarks were made on the implications of a random waste load. In this section we shall analyse a simple case in detail in order to interpret the meaning of effluent standards and effluent charges when the environmental quality vector Y is a random variable.

In order to simplify the discussion we shall restrict it to a single firm, which generates a single residual and which affects a single environmental quality.

We assume that the firm generates the flow ϕ of the residual. The firm can, however, decrease this flow by using a waste treatment process. The technology of this process is assumed to be extremely simple, so that it can be represented in the following way:

$$\begin{aligned} H + Ax &= \phi + e \\ H + Bx &= z \end{aligned} \tag{31}$$

ϕ is the flow of residuals generated, and e is a random variable with density $\gamma(e)$. The total flow of wastes entering the treatment process is thus $\phi + e$. This flow can either be disposed of into the environment directly, represented by H, or treated. The treatment process can be operated at any positive level. If it is operated at level x it can absorb the amount Ax of the residual. As a result of the treatment, Bx of the residual is generated ($B < A$). The total flow of residuals discharged into the environment is thus $H + Bx = z$.

If we set $D = A - B$, the treatment technology can be described by

$$z = \phi + e - Dx \qquad (32)$$

Assume that the variable average cost for the treatment process is constant and equal to c. Then the variable cost is cx.

As soon as the treatment plant has been constructed, it has a certain capacity $\bar{x}$, which means that waste loads in excess of capacity have to be discharged untreated into the environment. This capacity implies the following constraint: $x \leq \bar{x}$.

The average capacity cost, or the marginal cost of increasing the capacity, is given by $\bar{c}$. The total cost of operating the treatment process at the level x is then $cx + \bar{c}\bar{x}$.

Let us now define 'waste treatment policy' as follows:

Given the treatment capacity $\bar{x}$, and the waste load ϕ, a waste treatment policy is a function of the random variable e, $x(e)$, such that

$$x: (-\infty, +\infty) \to (0, \bar{x}) \qquad (33)$$

indicating the level at which the treatment process is operated for different random waste loads.

The firm can determine (i) the waste load ϕ, which depends on the output of the firm and the production processes it uses in producing this output, (ii) the treatment capacity $\bar{x}$, and (iii) the waste treatment policy $x(e)$. We shall assume that ϕ is given, and discuss the choice of treatment capacity and treatment policy when effluent standards are used and when effluent charges are used. We begin with effluent standards.

The effluent standard is formulated exactly as in the previous section:

$$\text{prob}(z > \bar{z}) \leq \rho \qquad (34)$$

The optimal treatment capacity and the optimal treatment policy for the firm is given by the solution to the following minimum problem:

$$\min\left[\bar{c}\bar{x} + \int_{-\infty}^{+\infty} cx(e)\gamma(e)de\right] \qquad (35)$$

where $\gamma(e)$ is the density of the distribution of e, subject to

$$x \leq \bar{x}$$
$$x \geq 0$$
$$\text{prob}(z > \bar{z}) \leq \rho \qquad (36)$$
$$z = -Dx + \phi + e$$

We have here assumed that the firm behaves as if it minimises its expected costs for waste treatment.

We shall now heuristically derive the optimal treatment policy.*

Let $\bar{e}$ be defined by $\bar{z} = \phi + \bar{e}$. When $e \leq \bar{e}$, the waste load is so small that even if discharged untreated, the relation $z \leq \bar{z}$ will not be violated. It therefore seems that $x(e) = 0$ is optimal for $e \leq \bar{e}$. Let the capacity be given for the moment. Then there exists a number $\tilde{e}$, such that $-D\bar{x} + \phi + \tilde{e} = \bar{z}$. When $e > \tilde{e}$, z will exceed $\bar{z}$ for each feasible capacity utilisation level x. With $\bar{x}$ given, the probability that $z > \bar{z}$ is thus independent of treatment of waste loads exceeding $\phi + \tilde{e}$. This leads to the conclusion that $x(e) = 0$ is optimal for $e > \tilde{e}$. For $\bar{e} \leq e \leq \tilde{e}$, let us study the policy $x(e) = (e - \bar{e})/D$. With this policy, we shall have $z = -e + \bar{e} + \phi + e = \bar{e} + \phi = \bar{z}$. The constraint $z \leq \bar{z}$ will thus never be violated with this policy so long as $\bar{e} \leq e \leq \tilde{e}$. It is obviously not profitable to use more treatment than what is given by this policy, because it will not affect the probability that $z > \bar{z}$. The policy is illustrated in Figs. 6.2 and 6.3.

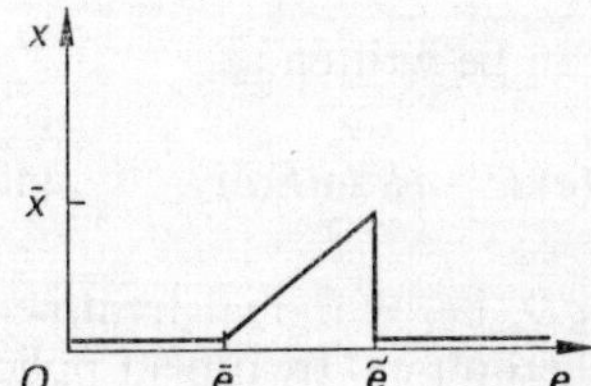

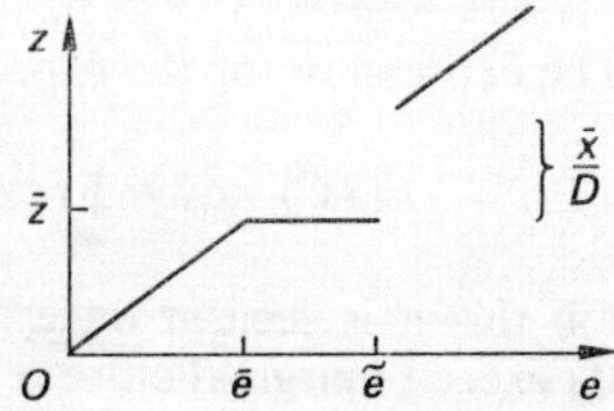

FIG. 6.2 Optimal waste treatment policy under effluent standards: treatment level

FIG. 6.3 Optimal waste treatment policy under effluent standards: residual level

The optimal policy can be summarised by

$$x(e) = \begin{cases} 0 & \text{if } e < \bar{e} \\ (e - \bar{e})/D & \text{if } \bar{e} \leq e \leq \tilde{e} \\ 0 & \text{if } \tilde{e} < e \end{cases}$$

and the capacity $\bar{x}$ is given by $\bar{x} = (\tilde{e} - \bar{e})/D$; $\bar{e}$ is determined from $z = \phi + \bar{e}$, and $\tilde{e}$ from $\int^{+\infty} \gamma(e)de = \rho$.

This waste treatment policy is very peculiar and does not seem consistent with usual environmental objectives. It implies a waste treatment plant which will be completely utilised with probability zero, and it implies that for large waste loads, no waste treatment will be undertaken. The source of this peculiar behaviour is of course the inflexible effluent standard. The standard does not take into account that even in those time periods when the waste load is larger

* In Mäler [7] a proof is given that the policy derived here is optimal. The general case with many residuals and many environmental qualities is also discussed in that book.

than the predetermined level, the region has an interest in the control of the discharge of residuals into the environment.

Let us turn now to effluent charges. Let the effluent charge be q. We assume as before that the flow ϕ is given and is constant, so that the firm behaves as if it minimised its expected waste disposal costs. The expected cost for waste disposal is

$$E(TC) = \bar{c}\bar{x} + c \int_{-\infty}^{+\infty} x(e)\gamma(e)de + q \int_{-\infty}^{+\infty} (e + \phi - Dx)\gamma(e)de \qquad (38)$$

where the first term is the capacity cost, the second the expected variable treatment cost, and the last term the expected payment for discharging residuals into the environment. The firm minimises this expected cost subject to the following constraints:

$$\begin{aligned} 0 &\leq x \leq \bar{x} \\ 0 &\leq e + \phi - Dx \end{aligned} \qquad (39)$$

The expression for the expected cost can be written as

$$E(TC) = \bar{c}\bar{x} + \int_{-\infty}^{+\infty} (c - qD)x(e)\gamma(e)de + q\phi + qE(e) \qquad (40)$$

From this it is seen at once that if $c > qD$, i.e. marginal treatment costs exceed marginal effluent charges, the optimal treatment policy is $x = 0$, $\bar{x} = 0$.

The interesting case is when $c < qD$. Assume for the moment that the capacity is given, $\bar{x}$. Then the minimising problem is reduced to

$$\min(c - qD)x(e) \qquad (41)$$

subject to
$$\begin{aligned} 0 &\leq x \leq \bar{x} \\ Dx(e) &= e + \phi \end{aligned}$$

The solution is obviously $x(e) = \min(\bar{x}, (e + \phi)/D)$. For small e we thus have $x(e) = (e + \phi)/D$. The policy is illustrated in Figs. 6.4 and 6.5.

The optimal capacity is determined from

$$\min\left[\bar{c}\bar{x} + \int_{-\infty}^{\infty} (c - qD)\min(\bar{x}, \frac{1}{D}(e + \phi)\gamma(e)de \right] \qquad (42)$$

We now have

$$\text{prob}((e + \phi)\frac{1}{D} < \bar{x}) = \int_{-\infty}^{Dx-\phi} \gamma(e)de \qquad (43)$$

and so (42) can be written as

$$\bar{c}\bar{x} + \int_{-\infty}^{Dx-\phi} (c - qD)(e + \phi)\frac{1}{D}\gamma(e)de + \int_{Dx-\phi}^{\infty} (c - qD)\bar{x}\gamma(e)de \qquad (44)$$

For a minimum, the derivative of (44) with respect to $\bar{x}$ must be equal to zero:

$$\bar{c} = (qD - c) \int_{Dx-\phi}^{\infty} \gamma(e)de \tag{45}$$

Both the policy and the capacity are now well determined. The policy has an intuitive appeal: when the waste load is small, the capacity utilisation is small, and when the waste load is large, all the capacity will be utilised. Moreover, for small waste loads all residuals will be treated, and no residuals will be discharged into the environment. Residuals will be discharged into the environment only

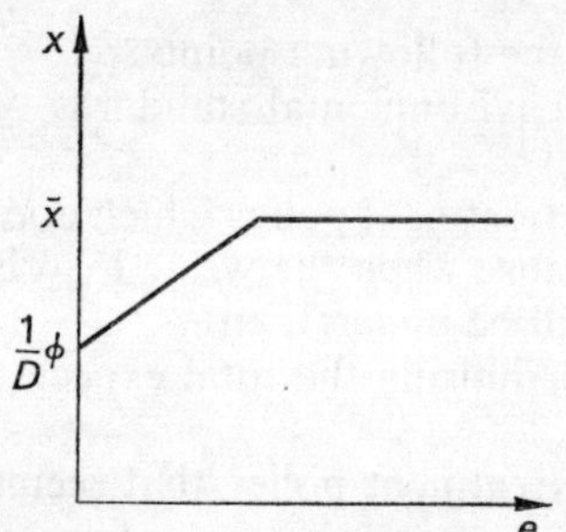

FIG. 6.4 Optimal waste treatment policy under effluent charges: treatment level

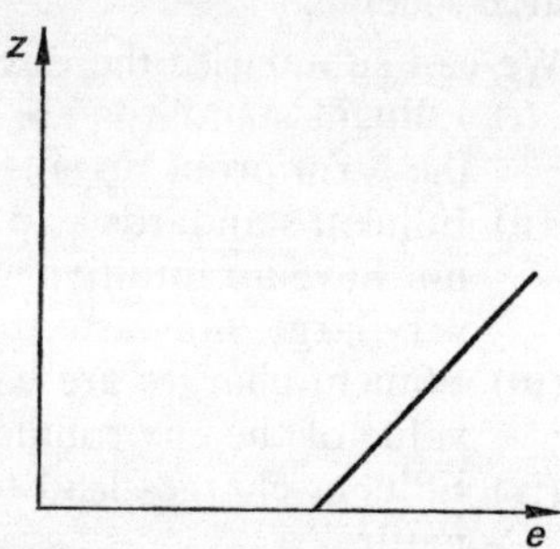

FIG. 6.5 Optimal waste treatment policy under effluent charges: residual level

when the capacity of the waste treatment is not large enough to treat all wastes. By choosing the appropriate effluent charge, it will be possible to satisfy the same effluent standard as before but with fewer harmful effects on the environment. But the effluent charge scheme does not satisfy the effluent standard at least cost, because the policy implications differ from those we derived for the effluent standard scheme. The difference between the two approaches is simply that they have different environmental objectives. We have seen that the effluent standard is based on an environmental standard $\text{prob}(Y < \bar{Y}) \leq \rho$. On what kind of environmental objective is the effluent charge based?

Assume for simplicity that the environmental interaction function can be written as

$$Y = -z = \sum_{k=1}^{K} z^k \tag{46}$$

We then have, in the previous notation, $q = \phi$. The total value to society from the environment can then be written as

$$-q \sum_{k=1}^{K} z^k - \sum_{k=1}^{K} TC^k \tag{47}$$

where TC^k is the treatment cost in firm k. Maximising this value is the same as minimising $\sum_{k=1}^{K}(TC^k+qz^k)$. But since z^k is a random value, this minimisation must be interpreted as minimising the expected value, or

$$\min \sum_{k=1}^{K} E(TC^k+qz^k) \tag{48}$$

We have already seen that if the effluent charge q is imposed on the firms, they will minimise $E(TC^k+qz^k)$, and so the total expected value of the environment will be maximised by using the effluent charge scheme.

We can summarise the discussion in the following points:

(i) Effluent standards are based on environmental standards of the form $\mathrm{prob}(Y \leq \overline{Y}) < \rho$.

(ii) Effluent standards lead to a waste treatment policy which does not have an intuitive appeal, because when the waste load is very large, no waste treatment will be undertaken.

(iii) Effluent charges are based on maximising the total expected value of the environment.

(iv) Effluent charges lead to a waste treatment policy that seems natural.

By introducing more complicated formulations of the environmental and effluent standards it is possible to decrease the peculiarity that characterises the waste treatment policy. Such more complicated standards* may be formulated as follows:

Determine a sequence of pairs $(\bar{z}(i), \rho(i))$, such that $\bar{z}(i) < \bar{z}(i+1)$, and $\rho(i) > \rho(i+1)$, and require that the discharge of residuals satisfy $\mathrm{prob}(z > \bar{z}(i)) < \rho(i)$, $i = 1, \ldots$ By using this more complicated but satisfactory form of the standards, the peculiar form of the waste treatment policy will be less dramatic, and it can be seen that if there are enough pairs which are close to each other, the optimal waste treatment policy will approximate that of effluent charges. On the other hand, it seems extremely difficult to enforce standards in this form. How does one prove in a court that an excess waste load is planned and not the result of a chance phenomenon?

REFERENCES

[1] *Air pollution across National Boundaries: The Impact on the Environment of Sulphur in the Air and Precipitation*, Sweden's case study for the United Nations conference on the human environment (Stockholm, 1971).

---

* The standards mentioned in fn.† on p. 199 are constructed in this way. It should be stressed, however, that these standards are only recommendations and the Environmental Protection Board has no legal means to enforce them.

[2] Baumol, W., and Oates, W., 'The Use of Standards and Prices for Protection of the Environment', *Swedish Journal of Economics* (1971).

[3] Feller, W., *An Introduction to Probability Theory and its Applications*, vol. 2. (Wiley, New York, 1971.)

[4] Hestenes, M. R., *Calculation of Variations and Optimal Control Theory* (Wiley, New York, 1966).

[5] *Hushållning med mark och vatten* (Civildepartementet, Stockholm, 1971).

[6] Kneese, A. V., and Bower, B. T., *Managing Water Quality: Economics, Technology, Institutions* (Johns Hopkins Press, Baltimore, 1968).

[7] Mäler, K.-G., *Studies in Environmental Economics* (forthcoming).

[8] *Riktlinjer för emissionsbegränsande åtgärder*, Statens Naturvårdsverk, Publikationer 1970: 2 (Stockholm, 1970).

APPENDIX

A COMPARISON BETWEEN BRIBES AND CHARGES

In order to compare bribes with effluent charges in environmental policy, the following simple general equilibrium model will be used.

The economy is assumed to be perfectly competitive. There are n different commodities and services. The set of all potential producers is denoted by A, and each producer is denoted by an index k. The net price vector is p. The set of producers actually operating on a market when the price vector is p is denoted by $K(p)$.

We assume that all individual production sets are strictly convex so that the supply functions are indeed functions and not correspondences. We follow the usual convention and regard outputs (including waste products) as positive quantities and inputs as negative quantities. The supply function of producer k (for all commodities) is written $x^k(p)$. The total supply vector is

$$x(p) = \sum_{k \in K(p)} x^k(p) \tag{A1}$$

The profit for producer k is $p^T x^k(p)$ (superscript T denotes transposition). There are H consumers in the economy. Each consumer owns a vector of resources, w^h, and a share in firm k's profit, α_k^h. The wealth of consumer h is then

$$R^h = p^T w^h + \sum_{k \subset K(p)} \alpha_k^h p^T x^k(p) \tag{A2}$$

Utility maximisation gives the individual demand functions (we assume strictly concave utility functions so that no correspondences will appear), $D^h(p, R^h, Y)$, where Y is a vector of environmental qualities (if environmental quality affects not only the satisfaction of consumers but also the production possibilities of the firms, then Y should also appear as an argument in the supply functions).

The total demand is then

$$D = \sum_{h=1}^{H} D^h(p, R^h, Y) \tag{A3}$$

Private equilibrium is defined by

(a) $z = \sum_{h \in K(p)} x^k(p) + \sum_{h-1}^{H} w^h - \sum_{h-1}^{H} D^h(p, R^h, Y) \geq 0$ (A4)

(b) $K(p) \leq \{k \in A;\ p^T x^k(p) \geq 0\},\ A - K(p) = \{k \in A;\ p^T x^k(p) \leq 0\}$

(c) $R^h = p^T w^h + \sum_{k \in K(p)} \alpha_k{}^h p^T x^k(p)\quad h = 1, \ldots, H$

The first condition is the short-run equilibrium condition that the excess demand for no service and no commodity must be positive. The second condition is that no firm operating on any market has a negative profit and that no potential producer can make positive profit by entering some market. The third condition is simply a definition of the wealth of consumer h.

For some commodities the excess supply will be strictly positive and this excess supply must be disposed of by discharging it into the environment. This is in particular true for waste products for which there is no demand at all. Use of the environment as a dumping ground will deteriorate the quality of the environment. We assume that the relation between the environmental qualities and the discharge of residuals (or excess supplies) is given by the environmental interaction function

$$Y = F(z) \tag{A5}$$

Next, assume that the society has as an objective to prevent the vector of environmental qualities from falling below certain ambient standards, Y, i.e.

$$Y \geq \bar{Y} \tag{A6}$$

In order to achieve this objective, the society may impose effluent charges given by the vector q. For most goods the corresponding components in q will be zero, but for residuals which are harmful to the environment the corresponding components will be positive. The effluent charges will change the net price vector from p to $p - q$, meaning that commodities which in private equilibrium have a zero price now have a negative price. The proceeds from the charges are assumed to be distributed in a lump-sum manner to the consumers.

The public equilibrium is now defined by

(a) $z = \sum_{k \in K(p-q)} x^k(p-q) + \sum_{h=1}^{H} w^h - \sum_{h=1}^{H} D^h(p-q, R^h, Y) \geq 0$ (A7)

(b) $K(p-q) = \{k \in A;\ (p-q)^T x^k(p-q) \geq 0\},\ A - K(p-q) = \{k \in A;$
    $(p-q)^T x^k(p-q) \geq 0\}$

(c) $R^h = (p-q)^T w^h + \beta^h q^T z + \sum_{k \in K(p-q)} \alpha_k{}^h (p-q)^T x^k(p-q) - h = 1, \ldots, H$

(d) $F(z) \geq \bar{Y}$

The first condition says as before that excess demand must be non-positive. The second condition is exactly as in private equilibrium. The third condition now defines wealth as the value of resources, the share of profits, and its lump-sum transfers of the proceeds from the effluent charges. The last condition is simply that the ambient standards must be met.

We know that this public equilibrium has certain normative properties, in particular that the ambient standards are met at lowest social cost.

Let us now introduce bribes or subsidies. We shall consider the following scheme for subsidising the polluters. For each agent a vector of commodities and services is determined, and the agent is paid in accordance with the amount by which he is able to reduce his supplies in comparison with this vector. For a firm k there is thus determined a vector $\bar{z}_f$, and the firm is paid $q^T(\bar{z}_f - x^k)$, where x^k is the firm's actual supply vector (q is equal to the vector of effluent charges considered above). For commodities that are not in excess supply the charge is zero, and consequently the firm is not subsidised for reducing the supply of such commodities. The same is true for commodities that are not harmful to the environment, even if they are in excess supply. It is natural to assume that $\bar{z}_f^k = 0$ for potential firms not operating on a market. If such a firm enters a market, it will not be subsidised but will have to pay effluent charges $q^T x^k$. It is possible, however, to imagine situations where new firms should also be subsidised. In such cases the firms must be subsidised before they start producing anything (which means that the owners must get a lump-sum transfer which continues after production has started), otherwise they will have incentives to start production due solely to the subsidy.

Similarly for consumers. The net supply of consumer h is $w^h - D^h$, and if the vector $\bar{z}_c^h$ is determined for him, he will be paid $q^T(\bar{z}_c^h - w^h + D^h)$.

The total expenditures for this scheme are

$$S = \sum_{k \in K'} q^T(\bar{z}_f^k - x^k) + \sum_{h=1}^{H} q^T(\bar{z}_c^h - w^h + D^h) = \tag{A8}$$

$$= q^T \left[\sum_{k \in K'} \bar{z}_f^k + \sum_{h=1}^{H} \bar{z}_c^h - \sum_{k \in K'} x^k - \sum_{h=1}^{H} w^h + \sum_{h=1}^{H} D^h \right] =$$

$$= q^T \left[\sum_{k \in K'} \bar{z}_f^k + \sum_{h=1}^{H} \bar{z}_c^h - z \right]$$

where K' is the set of firms actually operating on a market with this subsidy scheme (K' will be defined more precisely below). These expenditures are financed by lump-sum transfers from the consumers so that consumer h pays $\beta^h S$.

The wealth (or lump-sum income) of consumer h is then

$$R^h = p^T w^h + \sum_{k \in K'} \alpha_k^h \pi^k + x - \beta^h S \tag{A9}$$

where π^k is the profit of firm k:

$$\pi^k = p^T x^k + q^T(\bar{z}_f^k - x^k) = (p - q)^T x^k + q^T \bar{z}_f^k \tag{A10}$$

It is clear from this that if the firm is producing something, then its supply is determined from the same supply function as above, that is, $x^k(p - q)$. Obviously the same is true for the consumers, so that their behaviour can be described as maximisation of utility with prices $p - q$ and lump-sum income $R^h + q^T \bar{z}_c{}^h$.

The short-run equilibrium can now be characterised by

$$z = \sum_{h \in K} x^k(p - q) + \sum_{h=1}^{H} w^h - \sum_{h=1}^{H} D^h(p - q, R^h + q^T \bar{z}_c{}^h, Y) \geq 0 \quad \text{(A11)}$$

The set K' depends on the precise way the subsidies are administered. If the subsidy ends when firms stop production, then there are incentives for the firms to remain producing and K' will contain more producers than the corresponding set $K(p - q)$ when effluent charges are used. In this case the two approaches will differ with respect to resource allocation, and since the effluent charges scheme is efficient, bribes cannot be efficient.

On the other hand, if the subsidies continue after the firms have stopped production (as transfers equal to $q^T \bar{z}_f{}^k$ to the old owners), the subsidy will not have any effects on the incentives to stop production and the outcome is the same as when effluent charges are used, except that the income distribution is different (it is, however, possible to choose the shares β^h in such a way that the income distribution is the same in the two cases).

We have thus shown that it is possible to construct a scheme of subsidies that is equivalent to effluent charges. This scheme consists of a subsidy per unit of waste discharge reduction in comparison with a predetermined level for agents that already are polluters, and effluent charges for new firms. The subsidy must, however, continue to the owners of firms even after the firms have stopped their operations.

Discussion of the Paper by
Karl-Göran Mäler

Formal Discussant: Kneese. H. L. Mencken, a student of the American language and sarcastic commentator on the tendency of his times, once commented that 'for every problem economists have an answer, simple, neat and wrong'. One might paraphrase this statement in connection with the present paper and say that it concludes that effluent charges are neat and right but not so simple.

The contribution of the first part of the paper is to discuss, in a systematic way, the issues surrounding standards versus charges as a means of internalising costs. Its contribution is in providing a rigorous theoretical treatment of propositions already in the literature.

The second portion deals primarily with an analysis of stochastic aspects of the problem: this is a neglected area which consequently makes this part of the study of particular interest. Throughout the paper the central question being studied is how information requirements differ between the two approaches.

Concerning the deterministic part of the study, the general conclusion is that the determination of optimal effluent charges does not require more information than the determination of optimal effluent standards. Furthermore, in some important cases, effluent charges will require strictly less information than effluent standards.

The central conclusion in the portion dealing with stochastic aspects is that where the waste load is a random variable, effluent charges and effluent standards are not comparable from an efficiency point of view because charges are based on a maximisation of the expected economic value of the environment while effluent standards are based on a set of probabilistic environmental or ambient standards. But on their own terms optimal charges in some cases require less information than optimal standards.

One of the attractive features of the paper is the great care and precision with which the author states his assumptions. But some of them are open to question. For example, his assumption that there is a list of all possible residuals independent of any relative price configuration violates reality. In the real world, when effluent charges are imposed, a substance which was previously a non-marketable residual may be recovered and marketed because it is the most efficient way to respond to the new relative price structure. It would have been helpful if the author had provided us with a statement about the significance of this assumption, which runs contrary to observed fact.

Another assumption which invites discussion is that the objective of environmental policy is to attain a politically determined environmental objective at least cost. I agree that this is the most likely assumption to make, and the paper by Russell, Spofford and Haefele (Chapter 7) has quite a little to say about the political processes which might be used to make such political decisions. It does, however, rule out a class of cases where the effluent charges approach is clearly superior from an informational point of view. This is where a marginal damage function can be associated with the specific discharge of a particular source. In that instance the schedule of marginal damages can be imposed on the waste discharger and he will respond to it in an optimal manner. In this case the effluent charge requires no information concerning the cost of controlling residual discharges, whereas setting of discharge standards does require such information.

One should also note that most of the cases the author analyses are ones where the location of the discharge does not matter. These are situations where, for example, a pound of B.O.D. discharged anywhere affects the environmental standards similarly or, slightly more technically, where all the coefficients in the transfer matrix are identical. For the time being I follow this assumption in the present discussion.

For the static case the main conclusion is that information sufficient to set optimal standards is sufficient for optimal charges but not vice versa. When we confront a situation in which there are multiple dischargers we need know only the total benefit function for all dischargers and not the individual benefit function for each discharger. The intuitive explanation for this is that a single charge can be found which meets the standard if the total benefit function is known, and that charge will distribute residuals reduction optimally among firms. The reason is that all firms will equate their marginal benefits from discharge to this charge, which means that marginal benefits are equal everywhere. This in turn is a necessary condition for meeting ambient standards at least cost. With effluent standards this is not possible, since an amount of discharge must be assigned to each waste discharger so that equality of marginal benefits is realised. Therefore each individual marginal benefit schedule must be known. We may regard this as the basic conclusion for the static case, and clearly it is correct.

The author then studies a situation in which effluent standards may be superior to effluent charges in the sense that charges cannot assure the attainment of the ambient standard. As he demonstrates, this can occur when benefit functions are piecewise linear. One could even depart from his assumption of the concavity of benefit functions and discuss other cases in which an effluent charge would, except by accident, undershoot or overshoot the required ambient standard. The question then is whether this is a likely kind of situation. Empirical studies suggest that it is not in regard to industrial waste sources, which are the most important ones, but that it may be in connection with the treatment of municipal wastes. The usual municipal waste treatment cost function is such that it is rather flat over a large range of reduction and then rises rapidly as 100 per cent removal is approached. Consequently one might conclude that the situation Mäler analyses here could be a serious problem where municipal wastes are important. But the analysis, as he conducted it, neglects the results in larger systems of multiple sources of waste discharge. If benefit functions for individual dischargers are linear but lie at different levels, then the optimal solution requires that there be very high levels of removal at some sources and little or none at others. As the Delaware study, which was discussed in my paper (Chapter 3), illustrates, in such a situation the charges approach can come close to a least-cost solution even though the benefit functions are piecewise linear. I think that in this and other instances the author's paper suffers somewhat from not treating overall system effects in a very complete manner.

The author now abandons an assumption that he has carried on until this point and which has an extremely important result in terms of the information requirements of the two approaches. The impact of the result on the reader is rather muted because it is buried in considerable theoretical discussion, but it puts the comparison of the two approaches in an entirely new light. When the assumption is released that there is a fixed number of firms in the region, the conclusion is that to control effluent discharges to achieve an ambient standard will require that the regional authority must control entry of new industrial plants into the region. Effluent charges, on the other hand, may be determined from a knowledge of the aggregate benefit function for firms already in the region. If a firm plans to move to the region the calculated profitability of this move will depend, in part, on the charge, and it is up to the firm to decide whether the move is desirable. The point here is that while often effluent charges can be calculated from relatively limited information about firms already in the region, optimal effluent standards require that a full general equilibrium solution to the resource allocation problem in the entire economy be known to the authority. Since this is an impossibly high degree of knowledge, I think it may be concluded that optimal effluent standards cannot be implemented. This important conclusion should be highlighted to a greater extent than is done in the paper.

Rather incidentally, the author also comments that if the F-function is not known and policy is therefore defined in terms of aggregate flows of residuals, his conclusion still holds. In my view this conclusion should be stated more strongly also. In this case the superiority of the effluent charges scheme becomes particularly clear because by definition the location of discharges does not matter. If a reasonable estimate can be made of the average cost of controlling discharges, a near-optimal effluent charge can be set from this information alone.

Another point discussed is the comparative result when effluent standards and effluent charges are set on the basis of imperfect information and are therefore both likely to miss the optimum. The author points out that in this case there is no easily observable information by which this failure can be determined with respect to effluent standards. Assuming the standards are met, there is no way of knowing whether the solution is efficient, i.e. least-cost, without gathering additional information. In the case of charges, efficiency can be assumed but the ambient standard may not be met. In this case, however, the result is plain, for the failure of the standards to be met can be observed directly. Therefore the charges can be raised or lowered in a search for optimality without additional information. It is at this point that a discussion of the cost of dynamic adjustment would have been particularly appropriate. The problem is that if the optimal solution is missed to begin with, either in the case of charges or standards, investment decisions may be made which no longer permit the same path of adjustment which would have been possible if the charges or standards had been correct in the first instance. The paper could be improved by a consideration of this matter.

The remaining situation studied, where a random waste load prevails, is of particular interest. The case considered is where the waste load is random and the standard on effluent is set in a way requiring that a certain waste load not be exceeded more than a specified percentage of the time. The author rigorously demonstrates the result as far as the behaviour of a profit-maximising firm is concerned. But in general it is easy to see what happens. Given the plant the firm has designed to meet the probability of the standard being violated, once it becomes clear that the load will be so large that the standard is violated, it will cease any treatment altogether, thus leading to a very large violation. It may be noted that a similar result with respect to a reservoir system is discussed in my paper (Chapter 3) in connection with the Potomac case study. While it is common to set standards in this manner, the analysis convinces us that we really cannot believe that that is what society wants because it implies that the extent of violation of the standard is of no consequence.

On the other hand, with the charge set at the proper levels for meeting the standard, the firm will minimise its expected waste disposal costs. This implies that it will treat all the small waste loads and use its full capacity for large waste loads even when the standard is being violated. This means that the effluent charge will not meet the standard as stated at least cost, even though it does meet the standard. The author demonstrates that in fact the effluent charge meets another criterion which has more appeal, and that is that it maximises the total expected value of the environment. It would have been useful for him to compare the two cases he analyses with another which is common in practice. In this case the effluent standard is specified in the way indicated above, i.e. with both a level and a probability statement. Then a treatment plant for other measures is provided for meeting the standard, but with the difference that it is assumed that the treatment plant will operate continuously at a particular level. This is a third alternative with some interest in terms of real-world practice. Also, it would have been interesting to have some discussion of how charges versus standards may affect the *design* of the treatment facilities when the waste load is random. Many years ago George Stigler pointed out that when the production rate of a plant is variable there are benefits to designing flexibility into it. It would have been interesting to bring this consideration into the discussion.

Up to this point I have assumed, along with the author, that we are dealing with the situation in which the location of the waste discharge does not matter. There is only a little discussion in his paper of the situation where it does, and in general the conclusion is that in such a situation, given the same policy objective, the information requirements for the two approaches are the same. This is true in a strict theoretical sense, but in practice there may still be some advantages to an effluent charges scheme when it is compared with likely alternatives. In this connection I refer the reader back to the discussion of the Delaware estuary study presented in my paper (Chapter 3).

In conclusion, I should like to say that I believe this to be an excellent paper, one of the very best presented at the conference and one of the best treatments of the theoretical aspects of environmental economics I have seen. The general points I raise with respect to it are mostly in the nature of what might have been natural extensions to some further real-world issues. Let me repeat the main ones and perhaps add one or two others.

First of all, I think the dynamics of adjustment to a policy could have been discussed with benefit. As I see it, there are two main issues here. One is the question of how close one can come to an optimal solution through a process of iteration when there are costs of dynamic adjustment. The second relates to the processes of induced innovation. When one takes a longer view of the costs of controlling residuals discharges, one must take account of the processes of technological change. There is evidence to suggest that such change is induced by relative scarcity, i.e. relative prices of resources. Clearly, the incentive to innovation brought about by charges on the one hand and standards on the other must be somewhat different, but we are not enlightened on this issue.

As I noted, when the author releases the assumption that the number of firms in the region is fixed, spectacular differences in information requirements appear. It seems to me that this analysis could have been extended to consider the debate over bribes versus charges and the associated question of their potential equivalence which has been found in the literature over the years. I hope the author makes this extension because I believe that his models permit one to consider the question with great clarity.

Another aspect of adjustment processes which it seems to me requires additional discussion is the behaviour of large systems. This comes up particularly in connection with the study of adjustment processes when benefit functions are piecewise linear, but it also seems to have a bearing on the question of how close to optimality one can come through an iterative process.

The final issue that deserves theoretical consideration is the matter of the effluent charge considered as a tax. At first sight it seems that the discussions in the public finance literature relating to the excess burden argument with respect to excise taxation would be pertinent here. Cannot effluent charges and other taxes on externalities be considered to have an excess benefit? On the one hand they raise revenue and on the other they improve the allocation of resources.

None of these questions about other issues and extensions should be taken to diminish the excellence of the paper. It is first-rate.

Kolm interpreted the author's analysis as a defence of charges as a means of pollution control. However, the problem he considered happened to ignore the best case for charges (or subsidies). His objective was to limit pollution to a given overall level, and his tax simply distributed the creation of pollution between the polluting firms at the lowest total cost. But was the global level of pollution the right one? And if it was determined by a political process, was the authority initiating this process the one best suited to answering this question? Now, the determination of the optimum level of pollution must be examined under two headings: pollution abstention or abatement costs (which the author does take into account) and pollution damages of all kinds (which he does not). The best case in favour of the tax (or subsidy) occurred when the damage costs were known: a charge equal to this cost guarantees that polluters behave optimally, and the controlling authority has no need to know the polluters' production or utility functions. It likewise does not need to know what the level of pollution will be: it is sufficient to know that it will be optimal.

Of course, in many cases the damage cost was difficult to evaluate. However, in others it was quite easy (e.g. damages to industry). But when the evaluation was difficult, we had to make choices under conditions of uncertainty. We then faced a more general problem than that posed by the author: choose the optimum method of controlling pollution when both damage and abatement costs are not known to the policy-maker (perfect knowledge of one of them being a special case). The case when damage costs were known had just been mentioned. On the other hand, when the costs of reducing pollution were known, it could readily be shown that all control instruments were equivalent [1].

Uncertainty was furthermore only one of several criteria which must be taken into account when choosing a method of pollution control. Indeed there were several environmental *problems*, several controlling *instruments*, and several *criteria*. The general problem was to allocate instruments to problems (or the reverse) according to criteria. The instruments were: obligations or interdictions, taxes and subsidies, sale or buying of rights by the public sector, private markets in rights, direct agreements between concerned parties, internalisation through merger, education and moral persuasion, etc. The criteria were: social efficiency, distribution and justice, knowledge (of damage, costs, etc.), practicability (implementation, computation costs, etc.), the effect on public budgets, public and social organisation (degree of decentralisation, etc.), morality (for instance, 'you must not sell pollution rights – or tax polluters – if pollution is a crime'), etc. One sure thing about the result was that no tool was best in all cases, while every reasonably imaginable tool was the best one in some cases.

A further limitation of the author's model was that it could not account for dilution as a means of reducing pollution since, if we tried to extend his wording to include it, the functions could not have the assumed structures. Suppose the quantity of pollutant was as_1, where a was a constant and s_1 an input of the polluting production. Call t_1 the volume of dilutant (water, air, ground surface, etc.). Suppose that pollution is homogeneous in this volume so that the density of pollutant is $a(s_1/t_1)$. Finally, suppose that all that matters is the quantity of pollutant in a portion of volume b of the total dilution space. This quantity was $z_1 = ab(s_1/t_1)$. This should be a g-function in the author's analysis. But it was not convex as assumed. In fact the surface which represented it was a hyperbolic paraboloid: it was neither convex nor concave.

Another important consideration was whether the tax must, or need not, be a flat rate or, more generally, what the constraints were on its structure (and how much it costs). The author considered a flat-rate tax. In other words, his tax was similar to a uniform price in a competitive market. However, the main reason for this structure in markets was absent from the cases he considered. Uniform prices are imposed by the fact that each unit of the product can be resold by its buyer. All units are then necessarily sold at the same price and this prevents discrimination in relation to quantity or to buyer. But this assumption breaks down when the payment is a tax rather than a price. Discrimination is therefore possible in relation to both quantity and to taxpayer (incidentally, this permits the tax schedule to follow the cost-of-damage function when this is taken into account). However, there are other constraints on the tax schedule which sometimes require a flat-rate structure or which make the flat rate optimal, e.g. the high cost of information and the case of a large number of small substitutable polluters.

His own view on the comparison between taxes and subsidies was as follows. With the required convexities, especially with regard to the polluter, both led to efficient (Pareto-optimal) situations. But the two situations nevertheless differed. The difference consisted of a difference in distribution, which came from the implicit allocation of environmental rights which this choice of policy implied. To offer a polluter a subsidy if he agreed to reduce his pollution was equivalent to (a) granting him the right to pollute, and (b) buying a part of this right back from him. A tax, on the other hand, was equivalent to denying the right to pollute, and offering him the possibility of buying a part of it back by paying the tax.

There were also intermediate cases where the polluter was free to pollute up to a defined level: beyond this level he was taxed on the basis of the excess pollution created; if he polluted less he received a subsidy based on the difference between the actual and allowable levels of pollution. Again, with the required convexities, the outcome would be efficient.

In all these cases the environmental right denied to the polluter was granted to the public sector, supposedly acting on behalf of the victims, or directly to the latter when the situation involved direct exchange.

Of course, by assuming lump-sum compensation, or by considering partial equilibrium without income effects, one could construct models in which all these cases gave the same final result.

But non-convexities could make a difference between these measures from an efficiency viewpoint.

Consider a polluting firm. It produced quantity x of a certain product. The pollution created was a function of x, and we could therefore consider x as the tax base for a pollution tax. The market price p of x, and the tax rate t, are both assumed to be imposed on the firm. $C(x)$ represents the production cost of x. The choices available can then be portrayed diagrammatically as shown in Fig. 6D.1.

Without the pollution tax, the firm's maximum profit would be reached for production and cost represented by point A where the line AC of slope p was tangential to the cost curve. The result is a profit of OC.

We now consider social indifference curves in the plane (C, x). They represent all the reasons for being interested in these two variables (in particular, for x, they consider both its usefulness as a product and its

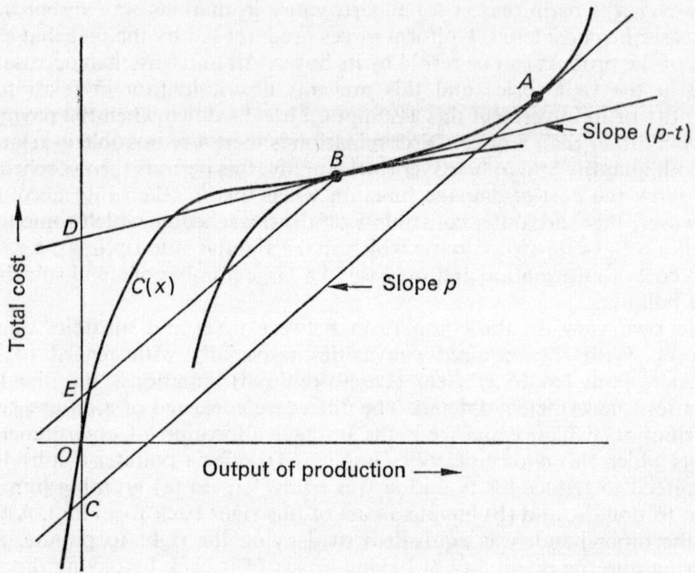

Fig. 6D.1   Policy choices available for deciding the optimum
pollution tax

noxiousness as a source of pollution). In this way we can determine the social optimum, with the cost curve representing our possibility frontier. Suppose this optimum is at B.

The tax rate t per unit of x must be such that the slope of $C(x)$ in B is $(p - t)$. But if the production is at B with this tax, the firm incurs a deficit of magnitude OD. Being a private firm, it would prefer to close down. But it is socially optimal that it should not. The firm must thus receive a subsidy at least as large as OD.

Finally, as a result of this whole operation, the financial relations between the firm and the public sector are twofold: the firm pays the tax tx and receives a subsidy. What is the overall result? If S is the subsidy, with $S \geq OD$, one has $OD = C(x) - (p - t)x$. It follows that

$$(S - tx) \geq (OD - tx) = (C(x) - px).$$

Since E is such that BE has a slope p (i.e. it is parallel to AC)

$$|C(x) - px| = OE$$

and $\{C(x) - px\}$ is positive or negative depending on whether E is above or below the origin at O. If E is above O, $(S - tx) > 0$, and the net result is a subsidy. In this case a tax, by itself, cannot induce the firm to behave optimally.

Of course, this situation was less interesting from the point of view of decentralising the necessity of knowing the utility of the possibility to pollute from the policy-maker to the polluter. And this was the essential interest in using a tax or subsidy method. But it would be a mistake to believe that it was an unrealistic case. On the contrary, this situation was probably fairly common, the increasing returns to scale being due to fixed costs. When we threatened to impose a pollution tax, polluters very often argued that this would force them to close down, thereby leaving local unemployment, etc. Sometimes it was not true: in the event they would both pay and stay. But at other times they would be forced to close down. And in this case there might be two situations. In some cases it was socially optimal to close down the plant – at least in the long run or after something had been done about labour immobility. But in other cases the social optimum required that the plant should stay in operation. Then it had to receive the subsidy described above.

But when the correct conclusion was that the firm should not pay tx, the measures which were usually taken were incorrect. In this case the authorities generally said: 'Since you cannot pay a tax of rate t, we shall accept that you pay only $t'(<t)$ on each unit of x'. Now, a previous result showed that the net payment from the firm to the authorities should be $(tx - S)$, which might be positive or negative. Defining the constant $x_0 = S/t$, this payment can be written as $t(x - x_0)$. If, at the optimum, $x > x_0$, this is a tax of rate t but on $(x - x_0)$, i.e. on the polluting activity in excess of a given constant x_0. And if at the optimum $x < x_0$, this is a subsidy to the firm, proportional at rate t to the amount $x - x_0$ of polluting activity which the firm abstains from at the given level x_0. This is the correct policy, in which we replace 'amount of a polluting activity' by 'amount of pollutant'. In brief, what has to be adjusted is the tax base and not the tax rate. (Note that it is this x_0 which embodies the allocation of environmental rights described above.) Clearly, non-linear taxes or subsidies could also be considered.

Up to now the discussion had focused on public action against polluters. If instead we considered direct exchanges between the polluter and his victims (charges or bribes), further problems arose, one being a necessary non-convexity in the victim's production or preference sets when we considered a dimension for the pollution. This non-convexity had the same effect as the previous one on the choice between a charge or a bribe, or of a starting-point x_0 describing allocation of rights, for an efficient exchange to be possible [2, 3].

However, this problem was not as serious as it seemed. If there was direct agreement between polluters and pollutees, there was no reason for the financial transfers to take the form of payments with prices imposed on these agents. And without this structure, convexity was irrelevant. If there were a great number of victims, the pollution was usually a 'public bad' to them and a public authority would have to act on their behalf: convexity of the victim's production or preference functions was thus again irrelevant; then the issue was not full decentralisation through markets but semi-decentralisation through taxation; the first one was prevented in the first place by the 'public good' aspect, which thus prevented the victim's non-convexity from being a problem.

Foster asked why one never contemplated charges for automobile pollution and suggested that the first reason was that the costs of metering automobile pollution were great while the task of discovering whether a given automobile was achieving the stated standard required periodic checks. In some cases, failure to achieve the standards could be observed by inspectors or police simply watching the traffic. But in general, enforcement was extremely difficult.

Another argument for pollution standards was that *over a period of time* there was no reason why automobiles could not become homogeneous in respect of the pollution function – that is, all makers should be able to achieve the same standard, at the same cost, in respect of the same type of vehicle. The costs of pollution control only began to rise substantially when the period allowed for adjustment was short relative to the gestation period of new technology, or if there were frequent changes in the standards set. In other words, the industrial benefit functions would only be identical in the long run.

On the question of 'bribes' he suggested that no one would suggest bribing *all potential producers*, since there was an infinite set of people prepared to accept something for doing nothing. One only considered bribes because they helped to offset the extremely high costs of adjustment faced by the existing set of agents.

Referring to one of Kolm's remarks, he likewise did not see the case for preferring standards to charges when one knew the direction of social costs in relation to the output of pollution but did not know their magnitude. This made it difficult to determine the level at which charges or standards should be set. However, it was still surely true, for an approach to efficiency, that there were still the same differences in the informational requirements postulated by the author.

Førsund noted that the author had stated that, if information on the aggregate benefit function had to be obtained from information about individual benefit functions, then there was *no difference* in informational requirements between the alternative approaches of charges versus standards. It was therefore of obvious importance to look into the possibility of estimating this aggregate function. But the author's reference to the estimation of consumption functions seemed ill-suited, since:

   (a) when cross-section data were used, *individual* data constituted the data base;

   (b) to interpret the average relationship established as an aggregate consumption function, one had to assume identical micro-functions as well.

Under these circumstances there was clearly no difference in informational requirements between the two approaches.

Thoss was pleased that some reference had been made to ambient standards. He felt that this was the one important standard that mattered. It kept the quality of the environment where we wanted it, regardless of whether or not it was optimal.

Müller observed that the author assumed, as was usual in this kind of analysis, that lump-sum compensation was possible. This was a reasonable first assumption, but it had to be relaxed at some stage because no real-world compensation could ever be of a lump-sum nature and, in any case, compensation was often not paid at all. We therefore needed to evaluate the allocative and distributional effects of different ways of reducing pollution. If charges had large undesirable distributional effects, and if regulations were no better because the private firm received the scarcity rent, we had to consider the alternative of socialising these firms.

REFERENCES

[1] Kolm, S.-Ch., 'Politique anti-pollution optimale en présence d'incertitude', *Kyklos*, vol. i (1971).

[2] ——, 'Possibilités et limites de la régulation des problèmes d'environnement et de nuisances par entente directe entre les intéressés', *Consommation* (1971).

[3] ——, 'Les non-convexités d'externalité' (mimeographed, CEPREMAP, 1971).

7 The Management of the Quality of the Environment

Clifford S. Russell, Walter O. Spofford, Jr., and Edwin T. Haefele

7.1 *INTRODUCTION*

One of the minor benefits of the recent wave of environmental concern is that arguments about technological externalities and unpriced resources are sufficiently familiar to require no reiteration or elaboration before a group of economists. Consequently we need not spend time explaining that there are interesting economic issues in the environmental field, but can go right to the background of our approach.

Let us rely on several more specific aspects of the received wisdom concerning residuals management and environmental quality and begin from the following propositions:

(i) By the very nature of the problem, free markets in residuals discharge or 'clean environment' rights do not, and probably cannot, exist.

(ii) In this circumstance it is hardly surprising that government intervention has become the prescription of choice (though not universally so, of course: see [7]).

(iii) The prospect and reality of widespread government action in this area has stimulated economists and others to work at the development of models designed to inform this action, to provide some technically rational basis for choosing one action over another.

(iv) The inconvenient fact that neither water nor airsheds are respecters of existing political boundaries has made it necessary to consider creating new units of government before carrying out public intervention.

These circumstances have, together with other broader influences, produced what we may call the classical economic and political models for environmental quality management. The classical economic models are designed to find the most efficient configuration of residuals discharge for a region, where efficiency is defined as the minimum sum of damages and abatement costs, when damages are known; or as the minimum abatement cost, when ambient quality standards are imposed. (See, for example, on the water quality side, [8], [37], [23], [43]. Similar work has been done on the air side: see [26] and [40].) These models are in the tradition of systems analysis, a field which grew up under the protective wing of U.S. Government executive agencies, particularly the military departments, in the years after the Second World War. The models follow the ancestral line; they implicitly or explicitly assume the existence of a decision-maker (a person or a 'board'), pursuing in a single-minded fashion the objective of economic efficiency, letting the distributional chips fall where they may, though occasionally applying his (their) wisdom to the assessment of side calculations showing one or another impact of the plan which even the most ingenious economists have been unable to measure in dollar terms.*

The assumptions of the economic models speak to a concern, common in the Western democracies, for businesslike efficiency in government decisions and operations. (That businesses, particularly large ones, are obviously inefficient themselves has never dampened the enthusiasm of the business-in-government advocates.) In the United States this concern has been manifest in the tremendous growth in the executive branch of government which has gone on since the Depression. The ideal is to 'get politics out' of decision-making, i.e. to concentrate on economic efficiency, though the result has simply been to allow the Congress to avoid doing its job, and to shove the 'politics' out into the new agencies, where the best-organised and financed special interest groups can nearly always win.

Given our romance with executive government, it is scarcely surprising that the 'design of institutions' for metropolitan (or other regional) environmental quality management has tended to produce appointive agencies. The most popular model – let us call this the classical political model – seems to be a commission or council to which each jurisdiction sends one man (the executive of the jurisdiction or his appointee). The voting is done as if by sovereign states, i.e. one vote per jurisdiction regardless of size, population, or degree of interest in the problem. This is the pattern for interstate river basin commissions (e.g. the Delaware River Basin Commission) and for metropolitan councils of governments (C.O.G.s). Variations on this theme are, of course, possible. The appointments may be made to achieve a 'balance' of special interest groups: for example, industry, municipalities and conservationists. The fundamental characters of

---

* This description is, of course, mildly overdrawn. In particular, some economists have recently been working to bring distributional considerations and the interplay of gaining and losing interest groups directly into their models. The most advanced work along these lines has been done at Harvard (see, for example, [7] and [13]).

the new institutions are, however, firmly in the executive mould. Politics, in the sense of voter preferences, appears to have been banished.

It is precisely here, at the exile of traditional politics and the apparent victory of efficiency, that the classical models stumble; and to their failure can be attributed some considerable portion of current disenchantment with our efforts at improving environmental quality. First, as we have mentioned, the problem of the choice of levels of environmental quality is not cleansed of politics by Congress's abdication of responsibility for choice. The new battleground is simply one in which the processes of representative democracy are not available to structure the political manœuvring. Choices are made, not by elected representatives trading votes on issues in order to reflect the direction and intensity of their constituents' concerns, but in the proverbial smoke-filled rooms, with technocrats and special interest lobbyists as the smokers (for a demonstration that a representative legislature, operating in this way, represents individual preferences, see [18]). Meanwhile the public is still encouraged to believe that all the standards and regulations issuing from the agency are somehow technically based. Admission that political consider- ations play a part is grounds for dismissal.

At the same time, economists and engineers offering their techni- cally reasonable models to these agencies find that their solutions are only acceptable when these solutions agree with the decisions already tacitly made. In these circumstances, the best of systems models can become simply a part of an elaborate 'snow job', with the inputs adjusted to produce the desired outputs. This role of systems analysis has been widely recognised and criticised in connection with military appropriations requests made to Congress. These abuses have stimulated some to suggest that Congress should have its own sys- tems analysis capability. We shall try to make clear that models structured for use by Congress should be different from those created to serve executive agencies. Indeed, one of the features of these executive-tailored models which makes them most vulnerable to such misuse is their concern with measuring damages, even those with a public bad character, in dollar terms. Since technical ingen- uity is required to force such functions out of real-world data, it is only through highly technical criticism that they can be discredited, once the appropriatness of the effort has been accepted.

Our position is implicit in the above discussion. We believe that the classical political and economic models are inadequate to the task of deciding on a socially desirable balance between residuals discharge (type, quantity, timing and location) and the quality of the air we all breathe, of the landscape we share and of the watercourses we use for so many purposes. We propose, most fundamentally, that legislative, not executive, bodies are the appropriate decision-makers in these social choice issues. This rests on a normative judgment that individual preferences should be reflected as accurately as possible in collective decisions. But it is a positive claim that, given this basic goal, legislative bodies can do the best job. This claim is not the subject of this paper. We take it as an axiom [18, 20]. This should not be construed as an endorsement of many present legislative practices (e.g. seniority, rigid committee structures) which seem designed to thwart individual preference aggregation.

In this paper we suggest that models designed for use by legislatures should be different in some respects from the classical efficiency models. In particular, a model designed to inform legislative decisions should not condense its 'answers' into a single, all-encompassing number, but should provide output showing each legislator what his constituents can expect by way of impacts from any particular policy. That is, the model should ideally show how environmental quality and costs (including such costs as the dislocation of unemployment) are distributed over legislative districts. By the same token, the model should not presume to include dollar values of damages (or changes in damages) attributable to *public* bads. This kind of evaluation is quintessentially the business of the legislature. We provide here an example of a model structured in this way and we show how it could be used by legislators. In addition, and on a different level, we claim that it is possible to say something about what kinds of legislative arrangements (district sizes, one-man-one-vote *v.* councils of government, etc.) are to be preferred in structuring the new institutions.

Our method of exploring these quite different questions has been to construct and experiment with a hypothetical region, modelling its residuals discharge activities and the natural systems which translate those discharges into ambient quality levels. This regional model has been linked to a model which simulates the vote-trading activities of a legislature assumed to be responsible for making the decisions about quality in that region. The regional model has been structured to give a great deal of information in physical as well as economic terms to each legislator. We hope to show that information in this form can be used in a systematic way by real legislatures to assist them in arriving at regional policy. Our legislative model is emphatically *not* an attempt to put real legislatures out of business, but is simply a device for allowing us to accomplish two things: first, to design the regional model for use in a legislative setting; and second, to compare the policies adopted by legislatures put together along different lines. The basic function of the simulation programme is to identify and accomplish the vote-trades which would take place in a real legislature. The details of the workings of both these models and of the links between them are described in the next section. The third and fourth sections of the paper describe some results of the research and draw from these the conclusions we have just anticipated.

7.2 *THE REGIONAL MODEL*

The Framework

In outline, the regional model discussed here is similar to the classical efficiency models mentioned above. It shares with these models a conceptual division into three parts: one in which the residuals discharges from man's production and consumption activities are determined; a second in which mathematical approximations of natural world processes are used to transform discharges into ambient environmental conditions;* and a third in which these ambient conditions are compared with exogenously imposed standards, or evaluated through damage functions. Our framework, as shown schematically in Fig. 7.1, differs from the classical models in three ways (for a more complete description, see [34]):

(i) It deals with air, water and solid waste (or land) problems simultaneously, reflecting not only the initial generation of residuals destined for all three receiving media, but also the exchanges among material and energy forms implied by various 'treatment' alternatives open to the dischargers. Thus, for example, fly ash removed from stack gases shows up as a solid residual requiring landfill volume, and standard sewage treatment alternatives generate sludge as they 'remove' biochemical oxygen-demanding organics (B.O.D.). This is nothing more than an application of the material and energy balance lessons taught to economists by Allen Kneese and collaborators [25].

We note, however, that ours is not a complete material and energy balance. Whether or not we are interested in a residual, and hence whether we include it in the model, depends on our time and space horizons. Thus, CO_2 is an important residual from the global point of

---

* This 'transformation' may simply involve dilution or dispersion or, at the other extreme, may require models of chemical and biological interactions which change residual forms and transmit effects through the ecological system. If these transformations are simple enough to be represented by linear functions, the three sections of the model may, in fact, be collapsed into a single linear program, as they are in the model used here. More generally, the functions will be non-linear and the problem retains its tripartite form for solution purposes, as we discuss briefly below.

view and in the very long run (perhaps a hundred or more years), but it is not of interest in regional models dealing with fairly short horizons. Thus, in our model we do not follow CO_2 (or heat) discharges to the atmosphere.

It is also worth noting that while this approach is conceptually correct, it may in practice prove to be an unnecessary refinement.

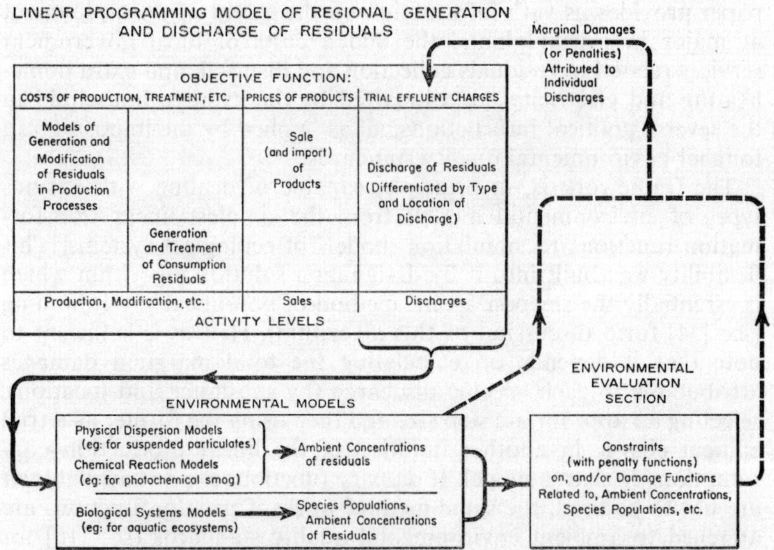

FIG. 7.1   Schematic diagram of the regional residuals management model

It may be possible to arrive at policies which are 'good enough', in some sense, without the complication and expense of an integrated model. It is, however, difficult to see how this question can be answered in the absence of any models using the integrated approach to provide the benchmark solutions.

(ii) The model of production, consumption and discharge activities, which in our model is couched as a linear programming problem, includes more than the traditional costs of residuals treatment or disposal. Instead of assuming that the generation of residuals is fixed per unit of sales, physical output, employment or what have you, our model includes options for decreasing generation through input substitution (low-sulphur for higher-sulphur fuel, capital for waste heat, etc.), process change (continuous *v.* batch diffuser in beet-sugar processing) and by-product recovery (sulphur from refinery gases). By the same token, these models are capable of reflecting the impact on residuals generation and discharge of changes in product mix, production technology, etc., caused by influences or policies exogenous to the regional residuals management problem (see [33]).

(iii) Third, we have modified the classical model framework by including, in the linear programming model of production, consumption and residuals discharge, information on the *distribution* of both the costs and benefits of possible environmental quality management policies. Thus, for example, the simple model we apply in this paper provides us with information on the extent of unemployment at major industrial plants, the added costs of local government services related to residuals collection and disposal, and extra home-heating and electricity costs incurred by the average household in the several political jurisdictions, all as implied by the imposition of tougher environmental quality standards.

The framework is, in principle, capable of dealing with various types of environmental models from the simplest linear transformation functions to simulation models of ecological systems. This flexibility was built into it by designing a solution algorithm which is essentially the steepest ascent method of non-linear programming (see [34] for a discussion of this algorithm). Here it is sufficient to note that it depends on calculating the total marginal damages attributable to each specific discharge (by substance and location), selecting an appropriate step size and then using the former as a trial effluent charge in another iteration of the linear production–consumption–discharge model. If damage functions are not available or are not to be used, the same method works if penalty functions are attached to ambient environmental quality standards (see [16] for a similar technique).

We gave the general framework this dimension of flexibility to allow for the future inclusion of more complex and meaningful environmental models, especially of aquatic ecosystems. These models will provide valuable information on the response to residuals discharges of socially important water quality indicators such as fish populations and algal concentrations. Unfortunately such models are not reducible to simple linear transformations. In fact, in general they must be solved using simulation techniques (see p. 233 for references to current work on these ecosystem models). In the model reported here, we use a very simple B.O.D./D.O. model based on the Streeter–Phelps [39] equations.

The Hypothetical Region and the Didactic Model

To carry out the research described in this paper, we constructed a small hypothetical region and then applied to it the modelling framework just described. This region, which we shall call Didactica, is shown in Fig. 7.2 and comprises a city, four suburbs and some hinterland through which flows a river and around which are located a number of major residuals dischargers. These latter, described in

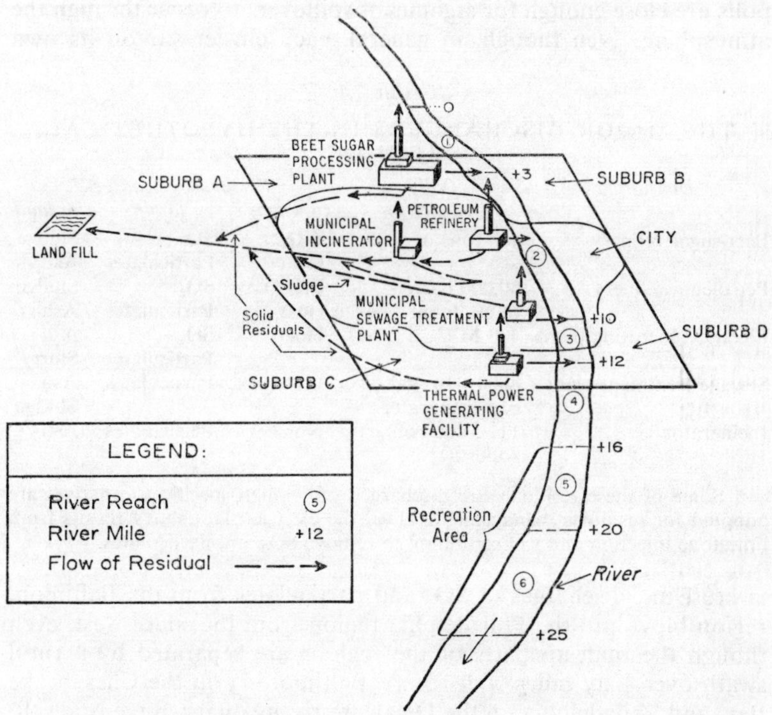

FIG. 7.2   Hypothetical region

a summary way in Table 7.1, consist of a beet-sugar refinery, a petroleum refinery, a coal-fired electric generating plant, a municipal sewage treatment plant (with sludge incinerator) and a municipal incinerator. We confine ourselves in this simple application to B.O.D. and heat residuals to water; SO_2 and particulates to the atmosphere; and ash, sludges and slurries to land. In addition, we have included consumption residuals (sewage, solid waste and emissions from home heating) from the households comprising the political sub-divisions.

It is worth pausing a moment to discuss the difficult matter of the choice of regional boundaries for residuals' management purposes. As we noted above, it would be most unlikely to find in the real world political boundaries conterminous with the limits of air or watersheds. The general rule is to draw the boundaries to internalise under the regional government or agency all the externalities. But this prescription is next to useless in an area like the eastern United States, since the series of population clusters which make up megalopolis are close enough for significant spillovers to occur through the atmosphere, even though, in general, each cluster sits on its own

TABLE 7.1

THE MAJOR DISCHARGERS IN THE HYPOTHETICAL REGION

Discharger	*Daily capacity*	*Residuals discharged**		
		to water	*to air*	*to land*
Beet-sugar refinery	400 tons of sugar	B.O.D. Heat	SO_2 Particulates	Sludge Ash
Petroleum refinery	50,000 barrels of crude oil	B.O.D. Heat	SO_2 Particulates	Sludge Ash
Electric generating plant	400 MW	Heat	SO_2 Particulates	Ash Slurry
Sewage treatment plant	8800×10^3 gal. of waste water	B.O.D.	Particulates	Ash Sludge
Incinerator	111·5 tons of solids	None	Particulates	Ash

* Some of these residuals are discharged only when specific alternatives are adopted for residuals management. Thus, the electric plant slurry results from limestone injection and wet scrubbing to remove SO_2 and particulates.

river. Thus, discharges of SO_2 and particulates from the Baltimore region blow into the Philadelphia region from the south-west, even though the built-up parts of the regions are separated by a rural swath over sixty miles wide. Since Baltimore is on the Chesapeake Bay, and Philadelphia on the Delaware river estuary, there is a well-defined boundary on the water side, and for practical purposes it seems wise to take advantage of these natural cuts to keep down the size of the region to be modelled and administered. (The Chesapeake and Delaware Canal does link the two bodies, but at present the stretches linked are both of fairly high quality.) One is left with the *ad hoc* choice of boundaries based on data availability (which is almost always based on political jurisdictions for economic data), the conformation of the river systems (trying to restrict the region to a single major watershed), and on the vagaries of annual wind patterns. Clearly, on the last item, one must be content with a regional boundary over which will come some quantity of residuals from neighbouring regions, and out of which will flow some residuals to cause damage elsewhere.

In the construction of Didactica we have begged these questions by assuming no significant population or pollution sources outside the square containing the city and its suburbs. We have allowed for a

B.O.D. load at the head of the river stretch of interest to us – perhaps from some upstream region. We justify inclusion of the stretch of river downstream from the city on the basis of its recreational importance to the residents of Didactica, but we do not impose a quality constraint on the river as it leaves the region.

The major features of all the large industrial discharges in Didactica (such as the thermal efficiency of the power plant) are assumed fixed over the run of interest, and only certain additions (such as cooling towers and higher-level treatment facilities) may be added. The municipal treatment plant is assumed to have installed only primary treatment. In Table 7.2 we summarise, for each of the major dischargers, the options for residuals management available in the model. The beet-sugar refinery alternatives are based on information in Löf and Kneese [27]. The petroleum refinery is a condensed version of a model developed at Resources for the Future and described in Russell [35]. The information for the electric power plant and its associated alternatives came largely from Delson and Frankel [10] and Cootner and Löf [6]. The municipal treatment plant vectors were constructed using the data pulled together by Smith and reported in [38]. The municipal incinerator characteristics are based on information contained in Combustion Engineering [5].

The generation of consumption residuals in the region has been estimated using coefficients related to the socio-economic characteristics we assumed for the several political divisions. The characteristics and related information on consumption residuals generation are summarised in Table 7.3. Thus, average daily solid residual loads per person were positively related to assumed per capita income and to the presence of commercial and small industrial firms. Home-heating requirements were estimated on the basis of degree days (using an average November/December day in the Philadelphia region), B.T.U. per degree day per room (from Ozolins and Smith [31], and average rooms per household (the last also related to per capita income, since we do not consider the possibility that the demand for a warm house is price-elastic). These requirements can be met using any of three sulphur grades of distillate fuel ('heating') oil: 0·5 per cent, 2·2 per cent or 3·8 per cent. Sulphur dioxide emissions are assumed to be twice the sulphur weight in the fuel burned. Particulate emissions are assumed to be 0·504 lb/bbl burned, independent of sulphur content (based on data in [42]). The heat content of distillate oil is taken as $5·84 \times 10^6$ B.T.U./bbl. The programme can then choose a policy for sulphur content of domestic fuel oil in each political jurisdiction. The extra costs to average households of going to lower-sulphur grades are reflected in the program as distributional constraints.

TABLE 7.2

RESIDUALS MANAGEMENT OPTIONS AVAILABLE
TO MAJOR DISCHARGERS

Discharger	Option	Residual affected	Secondary residual generated
Beet-sugar refinery	1. Partial or full re-use of flume water	B.O.D.	Sludge
	2. Secondary and tertiary treatment	B.O.D.	Sludge
	3. Cooling tower	Heat	*
	4. Burn lower-sulphur coal	SO_2	None
	5. Electrostatic precipitators (3 alternative efficiencies)	Particulates	Ash
	6. Digestion and landfill	Sludge	†
	7. Dewatering and incineration	Sludge	Particulates and ash
Petroleum refinery	1. Secondary and tertiary treatment and various re-use alternatives	B.O.D.	Sludge
	2. Cooling tower	Heat	*
	3. Burn lower-sulphur fuel	SO_2	None
	4. Refine lower-sulphur crude	SO_2	None
	5. Sell rather than burn certain high-sulphur products	SO_2	None
	6. Cyclone collectors on cat-cracker catalyst regenerator (3 efficiencies)	Particulates	Ash
	7. Sell rather than burn high-sulphur refinery coke	SO_2	None
	8. Digestion and landfill	Sludge	†
Electric power plant	1. Cooling tower	Heat	*
	2. Burn lower-sulphur coal	SO_2	
	3. Limestone injection and wet scrub	SO_2	Slurry
	4. Electrostatic precipitators	Particulates	Ash
	5. Settling pond	Slurry	Solid 'ash'
Treatment plant	1. Secondary or tertiary treatment	B.O.D.	Sludge
	2. Digestion and landfill	Sludge	†
	3. Dewatering and incineration	Sludge	Particulates Ash
	4. Dry cyclone	Particulates	Ash
Incinerator	1. Electrostatic precipitators (3 alternative efficiencies)	Particulates	Ash

* Heat is rejected to the atmosphere along with water vapour.
† The secondary residual here is sludge at a different location.

TABLE 7.3

CHARACTERISTICS OF CITY AND SUBURBS RELEVANT TO DOMESTIC RESIDUALS GENERATION

Jurisdiction	Population	Per capita income	Level of commercial activity	Average rooms per household	Solid waste* generation, domestic and commercial	Land area	Sewered waste water volume	Average B.O.D. load
					(lb./person/day)	(sq. km)	(10^3 gal./day)	(per 10^3 gal.)
City	45,000	Low to high	High	4 (reflecting apartments)	4·95	99	4,500 domestic 4,300 commercial/ industrial	2·5 lb. (0·25 lb./day/ person in domestic waste water)
Suburb A	10,000	Low to middle	Moderate	6	4·20	27	0	—
Suburb B	12,000	Very high	Very low	10	4·60	36	0	—
Suburb C	9,000	Low to middle	Moderate	6	5·00	36	0	—
Suburb D	12,000	High	Low	8	4·00	27	0	—

* Based on data in R. J. Black *et al.*, *The National Solid Wastes Survey: An Interim Report*, presented at the 1968 Annual Meeting of the Institute for Solid Wastes of the American Public Works Association (Miami Beach, Oct 1968).

Commercial and miscellaneous small industrial heating requirements for each jurisdiction were assumed to be some percentage of
that area's domestic heating requirement. These percentages varied
from 50 in the city to 5 in the wealthiest suburb. And while total
domestic heating was assumed to be spread uniformly over each
jurisdiction, commercial and small industrial establishments were
assumed clustered in one or two centres in each jurisdiction. These
heating requirements are satisfied by burning residual fuel oil, heavier
than distillate, but with the same emission rate for particulates and
a heat content of 6.59×10^6 B.T.U./bbl. Again, three sulphur contents
are available: 1.0 per cent, 2.3 per cent and 3.9 per cent. Both distillate and residual fuel oils in all three grades can be produced by the
refinery, but it is not required that all regional requirements be met
by the local refinery, or that the local production of these fuels be
used locally. Thus, it is quite possible that, in the absence of controls,
the local heating demand would be for higher-sulphur (cheaper) oils
which would be largely imported, while the local refinery produced
primarily lower-sulphur grades for export. (A situation very much
like this exists with respect to heating oils along the whole East
Coast of the United States today.)

For simplicity in handling domestic and commercial heating residuals as area sources (which represent the sum of discharges over
an area and are treated in the atmospheric models as though the
emissions came from a single point, usually in the middle of the
area, and usually with a low stack height assumed) and to make it
easier to interpret ambient concentrations in terms of the desires of
political subdivisions, we assumed that our city and its suburbs each
corresponded exactly with a different set of grid squares. This assumption gives us the rather odd-looking regional map of Fig. 7.3. Notice
that the region covered by the grid is 15 km on a side or 225 sq. km
in area. Each square is 3 km on a side, hence the subdivision areas
shown in Table 7.3.

The domestic waste water load for the city was simply assumed to
be 100 gallons per day per capita containing 0.25 lb. of B.O.D. The
suburbs, we assume, are still sufficiently sparsely settled to be able
to depend on septic tanks. The city's domestic sewage is augmented
by another 4.3 million gallons per day of commercial and industrial
waste water with a similar B.O.D. concentration. The city's waste
water is, of course, assumed sewered to the municipal treatment
plant.

The Natural Environment

Environmental models – air and water dispersion, chemical reaction,
and models of ecosystems – may be used to describe the impact on

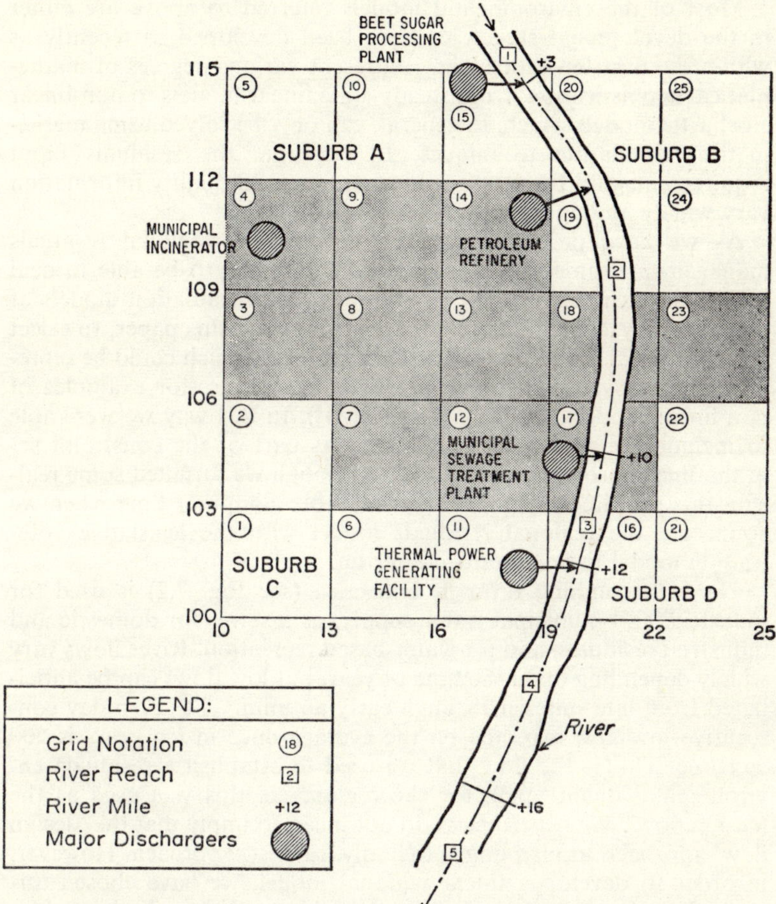

FIG. 7.3   Location of sources and receptors of gaseous residuals:
city and surrounding suburbs A, B, C and D

the environment of energy and material residuals discharged from the production and consumption activities of man. We use these models to predict the ambient concentrations of residuals and population sizes and species at various points in space throughout the environment, given (a) a set of residuals discharge levels from the linear programming model of regional generation and discharge of residuals, and (b) a set of values for the exogenous environmental parameters, e.g. stream flow and velocity, water temperature, wind speed and direction, atmospheric stability, and atmospheric mixing depth.

Most of the environmental models referred to above are either in the development stages, or have been developed as recently as within the past ten years. They represent various degrees of mathematical sophistication, from steady-state linear models to non-linear stochastic models which, in general, can only be solved using mathematical simulation techniques. In addition, the residuals input requirements and the outputs of environmental quality information vary widely among them.

As we have pointed out, even though our regional residuals management framework was originally designed to be able to deal with complex models of the environment, e.g. simulation models of aquatic ecosystems, we decided, for purposes of this paper, to select environmental models and quality parameters which could be represented by a system of linear algebraic equations (for examples of non-linear models, see [11], [4] and [24]). In this way we were able to include the environmental models as part of the constraint set in the linear programming model. Although we forfeited some realism, this simplification saved considerable computer time when we connected the regional residuals model with the legislative vote-trading model to be described later on.

The river running through Didactica (see Fig. 7.2) is used for industrial and municipal water supply, as a sewer for domestic and industrial residuals, and for water-based recreation. River flows vary widely depending upon the time of year, but low flows can be anticipated from late summer through early autumn. The seven-day consecutive low flow expected on the average once in ten years is 500 cu.ft/sec. This is the flow that we used to establish a residuals–environmental quality plan for the region, i.e. this was used as the 'design flow'. We nevertheless do not mean to imply that the 'design flow' approach to planning is the only, or best, approach. However, in order to develop a linear regional model, we have chosen this route. We plan to include the effects of time variations in the assimilative capacity of the natural environment in future models.

As shown in Figs. 7.2 and 7.3, the river within the limits of our hypothetical region is divided into six reaches. The boundaries of these reaches in river miles are also shown in these diagrams. Although the number of reaches selected was arbitrary, the river was delineated in this fashion in order to provide a small number of unique geographical sections of the natural world – in this case, the river – about which public choices regarding quality could be made. The ambient concentrations at the centres of these reaches are assumed to represent the water quality for the entire reach. Environmental models, given a set of residuals discharges, are used to provide these mid-reach concentrations. Data about the water resources of the region that are needed for these models are presented in Table 7.4. In Fig. 7.3 the positions of the dischargers are shown relative to reach boundaries.

TABLE 7.4

ASSUMED RIVER DATA FOR HYPOTHETICAL REGION

Hydrologic and water quality data

Design river flow*	500 cu.ft/sec.
Velocity	0·5 ft/sec. (8·2 miles/day)
Water temperature (river mile 0 +0)	68°F
Dissolved oxygen (D.O.) (river mile 0 +0)	7·2 mg/l
Dissolved oxygen saturation value at 68°F	9·2 mg/l

Value of parameters in water dispersion models

B.O.D. rate constant, K_1	0·20 days^{-1}
Reoxygenation rate constant, K_2	0·40 days^{-1}
Temperature decay constant, K_T	0·40 days^{-1}
Initial D.O. deficit (at the regional boundary), D_0	2·0 mg/l
Initial B.O.D. loading (at the regional boundary), L_0	20,000 lb./day

* Seven-day consecutive low flow expected on the average once in ten years.

The water quality indicators which we have selected for illustrative purposes in this paper are the dissolved oxygen levels and temperature in the river. The residuals from production and consumption activities in our region which affect these water quality measures are organic material (biochemical oxygen demand) and waste heat, respectively. Constraints on water quality are written in terms of limits on the dissolved oxygen deficit, D, and temperature rise, ΔT. The dissolved oxygen deficit, D, is defined as the difference between the dissolved oxygen saturation value and the dissolved oxygen level. Unfortunately the distribution of dissolved oxygen levels is not independent of the river temperature, since temperature affects the rate of biochemical oxidation (the oxygen-consuming mechanism) as well as the dissolved oxygen saturation level. For purposes of exposition we have assumed them to be independent, thus maintaining linear relationships.

(a) *The dissolved oxygen model.* To relate discharges of organic material – measured by its biochemical oxygen demand (B.O.D.) – to resulting dissolved oxygen (D.O.) levels in the river, we used the well-known Streeter–Phelps model [39].

Using this model and assuming steady-state conditions, the D.O. deficit due to B.O.D. discharges may be expressed as a linear function of these discharges. If we denote as X_{BOD} a vector of B.O.D. discharges (lb. per day), as A_{BOD} a matrix of transfer coefficients relating B.O.D. discharges to resulting D.O. deficits, and as D_X a vector of D.O. deficits at the mid-reach points along the river, the B.O.D./D.O.

deficit relationship due to discharges may be expressed in matrix notation as

$$A_{BOD} \cdot X_{BOD} = D_X$$

An initial D.O. deficit at the head of the river within the region, and an initial stream B.O.D. loading at this point, also impose D.O. deficits along the river. Hence, the total D.O. deficit at any downstream point along the river may be expressed as the sum of these three effects:

$$D_T = D_X + D_{DO} + D_L$$

where D_{DO} = a vector of D.O. deficits due to the initial D.O. deficit entering the region

D_L = a vector of D.O. deficits due to the stream B.O.D. loading entering the region

D_T = a vector of total D.O. deficits.

Because in this paper we did not choose to treat either the initial deficit or loading as management alternatives, D_{DO} and D_L are constants. For purposes of the model, the equation is rearranged as follows:

$$A_{BOD} \cdot X_{BOD} = D_T - D_{DO} - D_L$$

This relation, being linear in B.O.D. discharges, is placed in the constraint set of the linear programming model as $A_{BOD} \cdot X_{BOD} \leq R$. It must be recognised that this relation allows us to constrain D.O. deficit, not D.O. The latter must be evaluated separately, keeping in mind that the D.O. saturation value is also a function of river temperature.

(b) *The temperature model.* There are a variety of formulations in the literature for predicting the effects of man's activities on the resulting temperature of a body of water. The technique most widely used involves some form of energy budget analysis (see, for example, [2], [3], [9] and [14]). This method seeks to evaluate the net change in energy level of the river due to energy gains and losses via the following heat-transfer mechanisms: net thermal radiation, evaporation, conduction and convection. The most important source of energy in our model is waste heat added to the river as a residual of production and consumption activities.

By making simplifications similar to those necessary to linearise the B.O.D./D.O. relation (e.g. steady-state conditions, ignoring diurnal fluctuations), we can use the thermal energy balance equations for the river to derive another set of coefficients which translate discharges of residual heat (in B.T.U.) into temperature changes above natural for downstream reaches. If we denote as X_{BTU} a vector of heat discharges (B.T.U. per day), as A_{BTU} a matrix of transfer coefficients relating heat discharges to increased river temperatures, and as ΔT a vector of temperature increases along the river, this may be expressed in matrix notation as

$$A_{BTU} \cdot X_{BTU} = \Delta T$$

As this is a set of linear algebraic equations, it may now be incorporated within the constraint set of the linear programming model.

The Urban Atmosphere and its Quality

The portion of the hypothetical region in which air quality is of interest to us is shown in Fig. 7.3. This area, which includes a city and four suburbs, has been subdivided into 25 grid squares, each 3 km square. Each grid square is designated by a number from 1 to 25. Ambient residuals concentrations at the centres of these grids are assumed to represent the air quality for the entire grid. An air dispersion model is used to relate discharges of gaseous residuals within the region to mean seasonal ambient concentrations at the centres of the grids.

We designate two types of gaseous residuals dischargers within our hypothetical region: major dischargers which are treated as point sources, and minor dischargers which we handle as area sources. As noted previously, there are 5 major point sources in the region discharging gaseous residuals, and 25 area sources. Given the necessary x–y co-ordinates of each, distances in kilometres between all source mid-grid locations are evaluated automatically by the dispersal model.

The dispersion model, in turn, is based on certain assumptions about the meteorology of our hypothetical region: (i) a joint probability distribution for wind speed, wind direction and atmospheric stability (for the season of interest in this case, to correspond to the season of low river flows, we chose September data), typical of that of an East Coast (U.S.) city; (ii) a mean monthly maximum atmospheric mixing depth for this season of 1000 metres; and (iii) mean monthly temperature and pressure for this region of 68°F (20°C) and 1017 millibars (30·03 in. of mercury) respectively (see, for example, [21], [22] and [41]). As noted previously, we selected sulphur dioxide and suspended particulates to be representative of the air quality over this urban area. Because sulphur dioxide reacts chemically, and is otherwise lost from the atmosphere, we assumed for it a half-life of 3 hours.

The actual dispersion model we used is the one developed for the Federal Government's Air Quality Implementation Planning Program (I.P.P.) [40]. This model is based upon a diffusion model developed by Martin and Tikvart [29], which evaluates concentrations downwind from a set of point and area sources on the basis of the Pasquill [32] point source Gaussian plume formulation. Its output represents arithmetic mean seasonal concentrations based on the probabilities of occurrence of the 480 discrete meteorological situations implied by 16 wind directions, 6 wind-speed classes* and 5 atmospheric stability classes. The joint probabilities of occurrence for each of these 480 combinations are determined from actual meteorological data. Transfer coefficients are calculated on this basis for each source-receptor pair.

If we denote X_{SO_2} a vector of the sulphur dioxide discharges in the region (tons per day), as A_{SO_2} a matrix of transfer coefficients relating unit residuals discharges to mean seasonal ambient concentrations, and as Y a vector of mean seasonal ambient concentrations (micrograms per cu. metre) at the 25 mid-grid locations, the air quality of the region may also be expressed in the now familiar matrix notation as

$$A_{SO_2} \cdot X_{SO_2} = Y$$

A linear equation set similar to this may be used to evaluate ambient concentrations for suspended particulates as well, though the transfer matrices would not, of course, be the same for the two residuals. Because this equation set is linear, the discharge–concentration relationships for both sulphur dioxide and suspended particulates are placed in their entirety within the constraint set of the linear programming model.

7.3 *ENVIRONMENTAL QUALITY AND THE DISTRIBUTION OF COSTS*

It is clear that the model so far discussed, a slightly modified version of the classical model type to which we referred in the introduction, is designed to solve the following linear programming problem:†

$$\min c'X$$
$$\text{subject to} \begin{bmatrix} A_1 \\ A_2 \end{bmatrix} X \begin{matrix} \geq b_1 \\ \leq b_2 \end{matrix}$$

where X is the vector of activity levels of the various production alternatives, treatment and recirculation possibilities and discharges; c is

---

* The representative speeds associated with the six climatological wind-speed categories (0–3, 4–6, 7–10, 11–16, 17–21 and >21 knots) are given by the five mid-interval values of 0·67, 2·46, 4·47, 6·93 and 9·61 metres/sec., and by 12·52 metres/sec. (25·5 knots) for the >21 knots category [40].

† This is not to imply that cost minimisation is the only, or even the best, form in which to couch the residuals management problem. It is simply an easier form to discuss. All our comments below apply equally well to a formulation involving maximisation of net regional benefits subject to constraints on capacity, factor availability, etc., as well as E.Q. constraints. Indeed, this is the form our own didactic model takes.

a vector of the unit costs of these activities; b_1 is a minimum 'bill of goods' setting lower limits to production; b_2 is a set of ambient environmental quality constraints (e.g. upper limits on SO_2 in all grids, upper limits on the dissolved oxygen deficit in all river reaches); A_1 is the technology matrix (including residuals modification technology); and A_2 is composed of the several environmental model transfer matrices derived above. Thus, A_2 might be further partitioned as follows:

$$
\begin{bmatrix}
A_{BOD} & & & \\
& A_{BTU} & & \\
0 & & A_{SO2} & \\
& & & A_{SP}
\end{bmatrix}
\cdot
\begin{bmatrix}
\vdots \\
X_{BOD} \\
X_{BTU} \\
X_{SO2} \\
X_{SP}
\end{bmatrix}
\leq
\begin{bmatrix}
R_{BOD} \\
R_{BTU} \\
R_{SO2} \\
R_{SP}
\end{bmatrix}
$$

The right-hand side, b_2, would of course be similarly partitioned, as we show.

In essence, then, this model tells us the cost, to whomsoever these costs accrue, as they say in U.S. Government planning directives, of simultaneously meeting certain production and environmental quality requirements. In this form the classical models have been used to help in policy formulation, and it is likely that they will be so used again in the future. There is, however, mounting evidence, for both the public and its elected and appointed policy-makers to see, that aggregate regional costs (or net benefits) are not a very reliable guide to the political viability of a proposed regional policy. It makes a considerable difference to whom the costs and benefits do accrue. Thus, a city administration may, in principle, wish to clean up the river flowing past its waterfront, but since it will incur the clean-up costs and capture only a portion of the benefits, it may not support a regional plan calling for clean-up (see [36]). On another level, a working man *may* appreciate that tougher ambient standards on SO_2 and suspended particulates are likely to benefit his health in the long run, but he will almost certainly be opposed to a plan which involves the partial or complete shutdown of the plant at which he works.

Without at this point returning to the matter of what type of agency should be making environmental quality policy, it seems clear then that any decision process would be assisted by models which reflect the distribution of costs as well as the spatial distribution of quality. In addition, it may be useful to structure models in such a way that policy-makers can explore the implications of putting different constraints on the distribution of costs. It is a simple matter to include distributional information in the general model formulated in this section. All we have to do is to add a third part to the constraint matrix and a third set of values to the right-hand side. Thus, the problem becomes:

$$\min c'X$$
$$\text{subject to } \begin{bmatrix} A_1 \\ A_2 \\ A_3 \end{bmatrix} X \begin{matrix} \geq b_1 \\ \leq b_2 \\ \leq b_3 \end{matrix}$$

where A_3 is structured to allow the product $A_3 \cdot X$ to represent costs over and above the *status quo* (often the no-control) case incurred by various firms, political jurisdictions or other groups. Thus, one row of A_3 might represent increased costs of electricity to the customers of the regional utility(ies); another might be increased unemployment at a particular industrial plant implied by any cutbacks of production chosen to achieve overall production and quality goals at least cost. b_3 is a vector of acceptable levels for these costs: the largest acceptable increase in electricity costs or unemployment, for example. (It is possible to treat unemployment as an aggregate for the region, a sub-aggregate for a particular political jurisdiction, or for single important plants. The choice will depend partly on the particular regional economy and partly on the type of agency involved.)

The distributional constraints we have added to our model are summarised in Table 7.5. The technique of their insertion in the model is straightforward and is shown schematically in Fig. 7.4. For example, to deal with the electricity cost constraint, we add a row called *XTRCOSTE* to the linear constraint set. Each electricity production vector, EP_j, including the base (no residuals control) process, EP_1, has a positive entry, C_i, in that row giving the cost per megawatt-hour of using it. (This includes the power importation activity, EP_I, which is assumed to be more costly than domestic production.) Then, we add an activity which measures the percentage increase in cost, *assuming* a constant level of power consumption. Let us call this column *INCELCOST*. This vector has, as a negative entry in the *XTRCOSTE* row, a number equal to the total generation cost in the base case. In the simplest case this would be $EP_1 \cdot C_1 \equiv C_B$. The sum across the row is constrained to be zero. Thus, the value assumed by the activity *INCELCOST* in any solution is the fraction of the base-case cost represented by generation costs in the new solution. If costs are 10 per cent higher, for example, *INCELCOST* will be 'operating' at level 1·1.* To constrain the increase in electrical

---

* There are many ways to accomplish our end here. This method is a bit cumbersome for simple cases, but we used it because of complications involving inter-industry exchanges, treatment costs, etc.

TABLE 7.5

CONSTRAINTS ON DISTRIBUTION OF COSTS

Constraint on	*Notes and description*
1. Percentage increase in cost of electricity	Based on extra costs involved in adding cooling towers, precipitators, etc., compared with basic production process with no residuals control. Applies uniformly throughout region.
2. Percentage increase in costs of domestic residuals collection and disposal in city **M**, and suburbs **A**, **B**, **C** and **D**	For city: based on incremental costs of going to secondary or tertiary treatment; of added sludge disposal; of cleaning incinerator stack gases; and of using more efficient landfill methods. For suburbs: costs of using more efficient (higher compaction) landfill methods and/or transporting to more distant sites. Also share of increased incineration costs if incinerator is used by suburb.
3. Unemployment at beet-sugar and petroleum refineries	Beet sugar: based on degree of production cut-back only. Petroleum: based on process changes as well as total production level.
4. Increase in annual heating cost of average household implied by burning lower-sulphur fuel (city and each suburb)	Based on following price differential: High-sulphur D.F.O., \$4.35/b (base case) Medium-sulphur D.F.O., \$4.54/b Low-sulphur D.F.O., \$4.70/b. Annual figure obtained by blowing up average Nov/Dec day by annual average degree-day total for Philadelphia region.

FIG. 7.4   Distributional constraints: increase in cost of electricity

Row names (constraints)	Column names (activities)				Right-hand side
	Electricity production vectors $EP_1, \ldots \ldots, EP_n$	Power import	*INCELCOST*		
.					
.					
.					
XTRCOSTE	$+C_1, \ldots \ldots, +C_n$	$+C_I$	$-C_B$	$=$	0
ULXCSTE			$+1$	$\leq$	1·1 (or 1·05, or 1·2)
.					
.					
.					

TABLE 7.6

ILLUSTRATIVE RUNS OF DIDACTIC MODEL

	Run A	Run B (based on time unit of one day)	Run C
Objective function section			
Objective function	$89,250	$78,780	$75,150
Δ obj. fn. (cost) from Run A	—	$10,470	$14,100
Cost of changing distribution: B to C			$3,630
Environmental quality section			
Highest river temp.	77·9°F (Reach 4)	73·0°F (Reach 4)	73·0°F (Reach 4)
Lowest D.O.	1·73 p.p.m. (Reach 6)	5·87 p.p.m. (Reach)	5·87 p.p.m. (Reach 6)
Maximum SO_2 conc.	103 μg/m³ Grid 13)	60·0 μg/m³ (Grid 13)	60·0 μg/m³ (Grid 13)
Max. susp. particulates	63·4 μg/m³ (Grid 10)	36·6 μg/m³ (Grid 12)	30·4 μg/m³ (Grid 12)
Increased cost section			
Increase in costs of residuals disposal — City	—	46·3%	40·40%
A	—	—	—
B	—	—	17·6%
C	—	—	18·9%
D	—	—	
Increase in home-heating costs* — City	—	$11.10	$10.00
A	—	$15.40	$10.00
B	—	$25.80	$10.00
C	—	$15.40	$10.00
D	—	$9.25	$10.00
Increase in electricity cost	—	1·75%	9·70%
Sugar beet plant operating level (% capacity)	100	42·6%	75·0%

* Dollars per household per year.

Discharge information section

B.O.D.	Sugar refinery	33,420 lb./day	1,580 lb./day	692 lb./day
	Petroleum refinery	4,030 lb./day	764 lb./day	886 lb./day
	Municipal plant	15,400 lb./day	4,400 lb./day	5,330 lb./day
Heat	Sugar refinery	2,380 10^6 B.T.U./day	0 10^6 B.T.U./day	0 10^6 B.T.U./day
	Petroleum refinery	16,580 10^6 B.T.U./day	0 10^6 B.T.U./day	0 10^6 B.T.U./day
	Electric power plant	17,160 10^6 B.T.U./day	14,840 10^6 B.T.U./day	14,840 10^6 B.T.U./day
SO_2	Sugar refinery	6·86 tons/day	1·95 tons/day	3·43 tons/day
	Petroleum refinery	60·0 tons/day	46·5 tons/day	43·6 tons/day
	Electric power plant	28·8 tons/day	28·8 tons/day	4·25 tons/day
Part.	Sugar refinery	22·0 tons/day	2·55 tons/day	2·39 tons/day
	Petroleum refinery	4·80 tons/day	4·39 tons/day	4·31 tons/day
	Electric power plant	88·6 tons/day	58·9 tons/day	12·4 tons/day
	Municipal sewage plant	4·6 tons/day	6·1 tons/day	6·0 tons/day
	Municipal incinerator	2·0 tons/day	2·1 tons/day	1·3 tons/day
Solids to incinerator		101 tons/day	104 tons/day	65·4 tons/day
Solids to landfill		783 cu. yds/day	866 cu. yds/day	1,130 cu. yds/day

costs, we need only add another row called, say, *ULXCSTE*, and enter in it $a+1$ from the vector *INCELCOST*. The sum across this row would be constrained to be less than or equal to 1·05, 1·2 or whatever level was to be explored. Similar methods are employed in calculating and constraining each of the other distribution considerations.

The Capabilities of the Simple Regional Model
Before going on to discuss the linking of this regional residuals management model with a legislative vote-trading program, it will be worth while to examine the kinds of information the model can produce on its own. To give the reader some feel for this we include, as Table 7.6, a summary of the results of three runs of the model. These runs reflect both the efficiency and distributional considerations with which the model is capable of dealing.

In run A there are no constraints on environmental quality; we may think of it as the private optimum and we shall continue to use it as our reference point when we go on to the legislative solutions. For run B, we have constrained the annual average SO_2 concentrations in each grid square to be less than or equal to 60 micrograms per cu. metre ($\mu g/m^3$), the annual average suspended particulates to be less than or equal to 60 $\mu g/m^3$ (these are the proposed secondary standards from E.P.A. as reported in [15]), the D.O. deficit in the river to be less than 3 p.p.m. and the temperature rise to be less than 5°F. (Together, these imply a minimum D.O. level under our assumptions of 5·8 p.p.m.) Of particular interest in comparing the results of these two runs are:

(i) The overall cost, as given by the change in the regional objective function, is $10,470. This represents principally the control devices installed by the various industrial plants, of a higher level of sewage treatment by the municipality, and of the switch to lower-sulphur home and commercial heating fuels.

(ii) The distribution of these costs, as indicated by the forced cutback in production (and hence employment) at the beet-sugar refinery, the increases in home-heating bills of households in the various political jurisdictions, and the increase in the cost of municipal services.

(iii) The evidence of links between types of discharge. Thus, for example, in run B particulate discharges from sludge incineration at the municipal treatment plant are about 30 per cent higher than in run A. Solid residuals to landfill require a 10 per cent greater volume each day in run B. (The evidence of growth in solids load would be even more dramatic were it not that precipitated fly ash from the electric power plant is used as a sludge conditioner by the municipal treatment plant.)

The third run of the model (run C) demonstrates what happens when we require the same improvement in ambient environmental quality as in run B, but also require a different distribution of costs than is implied by the efficiency solution. Specifically, for run C, we require that the beet-sugar refinery operate at no less than 75 per cent of capacity; that home-heating bills be allowed to rise by no more than $10 per household per year on the average; and that the increase in the city's residuals disposal costs be kept to 40 per cent or less. From Table 7.6 we see that the cost of these distributional requirements, as measured by the incremental change in the value of the objective function, is $3630 per day, resulting from the forced use of more costly means of achieving the environmental standards. These costs show up, for example, in an increase of 9·7 per cent in the cost of electricity, and increases in the residuals disposal costs of suburbs B and D. Examination of the discharge section of the table reveals that in run C the electric power plant bears a very large share of the burden of reducing the SO_2 concentrations, while the beet-sugar refinery has to reduce its B.O.D. discharges even further in run C in order that the municipal cost constraint be met.

It is interesting to notice how large the required SO_2 discharge reduction is for the electric plant when the beet-sugar plant discharges are allowed to increase on run C. Thus, the same environmental effect *in the region* is achieved in run B by a total SO_2 discharge reduction of about 18·4 tons per day and in run C by a total reduction of 41 tons per day. This results primarily from the assumptions we made about stack heights, exit velocities, etc. The power plant's effective stack height is much greater than the beet-sugar refinery's and, as a result, a given reduction in SO_2 discharge at the power plant has a much smaller effect on regional ground-level concentrations. Note that the greater total SO_2 emissions under run B (which give a nearly identical pattern of ambient SO_2 concentrations within our region) imply greater transport of this residual out of the region. This raises again the question of interregional spillovers discussed above.

To give the reader some feeling for the distribution of benefits from these runs and to illustrate another useful form of the model's output, we include, as Figs. 7.5 and 7.6, maps showing the contours of concentrations resulting from the first two runs. For run A, the petroleum refinery contributes the major portion of the mean seasonal ambient sulphur dioxide concentrations at ground level. This is evident in Fig. 7.5. The two peaks of ambient concentrations

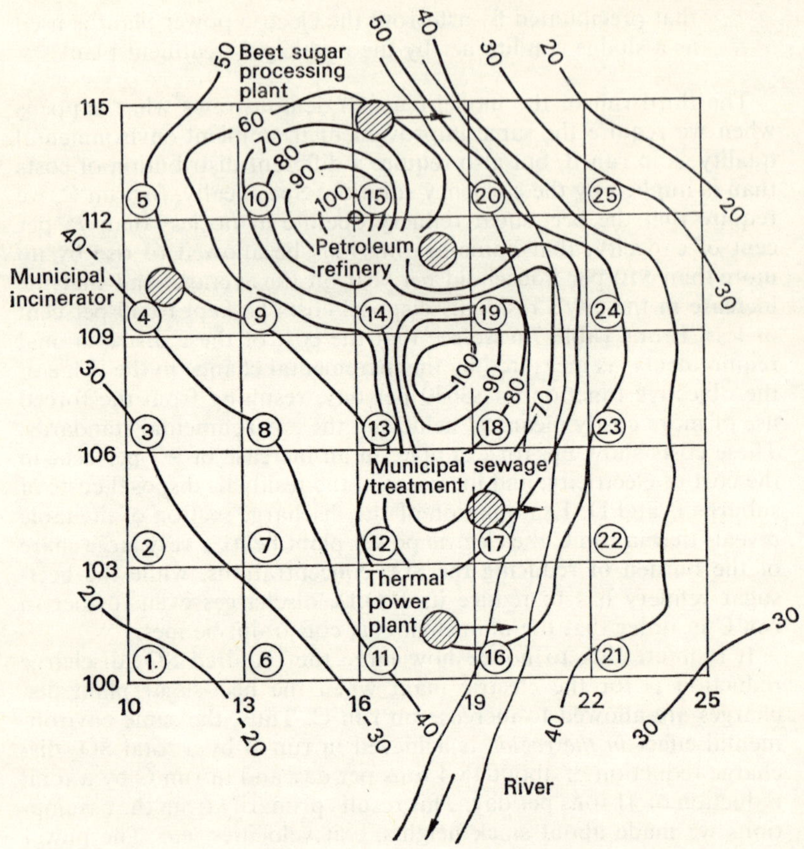

Fig. 7.5   Ambient concentration, run A: sulphur dioxide ($\mu g/m^3$)

associated with the petroleum refinery clearly indicate the two prevailing wind directions for this season. These are mean seasonal concentrations; the two peaks probably would not occur on the same day.

The tremendous variation in ambient concentrations over relatively short distances indicates that mid-grid concentrations (as used in Table 7.6) are not sufficient measures of air quality over the individual grids. Figs. 7.5 and 7.6 have both been prepared on the basis of concentrations at mid-grids, at mid-points along the sides of each grid, and at all grid intersections. Using mid-grid concentrations, Grid 13 exhibits the highest sulphur dioxide concentration (103 $\mu g/m^3$), as shown in Table 7.6. But with the finer receptor grid scheme, the maximum concentration (about 110 $\mu g/m^3$) was dis-

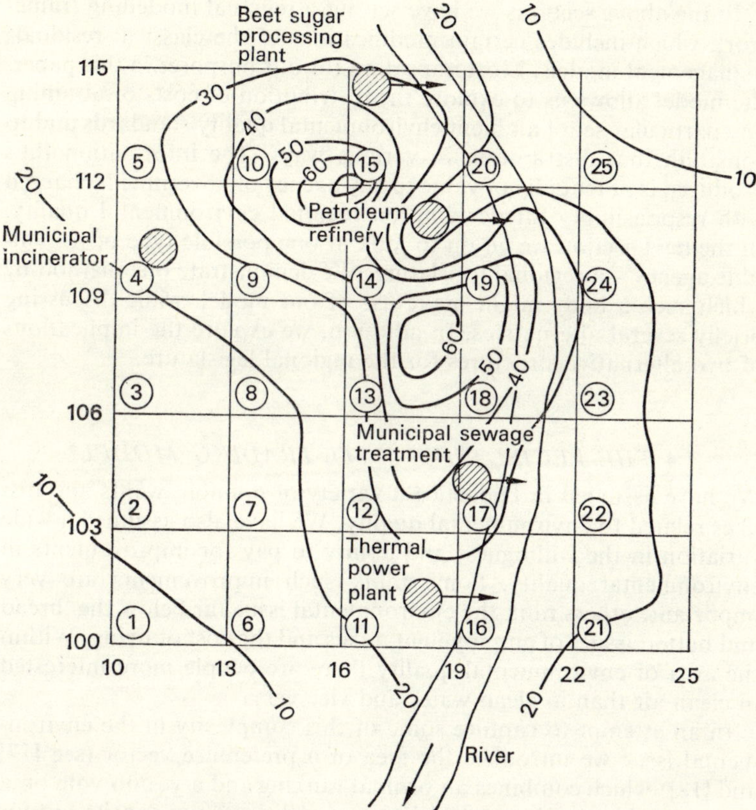

FIG. 7.6   Ambient concentration, run B: sulphur dioxide ($\mu g/m^3$)

covered to be at the intersection of Grids 9, 10, 15 and 14. This indicates that policies made on the basis of one ambient concentration prediction per grid, where the grids are larger than, say, 1 km on a side, may fail to allow for the peaks of the local concentration contours. Just how fine the grid needs to be as a practical matter almost certainly varies with local conditions.

A comparison of Figs. 7.5 and 7.6 indicates the improvement in ambient sulphur dioxide concentrations throughout the region due to a reduction in sulphur dioxide emissions from the petroleum refinery of 13·5 tons per day and to a reduction of approximately 4·9 tons per day from the beet-sugar refinery. Again we note that use of constraints on mid-grid locations is not sufficient to guarantee against higher levels in other parts of the grids. The pattern for run C is virtually the same as that for run B and is not included here.

In the above sections we have set out a regional modelling framework which includes certain modifications of the classical residuals management models. Most important for our purposes in this paper, the model allows us to explore the distribution of costs of attaining any particular set of ambient environmental quality standards and to constrain that distribution in various ways. The information thus produced is potentially of value to any agency of government charged with responsibility for deciding on regional environmental quality. In the next section we go on to look at one possible type of responsible agency – a regional legislature. We demonstrate one method by which such a body might make use of our model, while discussing briefly several alternatives. In addition, we explore the implications of two alternative structures for the regional legislature.

7.4 *THE LEGISLATIVE VOTE-TRADING MODEL**

We have assumed in Didactica a variety of opinion, wants and dislikes related to environmental quality. We have also assumed a wide variation in the willingness and ability to pay for improvements in environmental quality. Some think such improvements are very important; others rank the environmental issue far below the 'bread and butter' issues of employment, taxes and the cost of living. Within the area of environmental quality there are people more interested in clean air than in clean water and vice versa.

In an attempt to capture some of this complexity in the environmental issue we introduce the idea of a preference vector (see [17] and [18]) which combines an ordinal ranking and a yes/no vote on a given set of issues. In the legislative model the issues consist of four quality measures of the natural environment and the four measures of increased cost (Table 7.5) resulting from improvements in environmental quality. The quality measures of natural environment include (1) the level of dissolved oxygen in each reach of the river (calculated in terms of a dissolved oxygen deficit): D.O.D.; (2) increases in the temperature of each reach of the river, ΔT; (3) the level of suspended particulates in the air (measured in micrograms per cu metre), SP; and (4) the level of sulphur dioxide in the air (measured in micrograms per cu. metre), SO_2. We realise that these measures could probably not be used directly in a political process, although the experience in choosing levels of water quality in the Delaware shows a quick assimilation of technical information by laymen, particularly

* Special thanks are due to Elizabeth Mortland for developing an efficient program whereby this algorithm could be tied directly to the regional L.P. model.

if the technical measures are related to fish population, recreation potential and health hazards.

Arbitrary preference vectors (reflecting differences of tastes, incomes, etc.) were specified for each of the 25 grids. These are based on arbitrary upper limits on each of the eight measures previously outlined. Thus, Grid 1's upper limits vector is:

	Grid 1	
DOD	3·0	p.p.m. in reach 4
ΔT	5·0	°F in reach 4
SP	50·0	μg/m³ in Grid 1
SO_2	20·0	μg/m³ in Grid 1
Taxes	1%	increase of 1 per cent in suburb C
Unemp.	10%	increase of 10 per cent at the sugar refinery
Elec.	50%	increase of 50 per cent to each household in every grid
Heat	20	increase in dollars per year to each household in suburb C

If we allowed Grid 1 to be our decision-maker, these upper limits would be the constraints Grid 1 would put on the solution of the regional model. Since Grid 1 is only one of 25 grids, we wish to construct a social choice process to allow the upper limits vectors of all 25 grids to be expressed. Our preference vectors, one for each grid, are designed to do that. For example, we may display Grid 1's preference vector in response to run A (no constraints) of Table 7.6 in Table 7.7.

TABLE 7.7

GRID 1: PREFERENCE VECTOR

	Grid 1 *upper limits*	*Results of run A* *for Grid 1*	*Grid 1* *preference vector*
DOD	3·0	5·52	N
ΔT	5·0	9·89	N_1
SP	50·0	17·19	Y
SO_2	20·0	32·69	N_2
Tax	1%	0·0	Y
Unemp.	10%	0·0	Y_3
Elec.	50%	0·0	Y
Heat	$20	0·0	Y

N = No, Y = Yes

The numerical subscripts in the preference vector indicate the ordinal ranking of three (in this case) measures. Thus we are ranking, by assumption, an upper limit of 5°F heat rise in reach 4 of the river of first importance in Grid 1, an upper limit on SO_2 of 20 μg/m³ in Grid 1 as second in importance, and an upper limit of 10 per cent on unemployment at the sugar refinery as third most important to Grid 1. (For ease of computation we do not rank all measures: elements without subscripts are assumed to be all of equal importance but less important than any subscripted element.) Our ordinal ranking is based on the assumed socio-economic characterisations from which Table 7.3 was also derived.*

Grid 1's preferences may be summarised by saying that the citizens of Grid 1 are dissatisfied with an unconstrained solution (no constraints on present production and consumption activities in Didactica) in three out of four quality measures, while they are satisfied with the present tax and utility burdens on Grid 1. Since Grid 1 ranks the environmental measures above the financial ones, however, some additional financial burden would be accepted if necessary to achieve acceptable levels of water and air quality.

All other grids were assigned upper limits vectors and ordinal ranks to three or more measures. Using those vectors, we can display all 25 preference vectors in response to run A (Table 7.6). These are shown in Table 7.8. The Y votes on each row are tallied in the far right column. We see a unanimous approval of all financial measures but much disapproval of the present quality of the air and water. The stage is set, assuming our preference vectors are such that not all can be met simultaneously, for some sort of social choice process to be invoked.† Since the number of possible 'solutions', that is, technically feasible alternatives, may be said to be almost infinite, the social choice process cannot be simply a blind groping for a solution acceptable to some given percentage of grids, for the process would prove to be inefficient and the 'solution' ambiguous at best. Neither can we, without throwing away the concept of social choice, simply stand with run A or B or C without constraints reflecting the preference vectors, in the hope that the 'objectivity' of any one of these solutions will cow the residents of Didactica into acceptance.

In a real-world situation, whatever official has charge of the model (and a computation budget) might be tempted simply to meet the

---

* The ordinal rankings by real actors would clearly change as one or more upper limits were met and/or other upper limits were greatly exceeded. We have not investigated how such ordinal shifts would affect the algorithm, although it is clear that convergence problems might well occur.

† Two criticisms have been made of these vectors. The first, which we accept as probably valid, is that those items which can naturally be expressed in dollar terms (increase in cost of electricity and increase in taxes) should be so expressed. The second, which runs counter to the entire thrust of our argument, is that *all* the measures should be reduced to dollar terms because people (or their representatives) cannot trade apples for oranges. The Delaware estuary experience shows that people can think about, and argue over, trades of environmental quality indicators for dollars. This point is central to our critique of the classical models.

The unemployment rate is on the borderline. We feel that its valuation is best left to the legislators, but it can be argued that a dollar figure could be obtained for each district or political subdivision.

TABLE 7.8

VOTE MATRIX, RUN A (UNCONSTRAINED)

Issue	Grid: 1	2	3	4	5	6	7	8	9	10	11	12	13	14	15	16	17	18	19	20	21	22	23	24	25	Tally of Y votes
DOD	N	N	Y	Y	Y	N	N	Y	Y	Y	N	N_5	Y	N	Y_3	N	N	N_1	N	N	N	Y_1	Y_1	N_2	Y	11
ΔT	N_1	N	N	N	Y	N_3	N	N	N	Y	N_3	N_4	N	N	Y	N_3	N	N	N	N	N_2	N_2	N	N_4	N	3
SP	Y	Y	Y	Y	Y	Y_2	Y_2	Y	Y	Y_2	Y_1	Y_2	Y_1	Y	Y_2	Y_2	Y_2	Y	Y	Y_1	Y_3	Y	Y	Y	Y	25
SO_2	N_2	N_1	N	Y	N	N_1	N_1	N	N	N	N_2	N_1	N_3	Y	Y_1	N_1	N_1	N	Y	N	N_1	N_4	N	Y	Y_1	6
Tax	Y	Y	Y_1	Y_2	Y	Y	Y_3	Y_1	Y_2	Y	Y	Y_3	Y_2	Y	Y	Y	Y_3	Y_2	Y_2	Y_2	Y	Y_3	Y_2	Y_3	Y	25
Unemp.	Y_3	Y_2	Y_2	Y_1	Y_1	Y	Y_4	Y	Y_1	Y_1	Y	Y	Y	Y_1	Y_4	Y	Y	Y	Y_1	Y_3	Y	Y	Y_3	Y_1	Y	25
Elec.	Y	Y	Y	Y	Y_2	Y	Y	Y_2	Y_3	Y_3	Y	Y	Y	Y_2	Y	Y	Y	Y_3	Y_3	Y	Y	Y	Y	Y	Y	25
Heat	Y	Y_3	Y_3	Y	Y_3	Y	Y	Y	Y	Y	Y	Y	Y	Y	Y	Y	Y	Y	Y	Y	Y	Y	Y	Y	Y_2	25

first and second ranked upper limits of enough grids to ensure a majority (assuming there existed some political body through which these votes could be registered). Lacking any such political body, he might 'play around' with meeting a few more high-ranked upper limits (chosen on the basis of judicious knowledge of which areas could be most difficult if their limits were not met) and balance the additional limits against protests from special interest groups, industrial and environmental.

Indeed, even without any model, this procedure is an apt description of how agency personnel attempt, in their own words, to 'strike a balance' between different groups. In recent years, however, this procedure has worked less and less well, perhaps because a latent consensus on social goals has broken down. Whatever the reason, both the decisions being taken and the legitimacy of the decision-maker to take them are being questioned – not only questioned, but opposed, in the courts and in the streets.

It will be useful to explore in some detail what procedures could be used to find a solution based on the preferences of the 25 grids. The first method might be to see if all preferences (upper limits) of all the grids might be met simultaneously. If so, clearly it is Pareto-optimal to meet them. Since such a happy state is unlikely in the real world, we have set the upper limits vector so that it is not attainable in the Didactica regional model either. If all preferences cannot be met, then whose should be met, and in what order?

A second method is to meet each grid's upper limits by using each set separately as constraints on the regional model. This accomplishes two things: it allows us to make sure that each grid's upper limits preferences are internally consistent,* i.e. can be met simultaneously, and it allows us to see the kind of 'overlap' or complementarity between one grid's upper limits preferences and those of another grid. This step was taken in an earlier paper [19] and the results used to determine which one of the 25 solutions dominated the others (majority vote) under different vote-aggregation schemes, e.g. voting by grids or by municipality. Using this method tells us nothing about all other solutions and the relation of the 25 solutions to the unexamined solutions.

A third method is to pay more attention at the outset to the ordinal ranking of the different measures by each grid. For example, we might try to meet the first and second ranked preferences of all grids, or all first preferences, then all second preferences. Were we thinking about a strong party-oriented legislature or council, 'all'

---

* A grid's preferences do not have to be consistent since it could have 'if not this, then that' preferences in it. Identifying these beforehand will be useful, however, in a real-world situation.

might be replaced in the preceding sentence by 'majority party'. If 'all' were possible, the 'all' solution might have appeal over the 'majority party' solution, but would not necessarily be chosen by the majority party if meeting minority first and second ranked preferences meant giving up on majority party third and fourth ranked preferences. How the majority party acted would depend very much on what powers of retribution the minority party had, when the next election was coming up, and/or other political factors exogenous to our consideration here.

Any of the preceding methods could be employed in connection with the regional model presented in this paper, but they do not get us very far in terms of reality of conflicting preferences and the loose party structure characteristic of most American legislative councils. While replicating all the complexities of American legislative procedure and structure would be impossible, we can adopt a method which will replicate an important element in them, namely vote-trading. The essence of vote-trading is giving up on one issue to gain another issue you value more. It is distinguishable from bargaining, which involves a compromise, both sides giving up something in order to reach a position on one issue that both will accept. While both processes are used in legislative deliberations, they are logically distinct. Bargaining involves a loss to both parties on one issue but a minimum loss, i.e. not agreeing represents a greater loss to both parties. Vote-trading involves a gain to both parties but on different issues. Whether the bargaining path or the vote-trading path will be followed depends on the range of independent (tradeable) issues before the council and the heterogeneity of its members. If everyone thinks the same issue is most important, then obviously it will not be traded off by anybody, and differences of opinion will have to be compromised (bargaining) if a majority verdict is to emerge. Likewise, a council having only one matter before it must bargain because its members have no vote-trading possibilities.*

It will be seen, on reflection, that upper limits preferences and ordinal ranking of issues lend themselves easily to vote-trading, while bargaining could be simulated in a game-theoretic procedure with imputations assigned to outcomes. Since it is impossible to explore the almost infinite set of outcomes, there is little likelihood of us using our preference vectors to make sensible imputations on the feasible set. We have therefore chosen a vote-trading algorithm to explore the properties of different vote-aggregation patterns.

The basic idea of the vote-trading algorithm is to add constraints

---

* James Madison's famous defence of a large republic (in *Federalist Paper* No. 10) rests on the heterogeneity of its members and the broad range of issues that would come before its councils [28].

to our unconstrained run (run A in Table 7.8) such that N votes are converted to Y votes in some efficient, non-biased way. Vote-trading is efficient for this purpose because it focuses attention on high-ranked N votes (these are the upper limits violations of most concern to those grids). They are the upper limits the grids want most to be put in as constraints on the regional model. However, with vote-trading such a constraint can be put in only if the grid that wishes to put it in will accept (also allow to be added as a constraint) another upper limit on another issue that is desired by another grid. Constraints are put into the solution, therefore, in pairs. (There is nothing magical about 'paired' constraints. Three or more grids could agree on a constraint before it was put in. Paired constraints are simply easier, computationally, to use in the vote-trading algorithm.)

Since vote-trading was explained as giving something up for something of higher value, what is it that each grid gives up by this trade? To illustrate what is given up, let us pick out a vote-trade from Table 7.8 and see what happens.

	Grid 13 upper limits→	Grid 13 results from run A→	Grid 13 preference vector	Grid 24 preference vector	Grid 24 results ←from run A←	Grid 24 upper limits
DOD (Reach 2)	3·5	3·4	Y	N_2	3·4	3·0 (Reach 2)
SO_2	80	103	N_3	Y	38	40

The result of the vote-trade shown is to put in, as constraints, Grid 24's upper limit of $DOD \leq 3$ on reach 2 of the river and Grid 13's upper limit of $SO_2 \leq 80$ in Grid 13. What Grid 13 gives up is a 3·4 *DOD* outcome, an outcome it was happy with since it was below its own limit of 3·5. Grid 24 gives up an SO_2 outcome of 38 in its grid, an outcome it was happy with.

Now let us examine the reasoning behind this trade. Grid 13 says, 'I do not know what the effect will be of putting a tighter limit on *DOD* in reach 2 of the river, but I am willing to do it if I can get my SO_2 concentration down at least to my upper limit.' Grid 24 says, 'I want reach 2 of the river cleaned up more than it is now, and I'm willing to support an upper limit on Grid 13's SO_2 concentration if I can get support for a cleaner river'. Neither knows what effects these constraints will have on other things of concern to them, e.g. their tax bills.

Now, we could ask both grids to agree to accept whatever results come from this trade, i.e. to give up upper limits on, say, taxes entirely. We have not done that on the grounds that, should taxes go up dramatically, say 60 per cent increase, we could hardly find a way to enforce our rule. Moreover, since we are exploring the effects of putting in constraints, we feel it is more logical to assume only that the grids are giving up on run A outcomes, not giving up upper limits as such. Therefore, in the algorithm, if other upper limits are violated on a subsequent run, we allow the grids to vote N on those rows, and to try to trade on those issues if they can find a partner to trade with. In this way, higher-ranked limits are given priority but lower-ranked limits are allowed into the solution at a later stage (they may or may not be feasible at that time). Our method also means that changes of ordinal rankings are not considered. In a real-world situation, legislators and councilmen may well change their priorities as a result of the outcomes from a run of the model. Cycling and non-convergence could result from such switches, but the algorithm has not been extended to deal with that possibility.

We must make one more arbitrary rule before proceeding with the vote-trading algorithm. We shall assume that if a grid's first and second ranked measures are voted Y, then the grid is in favour of the solution; if they are not, then the grid votes N on the solution. These 'package' votes will be tallied in the bottom row of the trading tables (in Table 7.8 there would be 13 Y votes for run A, the unconstrained case). Since 13 is a majority of the 25 grids, we could say that the *status quo* (run A) is acceptable. However, the use of the vote-trading algorithm shows us that this majority is not a dominant majority. We start by solving the regional model a second time with the two constraints from the trade between Grid 13 and Grid 24, which was the only trade in run A.

The new solution generates a new vote matrix and hence new trading opportunities. The vote matrix resulting from the new solution is shown in Table 7.9. The results are easily summarised by comparing the two tables: we have improved the oxygen in the river (18 votes for instead of 11), lowered the ΔT also (8 for instead of 3), improved SO_2 substantially (22 for instead of 6), lost support on the tax rate (17 rather than 25) and still have unanimous satisfaction with the levels of *SP*, electricity and heating bills. What we have lost is our unanimous vote on the unemployment rate, which rate has jumped to 12·5 per cent, and only 7 grids will accept that. Moreover, for the whole package we have only 10 Y votes instead of 13.

As would seem likely, however, a number of trades are now possible, and when these are completed a new run of the regional model can be made, adding pairs of constraints to the constraints already in. The legislative model proceeds in this fashion until there are no further trades possible. In the instant case this occurs after three iterations with the following result, shown in Table 7.10.

The value of the objective function has gone from \$89,250 in the unconstrained solution to \$85,608 in the last solution, indicating

TABLE 7.9

VOTE MATRIX AFTER FIRST TRADE

Grid:

Issue	1	2	3	4	5	6	7	8	9	10	11	12	13	14	15	16	17	18	19	20	21	22	23	24	25	Tally
DOD	N	N	Y	Y	Y	N	Y	Y	Y	Y	N	Y_5	Y	Y	Y_3	N	N	Y_1	Y	Y	N	Y_1	Y_1	Y_2	Y	18
ΔT	N_1	N	N	N	Y	N_3	N	N	N	Y	N_3	Y_4	N	N	Y	Y_3	N	N	N	N	N_2	N_2	Y	Y_4	Y	8
SP	Y	Y	Y	Y	Y	Y_2	Y_2	Y	Y	Y_2	Y_1	Y_2	Y_1	Y	Y_2	Y_2	Y_2	Y	Y	Y_1	Y_3	Y	Y	Y	Y	25
SO_2	N_2	N_1	Y	Y	Y	N_1	Y_1	Y	Y	Y	Y_2	Y_1	Y_3	Y	Y_1	Y_1	Y_1	Y	Y	Y	Y_1	Y_4	Y	Y	Y_2	22
Tax	Y	Y	N_1	Y_2	Y	Y	N_3	N_1	N_2	Y	Y	N_3	N_2	Y	Y	Y	N_3	N_2	Y_2	Y_2	Y	Y_3	Y_2	Y_3	Y	17
Unemp.	N_3	N_2	N_2	N_1	N_1	N	N_4	Y	N_1	N_1	Y	N	Y	N_1	N_4	Y	N	Y	N_1	N_3	Y	N	N_3	N_1	Y	7
Elec.	Y	Y	Y	Y	Y_2	Y	Y	Y_2	Y_3	Y_3	Y	Y	Y	Y_2	Y	Y	Y	Y_3	Y_3	Y	Y	Y	Y	Y	Y	25
Heat	Y	Y_3	Y_3	Y	Y_3	Y	Y	Y	Y	Y	Y	Y	Y	Y	Y	Y	Y	Y	Y	Y	Y	Y	Y	Y	Y_3	25
Package	N	N	N	N	N	N	Y	Y	N	N	Y	Y	N	N	Y	Y	Y	N	N	Y	N	N	Y	N	Y	10

TABLE 7.10

VOTE MATRIX, NO FURTHER TRADES POSSIBLE

Grid:

Issue	1	2	3	4	5	6	7	8	9	10	11	12	13	14	15	16	17	18	19	20	21	22	23	24	25	Tally
DOD	N	N	Y	Y	Y	N	Y	Y	Y	Y	N	Y_5	Y	Y	Y_3	N	N	Y_1	Y	Y	N	Y_1	Y_1	Y_2	Y	18
ΔT	Y_1	Y	Y	Y	Y	Y_3	Y	Y	Y	Y	Y_3	Y_4	Y	Y	Y	Y_3	Y	Y	Y	Y	Y_2	Y_2	Y	Y_4	Y	25
SP	Y	Y	Y	Y	Y	Y_2	Y_2	Y	Y	Y_2	Y_1	Y_2	Y_1	Y	Y_2	Y_2	Y_2	Y	Y	Y_1	Y_3	Y	Y	Y	Y	25
SO_2	Y_2	Y_1	Y	Y	Y	Y_1	Y_1	Y	Y	Y	Y_2	Y_1	Y_3	Y	Y_1	Y_1	Y_1	Y	Y	Y	Y_1	Y_4	Y	Y	Y_2	25
Tax	Y	Y	Y_1	Y_2	Y	Y	Y_3	Y_1	Y_2	Y	Y	Y_3	Y_2	Y	Y	Y	Y_3	Y_2	Y_2	Y_2	Y	N_3	Y_2	Y_3	Y	24
Unemp.	Y_3	Y_2	Y_2	Y_1	Y_1	Y	Y_4	Y	Y_1	Y_1	Y	Y	Y	Y_1	Y_4	Y	Y	Y	Y_1	Y_3	Y	Y	Y_3	Y_1	Y	25
Elec.	Y	Y	Y	Y	Y_2	Y	Y	Y_2	Y_3	Y_3	Y	Y	Y	Y_2	Y	Y	Y	Y_3	Y_3	Y	Y	Y	Y	Y	Y	25
Heat	Y	Y_3	Y_3	Y	Y_3	N	Y	Y	Y	Y	N	Y	Y	Y	Y	Y	Y	Y	Y	Y	Y	Y	Y	Y	Y_3	23
Package	Y	Y	Y	Y	Y	Y	Y	Y	Y	Y	Y	Y	Y	Y	Y	Y	Y	Y	Y	Y	Y	Y	Y	Y	Y	25

reduction in economic activity. The small number of N's left in the matrix indicates that our initial upper limits preferences were not very demanding or in conflict. The initial starting-point and the rules for adding constraints did not, however, keep us from arriving at the point at which all 25 grids would accept the package. Had we accepted the original majority, we should clearly have been at a Pareto-inferior solution.

A slightly different set of rules for trading may be devised as follows. Instead of allowing trades to be made between any two grids, ask grids who vote Y on the run A package to try to trade with grids voting N on that package, the trades being, as before, agreements to add constraints to the regional model. This scheme was tried, using the same upper limits and ordinal rankings as before. One difference, inherent in the logic of this approach, is to allow a grid favouring the package to trade issues on which he is indifferent (no numerical subscript on either issue) if the grid opposing the package is made better off. Hence, under this rule a vector $\begin{bmatrix} Y \\ N \end{bmatrix}$ can be traded, whereas in the first set of rules only vectors such as $\begin{bmatrix} Y \\ N_1 \end{bmatrix}$ could be traded.

The use of this scheme brought the *DOD* row up to 24 Y votes, lost 3 Y votes on SO_2, gained the 25th vote on the tax row and lost 3 package votes (22 instead of 25 in favour of solution). The objective function dropped to \$81,532. We chose, for future runs, to use the unrestricted voting algorithm with no trades allowed when the grid was indifferent on the two elements.

To explore the possibilities of the vote-trading algorithm further, the upper limits vectors of each grid were tightened substantially. A further external constraint (analogous to a nationally imposed standard) was placed on water quality, *DOD*. Using these upper limits and our first set of trading rules, the trading sequence (nine iterations) stops well short of unanimity (15 Y package votes). This result is shown in Table 7.11.

Not only is the overall package opposed by 10 grids, but on one issue – taxes – only 8 grids favour the final outcome. It is now possible to test the dominance of the solution arrived at by the trading algorithm with some obvious alternatives. First, we ask the question, can we get a majority on the tax row without violating the constraints already imposed? In rough terms, is Table 7.11 a Pareto-inferior solution? The answer is no, the solution to the regional model goes infeasible when the tax constraint of the disaffected grids is added. Second, we ask, can we then start with a constraint that will guarantee a majority on the tax row, then work our trading algorithm

TABLE 7.11

VOTE MATRIX, TIGHT UPPER LIMITS

Grid:

Issue	1	2	3	4	5	6	7	8	9	10	11	12	13	14	15	16	17	18	19	20	21	22	23	24	25	Tally
DOD	Y	Y	Y	Y	Y	Y	Y	Y	Y	Y	Y	Y_5	Y	Y	Y_3	N	Y	Y_1	N	N	Y	Y_1	Y_1	Y_2	Y	22
ΔT	Y_1	Y	Y	Y	Y	Y_3	Y	Y	Y	Y	Y_3	Y_4	Y	Y	Y	Y_3	Y	Y	Y	Y	Y_2	Y_2	Y	Y_4	Y	25
SP	Y	Y	Y	Y	Y	Y_2	Y_2	Y	Y	Y_2	Y_1	Y_2	Y_1	Y	Y_2	N_2	Y_2	Y	N	N_1	N_3	N	Y	Y	Y	20
SO_2	Y_2	Y_1	Y	Y	Y	Y_1	Y_1	Y	Y	Y	Y_2	Y_1	N_3	Y	Y_1	Y_1	Y_1	Y	Y	Y	Y_1	Y_4	Y	Y	Y_2	24
Tax	Y	Y	N_1	N_2	Y	Y	N_3	N_1	N_2	Y	Y	N_3	N_2	N	Y	N	N_3	N_2	N_2	N_2	N	N_3	N_2	Y_3	N	8
Unemp.	Y_3	Y_2	Y_2	Y_1	Y_1	Y	Y_4	Y	Y_1	Y_1	Y	Y	Y	Y_1	Y_4	Y	Y	Y	Y_1	Y_3	Y	Y	Y_3	Y_1	Y	25
Elec.	Y	Y	Y	Y	Y_2	Y	Y	Y_2	Y_3	Y_3	Y	Y	Y	Y_2	Y	Y	Y	Y_3	Y_3	Y	Y	Y	Y	Y	Y	25
Heat	Y	Y_2	Y_3	Y	Y_3	N	Y	Y	Y	Y	N	Y	Y	Y	Y	Y	Y	Y	Y	Y	Y	Y	Y	Y	Y_3	23
Package	Y	Y	N	N	N	Y	Y	Y	Y	N	N	Y	Y	Y	Y	N	Y	Y	Y	N	Y	N	N	N	Y	15

forward to a new overall majority? This question addresses the issue raised earlier: are we foreclosing some solutions by our starting-point? We found it possible to solve the model with a bare majority on the tax row (13) as a starting-point, but no majority could be found for the solution as a package. Even after all trades were accomplished, the solution garnered only 7 Y votes. We further tested the extent of this infeasibility by raising the upper limits on those bounds responsible for the infeasibility, stopping only when it was clear that no feasible solution was in sight when the original upper limits were slacked off by a factor of 8.

Such tests on particular cases are not sufficient for us to assert that the trading algorithm always finds the dominant solution, but they are at least reassuring in the instant case.

The preceding discussion assumed that a legislature existed which allowed the preferences of all 25 grids to be represented in the decision process. Rarely does such a mechanism exist in a metro-politan area (an exception is in the new Metropolitan Council of Minneapolis–St. Paul, organised from 14 equal population districts). A typical pattern in an American metropolitan area is a council of governments (C.O.G.) in which each municipality is represented. This pattern of preference aggregation can be duplicated in the model, working with the city and four suburbs as follows:

    Suburb A = Grids 5, 10, 15
    Suburb B = Grids 19, 20, 24, 25
    Suburb C = Grids 1, 2, 6, 11
    Suburb D = Grids 16, 21, 22
    City   M = Grids 3, 4, 7, 8, 9, 12, 13, 14, 17, 18, 23

We have only to aggregate preference vectors (assuming the orig-inal slack grid upper limits and ordinal rankings) on some basis (say simple majority) and use the resulting vector* as representative of the municipality. For example, starting from the run A case (Table 7.8) and combining Grids 5, 10 and 15, we have the results shown in Table 7.12.

The process of aggregating by municipality has masked the trading possibilities inherent in the grid voting vectors. However, the C.O.G. aggregation correctly reflects that the unconstrained solution has majority support. We know, however, that the unconstrained case is not a dominant solution. Moreover, C.O.G.s must often operate on a consensus basis. The C.O.G. staff could point out, on the basis of the air pollution pattern, that putting an SO_2 upper limit on Grid 13 of,

---

* These vectors are composites of the grid vectors making up the municipality, constructed by majority vote but allowing for vote-trades within each jurisdic-tion. No trades exist in this example.

TABLE 7.12

CITY-SUBURBAN VOTE MATRIX BASED ON RUN A

Municipality

Issue	*a*	*b*	*c*	*d*	*m*	*Tally*
DOD	Y	N	N	N	Y	2
ΔT	Y	N	N_3	N_2	N	1
SP	Y_2	Y_3	Y_2	Y_3	Y_4	5
SO_2	N	Y	N_1	N_1	N_3	1
Tax	Y_1	Y_2	Y	Y_4	Y_2	5
Unemp.	Y_3	Y_1	Y_4	Y	Y_1	5
Elec.	Y	Y	Y	Y	Y_5	5
Heat	Y	Y	Y_5	Y	Y	5
Package	Y	Y	N	N	Y	3

say, 80 $\mu g/m^3$ would probably result in making more grids satisfied with the SO_2 level. Likewise a limit of $\Delta T \leq 4°F$ in river reaches 2 and 4 could be expected to improve the heat situation in the river.

In the absence of trades, the C.O.G. (like the authors) might well take this advice.

This result, presented again by municipality, results in a voting matrix as shown in Table 7.13.

We now have four municipal jurisdictions in favour of the package (although on a grid-by-grid count five grids are still opposed). If a two-thirds majority were the C.O.G. decision rule, this would be the final solution (there again being no trades possible) – a solution Pareto-inferior to the grid trading solution to the same problem (Table 7.10). However, we still do not have unanimity; therefore we continue to add constraints, on the advice of staff, to ease the pain in suburb C, the only N package vote remaining. The three easiest-to-meet SO_2 limits in suburb C were then added to the regional model. These result in adding 3 more Y grids (total 22), and when the grid voting sections are aggregated again by municipality, they reveal that

TABLE 7.13

CITY-SUBURBAN VOTE MATRIX BASED ON RUN A, WITHOUT TRADES

Municipality

Issue	*a*	*b*	*c*	*d*	*m*	*Tally*
DOD	Y	N	N	N	Y	2
ΔT	Y	Y	Y_3	Y_2	Y	5
SP	Y_2	Y_3	Y_2	Y_3	Y_4	5
SO_2	Y_3	Y	N_1	Y_1	Y_2	4
Tax	Y	Y_2	Y	Y	Y_3	5
Unemp.	Y_1	Y_1	Y	Y	Y_1	5
Elec.	Y	Y	Y	Y	Y	5
Heat	Y	Y	Y	Y	Y	5
Package	Y	Y	N	Y	Y	4

we have achieved a unanimous Y vote. All five municipalities favour the package, although only 11 grids favour the DOD level and 3 grids disapprove of the package. This solution is also Pareto-inferior to the Table 7.10 solution. The three grids that disapprove of the solution are, unfortunately for them, each in a different municipality.

If we try a C.O.G. aggregation with our tighter upper limits vectors, we find it impossible to move at all. This results from being confronted with the initial voting matrix shown in Table 7.14.

TABLE 7.14

Municipality

Issue	a	b	c	d	m	Tally
DOD	Y	N	Y	Y	Y	4
ΔT	Y	Y	N_3	N_2	N	1
SP	Y_2	Y	Y_2	N_3	N	3
SO_2	Y	Y	N_1	N_1	N_3	2
Tax	Y	Y_2	Y	Y	N_1	4
Unemp.	N_1	N_1	N	Y	N_2	1
Elec.	Y_3	Y	Y	Y	Y	5
Heat	Y	Y	Y	Y	Y	5
Package	N	N	N	N	N	0

Here, to convert municipalities from voting N to Y, we have either to treat the N_1's and N_2's of each jurisdiction, which is infeasible (no solution of the regional model is possible with these upper limits as constraints), or we must be told which majority's limits we should be concerned with, thus violating our non-biased approach. Yet we know, by the vote-trading algorithm, that a majority solution exists (Table 7.11) which also converts a majority of the jurisdictions to Y.

It is possible, of course, to choose preference vectors in the grids which can be aggregated to have the same solution by municipality as by grid trades. One need do one of two things: either make sure grid trades occur within a jurisdiction, or that the desire for such trades represents a majority opinion in two jurisdictions. Thus, either heterogeneous communities within which trades can take place, or communities homogeneous within but heterogeneous as a metropolitan whole, could have trading potential. The case we investigated possessed, however, neither characteristic to a sufficient degree to trade itself to a Pareto-efficient solution through a C.O.G. mechanism.

It may be useful to compare the solutions of the regional model under our various assumptions. Table 7.15 shows the results of run A with the results of three of our political solutions. The shifts in the distribution of costs as different environmental constraints are imposed are noteworthy. So too is the extent to which the C.O.G.

TABLE 7.15

POLITICAL SOLUTIONS, DIDACTIC MODEL

	Unconstrained case (*run A*)	*Legislative trades slack constraints* (*Table 7.10*)	*Legislative trades tight constraints* (*Table 7.11*)	*C.O.G. solution slack constraints* (*Table 7.13*)
Objective function	$89,250	$85,608	$77,481	$87,960
Δ obj. fn. (cost) from run A	—	$3,642	$11,769	$1,290
Highest river temperature	77·9°F (Reach 4)	72°F (Reach 4)	72°F (Reach 4)	72°F (Reach 4)
Lowest D.O.	1·73 p.p.m. (Reach 6)	3·43 p.p.m. (Reach 6)	5·48 p.p.m. (Reach 6)	2·07 p.p.m. (Reach 6)
Maximum SO_2 concentrate	103 μg/m³ (Grid 13)	80 μg/m³ (Grid 13)	75 μg/m³ (Grid 13)	80 μg/m³ (Grid 13)
Max. susp. particulates	63·4 μg/m³ (Grid 10	34 μg/m³ (Grid 12)	32 μg/m³ (Grid 12)	63 μg/m³ (Grid 10)
Increase in costs of residuals disposal — City A	—	0·01%	12·8%	—
B	—	—	—	—
C	—	—	17·6%	—
D	—	—	—	—
E	—	10%	18·9%	—
Increase in home-heating costs* — City A	—	$10.38	$8.63	$9.59
B	—	—	—	—
C	—	$14.65	$14.65	$14.65
D	—	$15.44	$15.44	$8.81
E	—	$21.12	$21.12	$14.83
Increase in elec. cost	—	2·7%	4%	2·6%
Sugar beet plant op. level (% capacity)	100%	90%	90%	100%

Selected Discharge Information

B.O.D.	Sugar refinery	33,420 lb./day	20,040 lb./day	964 lb./day	33,420 lb./day
	Petroleum refinery	4,030 lb./day	1,820 lb./day	395 lb./day	4,030 lb. day
	Municipal plant	15,400 lb./day	15,400 lb./day	11,882 lb./day	15,400 lb./day
Heat	Sugar refinery	2,380 10^6 B.T.U./day	266 10^6 B.T.U./day	0	892 10^6 B.T.U./day
	Petroleum refinery	16,580 10^6 B.T.U./day	0	0	
	Electric power plant	17,160 10^6 B.T.U./day	11,698 10^6 B.T.U./day	11,869 10^6 B.T.U./day	11,295 10^6 B.T.U./day
So_2	Sugar refinery	6·86 tons/day	6·79 tons/day	6·79 tons/day	6·86 tons/day
	Petroleum refinery	60 tons/day	54 tons/day	48·6 tons/day	52·4 tons/day
	Electric power plant	28·8 tons/day	28·8 tons/day	28·8 tons/day	28·8 tons/day
Part.	Sugar refinery	22 tons/day	4·2 tons/day	7·58 tons/day	22 tons/day
	Petroleum refinery	4·80 tons/day	4·63 tons/day	4·48 tons/day	4·7 tons/day
	Electric power plant	88·6 tons/day	77·4 tons/day	24·5 tons/day	89 tons/day
	Municipal sewage plant	4·6 tons/day	4·6 tons/day	5·08 tons/day	4·6 tons/day
	Municipal incinerator	2·0 tons/day	2·0 tons/day	1·5 tons/day	2·0 tons/day
Solids to incinerator		101 tons/day	102 tons/day	77 tons/day	101 tons/day
Solids to landfill		783 cu. yds/day	869 cu. yds/day	1,207 cu. yds/day	783 cu. yds/day

* Dollars per household per year.

solution (column 4), while it does not produce the same quality environment as is reflected in column 2, manages to keep costs to industry, governments and individuals down. Another contrast can be seen between column 3 of this table and run B (Table 7.6), two runs having roughly similar objective function values. Run B, having to meet no constraints on increases in the city's costs of residuals disposal (which go up 46·3 per cent), handles residuals in a very different way than does the solution in column 3 of Table 7.15, a solution which had to be approved by a majority of the 25 grids (city costs up only 12·8 per cent).

We also note here that should we delete one or more of the issues (not allowing our mythical C.O.G. or legislature to concern itself with the issue), a very different result would occur. Most noticeably, if decisions about environmental quality (our four quality measures) were to be made apart from decisions of what costs to bear, then the group deciding environmental quality will opt for higher levels than it would if the costs (and their distribution) were faced explicitly.

Not presented in detail is the fact that, in our example, were the increases in costs of handling solid residuals shared equally throughout the metropolitan region, rather than charged directly to the jurisdiction affected, then a different set of trades would have occurred in all runs. Using the regional model to experiment with different tax-sharing devices, with given preferences, may reveal ways whereby the cost of achieving a desired level of environmental quality will be acceptable.

Use of the Regional Model

There is no reason to believe that any existing legislature or council would use the trading routine we outlined in the preceding section. As we said earlier, there is no necessity for them to do so since the distributional information may be adapted to a variety of decision paths.

Let us take some examples. In real councils there is often a desire, sometimes even a necessity, to let everybody win. A chairman might ask each member to write down one constraint that the member really wants, or needs, in his district or area. These constraints could be collected and used as the first set of constraints for the model. If the regional model can be solved with these constraints, the effects of the solution could again be examined and one additional constraint added by each member. This process could be continued until the regional model fails to solve. The last solution could be adopted or used as a starting solution to trades or bargains.

Bargaining could occur either after the process described in the preceding paragraph was completed or *ab initio*. In general terms, bargaining can occur whenever two or more council members perceive that some constraints are in direct conflict. In such instances, real council members may wish to bargain, that is, to agree to slack off their constraints slightly if the other side will slack off its constraints. In doing so, they may wish to use the regional model to help them find the most agreeable bargain – one with minimum change in both constraints, one with minimum variation in percentage rise between the two, or whatever other criterion is agreed upon by the pair of bargainers.

It will quickly be seen that such constraint slackening will probably come late in the game, since no one will be willing to change his upper limits until it is clear that they either cannot be met as they stand, or that a majority favours a solution in which one's upper limits are greatly exceeded.

The actual use of the regional model, therefore, may vary considerably from one group to another. Some strictures on use are, however, apparent. For example, the distributional information has potentially explosive political content. That fact can be used to advantage for good or ill. Suppressing it may be politically expedient to some but can rarely, if ever, be either ethically or legally justified when taxpayers' moneys are being used. It may not be desirable, however, to have all experimental runs made public so long as they are shared by all the council members. (The potential for mischief by a clique within the council or legislature, a committee for instance, would be very large. Such a group could construct a very biased solution.)

Another stricture on use relates to the choice of a method to proceed once unanimity has broken down, as it inevitably will early on in any real situation. One can imagine, in a strong party legislature or council, that the majority party might well wish to give priority to meeting the constraints of its members, with minority preferences being met later if at all. This method is not so bad as it might appear, so long as the distributional impact of their deliberations is known to all council members, *and* the minority party has equal access to the regional model to design counter-solutions. These latter solutions will, of course, attempt to meet minority preferences *plus* improving the lot of a sufficient number of the majority members to place the majority solution in jeopardy. That process will, in practice, turn out not too differently from our second set of trading rules.

In more non-partisan situations the idea of coalition building (as in our first set of rules) may have appeal as an equalitarian procedure which may be able to reach a dominant solution in the absence of cycling.

In any event there will be some pressure on the operators of the model, who must be like Caesar's wife. Sometimes both majority and minority party staff will be needed to ensure ease of access, shared results and general trust. There will be opportunities for technicians to facilitate a solution, or to obstruct one by purely technical means which go unnoticed by the members of their staffs. These opportunities and dangers are present in many public service posts and cannot be completely eliminated. Just as both sides in disputes have for centuries employed lawyers, they now must employ programmers, economists and systems analysts if the technician's temptation to play God (always for the public good, of course) is to be minimised.

REFERENCES

[1] Black, R. J., *et al.*, *The National Solid Wastes Survey: An Interim Report*, presented at the 1968 Annual Meeting of the Institute for Solid Wastes of the American Public Works Association (Miami Beach, Oct 1968).

[2] Brown, G. W., 'Predicting Temperatures of Small Streams', *Water Resources Res.*, vol. 5, no. 1 (1969) p. 68.

[3] Burt, W. V., 'Heat Budget Terms for Middle Snake River Reservoir', in 'Water Temperature Studies on the Snake River', *U.S. Fish and Wildlife Service Technical Report*, no. 6 (1958).

[4] Chen, C. W., 'Concepts and Utilities of Ecologic Models', *J. Sanitary Eng. Div., Proc. A.S.C.E.*, vol. 96, SA5, Proc. Paper 7602 (Oct 1970) pp. 1085–1097.

[5] Combustion Engineering, Inc., *Technical Economic Study of Solid Waste Disposal Needs and Practices*, IV (U.S.D.H.E.W., Bureau of Solid Waste Management, Rockville, Md., 1969).

[6] Cootner, P., and Löf, G. O. G., *Water Demand for Steam-Electric Generation* (Resources for the Future, Washington, D.C., 1965).

[7] Dales, J. H., *Pollution, Property and Prices* (Univ. of Toronto Press, 1968).

[8] Delaware River Basin Commission, *Final Progress Report: Delaware Estuary and Bay Water Quality Sampling and Mathematical Modeling Project* (May 1970).

[9] Delay, W. H., and Seaders, J., 'Predicting Temperatures in Rivers and Reservoirs', *J. Sanitary Eng. Div., Proc. A.S.C.E.*, vol. 92 (1966) pp. 115–34.

[10] Delson, J. K., and Frankel, R., 'Residuals Management in the Coal-Energy Industry', unpublished MS (Resources for the Future, Inc., Aug 1972).

[11] DiToro, D. M., O'Connor, D. J., and Thomann, R. V., 'A Dynamic Model of the Phytoplankton Population in the Sacramento–San Joaquin Delta', in *Advances in Chemistry Series*, no. 106, 'Nonequilibrium Systems in Natural Water Chemistry' (American Chemical Society, 1971).

[12] Dorfman, R., and Jacoby, H., 'A Model of Public Decisions Illustrated by a Water Pollution Policy Problem', in *The Analysis and Evaluation of Public Expenditures: The P.P.B. System* (Joint Economic Committee, Washington, U.S.G.P.O., 1969) I, pp. 226–74.

[13] ——, ——, and Thomas, H. A. (eds.), *Models for Managing Regional Water Quality* (Harvard Univ. Press, Cambridge, Mass., 1973).

[14] Duttweiler, D. W., *et al.*, 'Heat Dissipation in Flowing Streams', *Report of the 1960–1961 Advanced Seminar* (Johns Hopkins Press, Baltimore, 1962).

[15] *Environmental News* (E.P.A., Washington, D.C.) 30 Jan 1971.

[16] Fiacco, F. A., and McCormick, G. P., *Nonlinear Programming: Sequential Unconstrained Minimisation Techniques* (Wiley, New York, 1968).

[17] Haefele, E. T., 'Coalitions, Minority Representation, and Vote-Trading Probabilities', *Public Choice*, vol. 8 (spring 1970) pp. 75–90.

[18] ——, 'A Utility Theory of Representative Government', *Amer. Econ. Rev.*, vol. 61, no. 3 (June 1971) pp. 350–67.

[19] ——, 'Social Choice in Residuals Management: Politicising an Economic Model', *Proc. Amer. Pol. Sci. Assoc.* (Chicago, Sep 1971).

[20] ——, 'Environmental Quality as a Problem of Social Choice', in Kneese, A. V., and Bower, B. T. (eds.), *Environmental Quality Analysis: Theory and Method in the Social Sciences* (Johns Hopkins Press, Baltimore, 1972).

[21] Holzworth, G. C., 'Estimates of Mean Maximum Mixing Depths in the Contiguous United States', *Monthly Weather Review*, vol. 92, no. 5 (May 1964) pp. 235–42.

[22] ——, *Mixing Heights, Wind Speeds, and Potential for Urban Air Pollution throughout the Contiguous United States*, Publ. No. AP-101 E.P.A (Office of Air Programs, Research Triangle Park, N.C., Jan 1972 (also available from U.S. Government Printing Office, Washington, D.C. 20402).

[23] Johnson, E., 'A Study in the Economics of Water Quality Management', *Water Resources Res.*, vol. 3, no. 2 (1967) pp. 291–305.

[24] Kelly, R. A., 'Conceptual Ecosystem Model of the Delaware Estuary' (mimeo, Resources for the Future, Washington, D.C., 1972).

[25] Kneese, A. V., Ayres, R. U., and D'Arge, R. C., *Economics and the Environment* (Resources for the Future, Washington, D.C., 1970).

[26] Kohn, R., 'Linear Programming Model for Air Pollution Control: A Pilot Study of the St. Louis Airshed', Paper 69–64, 62nd Annual Meeting, Air Pollution Control Association (New York City, June 1969).

[27] Löf, G. O. G., and Kneese, A. V., *The Economics of Water Utilisation in the Beet Sugar Industry* (Resources for the Future, Washington, D.C., 1968).

[28] Madison, James, *The Federalist Papers* (1787).

[29] Martin, D. O., and Tikvart, J. A., 'A General Atmospheric Diffusion Model for Estimating the Effects on Air Quality of One or More Sources', *A.P.C.A. Journal* (June 1968) pp. 68–148.

[30] *New York Times*, 22 Jan 1972, p. 27; 27 Jan 1972, p. 18.

[31] Ozolins, G., and Smith, R., 'A Rapid Survey Technique for Estimating Community Air Pollution Emissions', P.H.S. Publ. No. 999-AP-29 (National Air Pollution Control Administration, Raleigh, N.C., Oct 1958) App. C, pp. 43–5.

[32] Pasquill, F., *Atmospheric Diffusion* (Van Nostrand, London, 1962).

[33] Russell, C. S., 'Models for the Investigation of Industrial Response to Residuals Management Action', *Swedish J. Econ.*, vol. 73, no. 1 (1971) pp. 134–56.

[34] ——, and Spofford, W. O., Jr., 'A Quantitative Framework for Residuals Management Decisions', in Kneese, A. V., and Bower, B. T. (eds.), *Environmental Quality Analysis: Theory and Method in the Social Sciences* (Johns Hopkins Press, Baltimore, 1972) chap. 4.

[35] ——, 'Models of Response to Residuals Management Action: A Case Study of Petroleum Refining', to be published by Resources for the Future.

[36] ——, 'Municipal Evaluation of Regional Water Management Proposals', in Dorfman, R., Jacoby, H., and Thomas, H. A. (eds.), *Models for Managing Regional Water Quality* (Harvard Univ. Press, Cambridge, Mass., 1973) chap. 4.

[37] Schaumberg, G., *Water Pollution Control in the Delaware Estuary* (Harvard Water Program, Cambridge, Mass., 1967).

[38] Smith, R., 'Costs of Conventional and Advanced Treatment of Waste Water', *J. Water Pollution Control Fed.* (Sept 1968) pp. 1546–74.

[39] Streeter, H. W., and Phelps, E. B., 'A Study of the Pollution and Natural Purification of the Ohio River', *Public Health Service Bulletin*, no. 146 (Washington, D.C., 1923).

[40] T.R.W., Inc., *Air Quality Implementation Planning Program* (Environmental Protection Agency, Nov 1970) I and II (also available from National Technical Information Service, Springfield, Va. 22151, PB 198 299 and PB 198 300 respectively).

[41] U.S. Department of Commerce, *Climatic Atlas of the United States* (U.S.G.P.O., Washington, D.C., June 1968).

[42] U.S. Environmental Protection Agency, Office of Air Programs, *Compilation of Air Pollution Emission Factors* (Research Triangle Park, N.C., Feb 1972) pp. 1–4 to 1–7.

[43] Wright, W. A., 'Alternative Approaches to Water Quality Management in the Trent River Basin', unpublished dissertation (University College, Oxford) 21 Mar 1972.

Discussion of the Paper by Clifford S. Russell et al.

Formal Discussant: Prud'homme. The first model presented in this paper, which I shall call the economic model, has a number of interesting features which are rarely found in similar models:

(i) it incorporates relationships between different kinds of residuals (air pollution, water pollution, solid waste disposal);
(ii) it yields data at a fine geographical level (grid);
(iii) it incorporates substitution possibilities.

On the whole, however, this model is not very different from the Delaware and Potomac models (which it refers to), and I shall make only two points.

My first point relates to the use and meaning of linear programming. Minimising an objective function is a legitimate procedure for a normative model like the Mills model [1]. But this is not a normative model; it is a descriptive one. It does not tell us what *should* be done, but what *can* be done, so that politicians can make choices. As such, it may be quite misleading. The price tags (taxes, unemployment) attached to the quality contraints are not the likely prices, but the lowest possible prices. Are we sure that the two sets of prices are equal, or, if you prefer, that the economy of Didactica is efficient?

This matter is of considerable importance to a politician (and even for a citizen), since a tax increase of *about 10 per cent* is quite different from an increase of *at least 10 per cent*. Beckerman touched upon that problem in his discussion of the Kneese paper (Chapter 3). He raised the objection, but quickly dismissed it, asserting that firms do behave as we say they do. Two observations are relevant. The first is that the efficient state is only arrived at when all the relevant information is available. And we have all agreed that the costs of obtaining information are high.

The second is that the efficient state is only arrived at in the long run, and that does not help us very much. We want to know, and to predict in the short run, what is likely to happen when a given set of constraints are applied. Politicians are not re-elected in the long run.

Let me mention a case in point. In France, as in a number of other countries, unemployment in depressed areas is tackled by means of investment subsidies. This does not seem to be very logical, since it lowers the cost of capital relative to that of labour in the areas where we want labour-intensive developments to take place. I therefore asked some students to look at this problem to see whether investments in subsidised-investment areas really had a capital–labour ratio higher than investments in non-subsidised areas (making due allowance for wage differentials). The answer seems to be that they are not. Businessmen apparently build exactly the same plants in both kinds of areas. It is true that there are serious data problems involved, and that these findings can only be considered as indications rather than as evidence. But they are indications of the fragility of our assumption.

In short, and this is my first point, the predictive value of a linear programming model seems open to discussion.

My second point is about the usefulness of the model. As it stands, the model is not a theoretical Uzawa-type model (Chapter 1) that yields theoretical, but significant, insights; nor is it an applied (Delaware) type of model that yields limited but relevant information. It is in between. It is a hypothetical example. And I am not sure that there is room for that kind of model. I do not really know. But I feel that to be useful it should develop into either a theoretical or an applied model. The trouble is that both alternatives appear to be equally difficult.

To make a theoretical model out of it, one would have to simplify to such an extent that all the interesting and unique features of the model would probably be lost in the process. I should appreciate it if people like Uzawa, or Kolm, or of course Sir John Hicks, who have demonstrated such a great talent at theoretical model-building, would care to comment on this point.

To make an applied model out of it seems to be a more natural development. But the amount of information and calculation it would require is so enormous that it does not seem to be a feasible undertaking. There again I may hopefully be wrong, and I should be glad to hear more about planned applications of the model to real-life situations.

In short, my second point is that the authors have struck a very nice balance between the conflicting goals of simplicity and applicability, capturing enough elements to make the model relevant while at the same time remaining simple enough to make it manageable, but that such a combination may unfortunately not be very fruitful.

The second model presented in the paper is a model of social choice. Everybody will realise how interesting and original it is. It is furthermore quite general and does not only apply to environmental problems. It raises a number of questions:

(i) It is a model of social choice for groups of individuals. In the paper it is applied to a legislature in which one man represents each grid. An attractive feature of this application is that votes and vote-trading do take place in legislatures. On the other hand, it assumes that grids are homogeneous, and that all the inhabitants of a grid feel the same way about economic and environmental problems. I wonder whether non-spatial groupings (e.g. income, age, race, political persuasion, etc.) might not be more useful.

(ii) The model does not get rid of the free-rider problem, or does it? The grids seem to have no interest in revealing their preferences, since this only makes it more difficult for them to find trading partners. But I am not clear on that point, and I am not sure that they stand to lose something by revealing their preferences.

(iii) The model is not a satisfaction-maximising device; it is a dissatisfaction-minimising one. The 'best' solution is the one that will 'satisfy' the largest number of grids, not the one that will most satisfy grids. The old distinction between efficiency and distribution is thrown out. The model does not even try to maximise efficiency under distributional constraints. It does something different. This might be exactly the kind of model we need; but we may wish to think twice about it.

(iv) The role of the vote-trading concept is not entirely clear. Is it trying to describe what actually takes place in a legislature? In this case it is a hypothesis, quite an interesting one, but one which requires testing.

Or does it seek to prescribe what should take place in a legislature? In this case it should help us to find better institutional arrangements that will facilitate vote-trading.

As we can see, this paper raises more questions than it provides answers; and this is certainly what a good paper ought to be about.

Rothenberg argued that the 'free-rider problem' in vote-trading was not nearly so great as in many other problems of public choice. This was because vote-trading could be multilateral and there was a basic symmetry in the interests of every potential trader to be part of a multilateral trading arrangement. This happened because his presence made a difference to the nature of the trade being carried out by other members and, since everybody's trade in a legislature was intrinsically blessed with the generation of external effects upon everybody else in the legislature, everybody could ensure a trade more appropriate to his preferences only if he was willing to be part of the agreement. It was like an *n*-person strategic problem in which no person had intrinsically more bargaining power than anyone else and must thus be willing to give up in order to be able to gain.

It clearly did make a difference in these games in which order the bargains were arranged.

Bolwig wondered how realistic these vote-trading models were. He suspected that low-income groups did not count sufficiently in local government affairs because they did not vote as regularly as middle-income groups. The most important bias seemed to be that the leaders of public opinion, mainly representing the views of middle- and higher-income groups, only spoke to a small proportion of the population. Any vote-trading model should thus recognise this bias and allow for its effect on any simulated policy choices. The net effect was clearly a function of the degree of centralism/federalism present in the voting system and the degree of central government interference in the local decision process.

Foster wondered whether the authors' second model was quite what it appeared. If it was a realistic vote-trading model it would have to recognise the effects of coalitions and the sequence of bargains. On reflection, the value of the model could be enhanced if it was viewed in another administrative context. For example, imagine that there is a conscientious administrator weighing the interests of the different grid squares. He is not merely interested in an acceptable solution. He would prefer a more acceptable one defined as one in which, through trade, every grid was able to increase its utility. He thus examines the possible tradeoffs and finds such a solution, without simulating the behaviour of coalitions. By definition, this will be a superior solution.

However, one should be able to reach the same solution by cost–benefit analysis. If each preference vector was transferred into a preference scale ranking elements on the basis of ability of pay, the administrator could also arrange trading so as to reach a superior solution. He could also simulate the principle of one grid one vote by constraining the total value revealed by each grid to unity, so that each grid had equal weight. This would imply, for example, that if grid M felt more strongly than grid N about the choice, as evidenced by higher willingness to pay, or had a larger population or higher income per head, this would not be allowed to influence the outcome.

If this is a correct reconciliation of the second model with cost–benefit analysis, then the second model does not simulate the behaviour of a legislature, but represents a hypothesis about the weight to be given to areas in coming to a recommendation. The question is therefore why the peculiar simplifying assumptions governing the expressions of preference in the model should be preferred. If preference can be expressed in money terms, or even as a scalar, it would be both computationally easier and probably more intelligible to use money valuations as the basis for decisions.

Thoss, referring to the objection that linear programming was a normative device, argued that this did not limit the value of applying it in this situation. After all, you could model all the predictable and descriptive parts of the model in its constraints. The model thus dealt with inequalities and sought to find those that were binding. The optimising process thus simply identified the binding constraints.

Tulkens noted that (i) the paper did not mention whether the vote-trading algorithm proposed did lead to a Pareto optimum (although certain outcomes were called 'Pareto superior' to others). Could the authors characterise their solution with respect to that criterion? (ii) Since the algorithm was applied to the choice of public goods, it might be interesting to compare its properties and performance with other procedures developed recently for the optimal allocation of such goods, namely by Drèze and de la Vallée Poussin [2] and by Malinvaud [3]. In particular, the game-theoretic problem of free-rider behaviour might have a quite different impact under each procedure.

Kolm asked why we used votes instead of a money valuation. He suggested that one reason was because of the distributional effects of environmental policies. One man one vote implied a higher weighting to the lower-income groups. This was often done in relation to particular policies designed to achieve specific distributional objectives, and this was why vote-trading was often forbidden or thought to be immoral.

Uzawa observed that the problems of 'free riders' were usually mentioned in the context of public goods. However, this problem also applied to the standard exchange of private goods.

Russell, Spofford and Haefele replied that the comments by Prud'homme captured many of their own concerns about the models. On a couple of points, however, they felt that there was a failure to communicate and they wished to correct this.

It was not true that the inhabitants of any one grid were assumed to have homogeneous tastes. Their legislators did 'represent' their grids in roughly the same way as real legislators represented districts, taking account of intense minorities and so forth [4].

On a related point, they agreed with Rothenberg on the nature of vote-trading. Examples could be constructed to show that the sequence of trades was critical and such cases might well be important in the real world.

The ordinal rankings could not be translated into the scalar measure proposed by Foster except by vitiating the whole exercise. Certainly the process of exogeneously giving weights to grids could be done, just as prices on goods could be assigned by someone on some kind of 'evidence' about how strongly people wanted different things. But markets and competitive elections had both theoretical and practical advantages over such administered devices.

Finally, their vote-trading model traded *among* Pareto-optimal states, since the use of the linear programming model guaranteed that no Pareto-inefficient states were considered.

REFERENCES

[1] Mills, E., 'Sensitivity Analysis of Congestion and Structure in an Efficient Urban Area', in Rothenberg, J., and Heggie, I. G. (eds.), *Transport and the Urban Environment* (Macmillan, London, 1974).
[2] Drèze, J., and de la Vallée Poussin, D., 'A Tâtonnement Process for Public Goods', *Rev. Econ. Studies*, vol. 37 (1971) pp. 133–50.
[3] Malinvaud, E., 'A Planning Approach to the Public Good Problem', *Swedish J. Econ.*, vol. 73 (1971) pp. 96–112.
[4] Haefele, E. T., 'A Utility Theory of Representative Government', *Amer. Econ. Rev.*, vol. 61, no. 3 (1971) pp. 350–67.

Part III

Evaluation and Consolidation Panel

8 Urbanisation and Environment: Retrospective and Prospective Views

Ian G. Heggie, Henry Tulkens, Rainer
Thoss, Karl-Göran Mäler and
Edwin S. Mills

A Retrospective Summary: Ian G. Heggie

The main theme of this conference was 'Urbanisation and Environment' and it followed close on the heels of the U.N. Conference on the Environment held in Stockholm. However, it would be misleading to pretend that the collected papers presented at this conference represented an organised attempt by economists to become involved in the present controversy surrounding the relationship between man and the environment. Nor should they be interpreted as a definitive statement of the role the economist feels he can, and should, play in managing the environment. Rather, it must be interpreted as a series of separate statements linked by the common themes of Urbanisation and Environment.

The papers presented at the conference can be divided into four main groups:

(1) The macro-economic approach to understanding urban and regional structure.
(2) The macro-economic approach to managing water resources.
(3) The role of transport in the urban environment.
(4) The management of the environment by means of charges and regulations.

Although each topic represents a separate field of study, all are linked by their emphasis on the environment. This immediately raises one of the recurrent themes of the conference: what is the role of the economist in planning man's environment? In simpler terms, what can economists, as a profession, contribute towards the achievement of an optimum urban and regional environment: the term 'optimum' being interpreted in terms of human satisfaction. The answer is partly given in the papers we have considered. So many of the issues discussed at this conference have either required, or have pointed to, the need for a multi-disciplinary approach. The discussions on urban structure, on transportation, and on water resources relied, to an important extent, on material contributed by physical planners, psychologists, sociologists, natural scientists and engineers. It was likewise apparent that much prospective work in these areas – I have in mind the references to the Streeter–Phelps models – can only be effectively prosecuted in a multi-disciplinary context. As economists, we should therefore ponder what role we can best play in initiating such research.

The discussion following each of the 15 papers presented at this conference generated several recurring topics from which the panel selected eleven for inclusion in this summary. The mode of presentation will be for me to summarise briefly each topic after which Tulkens, Thoss, Mäler and Mills will sequentially introduce a few brief comments on the further research they think is required in each area. The first three discussants will introduce three topics each; Mills will introduce the remainder. After introduction the topic will be thrown open for general discussion.

1. *Preferences*. The first recurrent topic was that of preferences. It is clearly desirable to have a working knowledge of people's preferences to provide the basis for optimum decisions. However, we are not only interested in the displayed preferences of people and firms explaining why and how they presently behave; we must know enough about the fundamental determinants of behaviour to frame a set of incentives capable of achieving stated policy goals. It is not enough simply to explain why people migrate; if the optimum solution – including externalities – requires that they should not migrate then how do we persuade them not to?

2. *Distribution*. The second recurrent issue was that of distribution. This is clearly important in all studies of the environment and has to consider the regional distribution of environmental variables as well as the distribution of costs and benefits between individuals and its distribution over time. The central question is, who is doing what, to whom and by how much? It is also important to distinguish between the distributional effects of policies affecting existing, as opposed to potential, firms and individuals. In other words 'sunk costs' should feature in all our distributional calculations.

3. *Objective Functions*. At first glance this seemed a fairly simple issue and was initially discussed in terms of efficiency versus distribution. This clear distinction disappeared, however, once further complications were introduced. It was suggested that 'policy statements' should be clearly distinguished from 'policy intentions': in some countries policy statements were issued, and not implemented, to achieve maximum political impact; in others they were explicitly not stated to avoid the divisive effects that such statements frequently

induced. Another suggestion referred to the objective of preserving a unique environment. Although this was light-heartedly characterised as simply another way of providing 'outdoor relief' the objective does gain credibility when related to the preservation of unique cultures in developing countries (e.g. the Masai) or to the preservation of the historic centres of so many European cities.

4. *Environmental Effects.* A recurrent theme under this heading was 'what is the environment?' and 'how can it be measured?' It was pointed out that the concept of environment cannot stop short at the statement of residuals. Environment is a question of quality as well as of sewerage, and one must avoid allowing a clinical definition to result in cultural suffocation. At a more technical level there was the question of measurement: D.O.* concentrations cannot readily be translated into degrees of satisfaction. The question of reversible versus irreversible effects was also raised; as was that of the durability of residuals and the possible catastrophic breakdown of systems. The malleability of input–output coefficients was also raised and this tied in with the questions of the sequence in which decisions were taken, the durability of capital assets and their adaptability over time.

5. *Micro versus Macro Models.* There was some genuine conflict on this subject. The micro modellers argued that macro models remain abstract, and while such models have their utility and aesthetic appeal, it is only infrequently that they can be used to solve genuine decision problems. The plea was made that the economic modeller should try to capture in his model as much of the real world as possible. Input–output models came in for considerable criticism because they are usually estimated by sector and thus dismiss the problem of space which is so important in most residual studies. In return it was argued that different types of models serve different purposes and part of the skill of the research worker was related to choosing the best model to suit the problem being analysed. International trade models and models simulating the effect of the new auto anti-pollution legislation were cited as examples supporting the usefulness of the macro approach.

6. *Growth Models and the Environment.* In spite of its importance, this topic generated very little discussion. The problem with most macro-economic growth models is that they do not accurately replicate the growth process over extended periods of time because they do not systematically allow for the effect of technological innovation in their coefficients. Very few incorporate a finite set of

---

* D.O. = Dissolved oxygen content and is a measure of a certain type of water pollution.

natural resources and they likewise do not allow for the cumulative effect of residuals on the environment. Because this topic has only received limited attention at the conference, I hope that participants will feel free to contribute further ideas in the discussion this morning.

7. *Pollution Controls*. This resulted in a very wide-ranging discussion related to the choice between charges and/or subsidies versus regulations and bribes versus charges. It was difficult to disentangle the questions of efficiency from those of effects. The degree of generalisation possible was also limited. Effluent charges in a river basin appeared to be efficient and easy to administer; charges for auto emissions, though possibly efficient, appeared extremely difficult to administer. Even parking charges – in Paris at least – seemed likely to have a problematic effect. There is likewise no guarantee that congestion charges in urban areas will necessarily lead to everyone being better off. It was suggested that if pollution costs were included in prices then such charges would become regressive. It was also doubted whether congestion charges would have much effect on city size. It might lead to a reallocation of resources between cities of different size, but may result in little decentralisation whilst at the same time fortuitously increasing the rental income of Central Area landlords. The dual problems of efficiency versus effects thus loomed large in the discussions.

8. *Compensation*. This topic produced a great deal of discussion. The central issue was, who is entitled to compensation and what implied rights do, and should, the public enjoy? Everyone is affected by some aspects of public policy, but why do we intuitively feel that people are entitled to be compensated for some effects and not for others? A great deal seems to hinge on 'reasonable expectations' and hence on the degree and type of uncertainty associated with the loss (or gain) of amenity. A further point, which did not clearly emerge from the discussion, was the relationship between the rights to compensation and Hicks' equivalent and compensating surplus. Amenity values generally constitute a large part of a person's real income. When he suffers an amenity nuisance, or when he gains added amenity, the effects are therefore often non-marginal and the difference between the compensating and equivalent surpluses can no longer be ignored.

9. *Economists and Public Policy*. This topic was less well discussed. There were several references to what economists thought policy-makers should do, but none to the specific role of the economist in the decision-making process and to the mode of communication most appropriate to our status and profession. Attempts to change

the world will never succeed as long as they are directed at the wrong people, in the wrong place and in the wrong way. If we wish to influence policies and if we wish to improve the urban and engineering planning process, it is essential that we direct our ideas, in a fluent and coherent form, to the most responsive parts of the relevant decision process.

10. *Empirical Studies.* Apart from those examples embodied in the written papers, nearly all the references to empirical studies have come from the U.S. True, we have been treated to some anecdotal references to the anti-social behaviour of Parisian parkers, but this has not measured up to the great weight of empirical data turned out by our American colleagues. There are clearly many areas in which further empirical studies are required. There is clearly scope for a detailed study of personal preferences; why do firms choose one location instead of another (they never seem to behave as economists want them to); what incentives would persuade them to move; how are people in different situations likely to respond to parking charges; how would this response be affected by complementary improvements in public transit services; and so on. In other words, can we not tabulate, on an empirical basis, how people react to different types of environmental variables?

11. *Urban Structure.* The central issue discussed in this area was what are the causes and consequences of changes in urban structure. These two aspects were not always clearly distinguished. At times transport facilities were said to 'cause' urban migration; at others it was the urban migration that led to the 'need' for new transport facilities. Both are consistent, but it means that many of the phenomena observed in urban environments can only be interpreted against the backdrop of the way in which the particular urban centre has developed and the natural legacy of capital (sunk costs) and the established patterns of behaviour that it has inherited.

That concludes my brief summary of the main issues that have been discussed at this conference. I shall now hand over to the four panellists and ask for their ideas on what prospective research they now think is desirable in each area.

Preferences, Distribution and Objectives: Henry Tulkens

I have been assigned the risky task of raising the three following types of problems: (i) the choice of objectives in environmental analysis; (ii) the determination of society's preferences in that domain; and

(iii) the treatment of the distributional issues involved.

As regards the *choice of objectives*, it appears from the set of papers presented at the conference that a certain division of labour is taking place among economists. If we draw a production possibility curve (see Fig. 8.1), we can summarise the final objectives of all economic analysis in the environmental field as those of *determining*

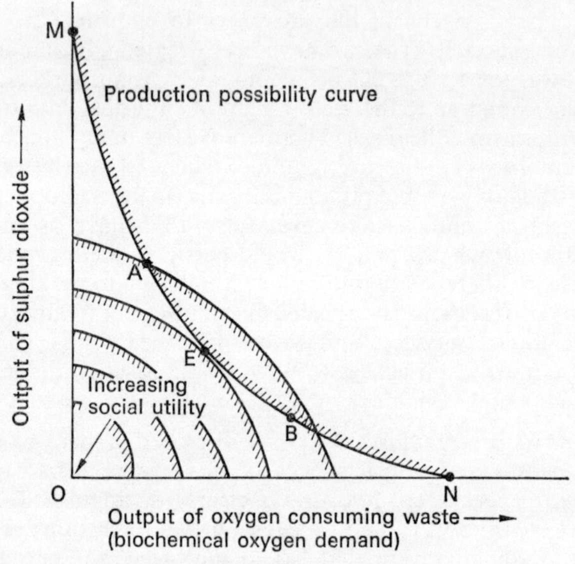

Fig. 8.1   Effect of pollutants on economic choices

a point such as *E*, and of *finding procedures* to help society to get there. At least four of the papers, which I would call 'efficiency-oriented', choose as objective how to reach some point, say *A* or *B*, lying on the boundary *MN* of the production set. In three other papers, on the other hand, the objective is exactly point *E*, i.e. they are 'welfare (or optimality)-oriented'. Given the complexity of the problem involved, such a division of labour is surely a good thing, provided communications continue between the two types of studies.

As far as efficiency studies are concerned, I find it comforting to notice that existing general techniques of economic analysis, such as mathematical programming or input–output analysis, prove to be applicable to this new field; no genuinely new techniques seem to have been devised or are needed. Moreover, these techniques appear to be precise enough to allow for a genuinely multi-disciplinary approach to be adopted. Clearly, for example, the Kneese [1],

Russell *et al.* [2] and Thoss and Wiik [3] programming models appear as bridges between engineering knowledge and economic calculus. This suggests that both the stochastic and dynamic components of pollution phenomena might now usefully be incorporated into the optimisation process.

When it comes to the welfare-oriented studies, we clearly confront one of the most basic problems of economics, namely the *determination of society's preferences*. The papers by Kolm [4] and Uzawa [5] have shown the static and dynamic implications of the well-known Pareto-optimality principle. But, even within that rather simple framework, some essential elements of social preferences with respect to the environment remain obscure: for instance 'merit wants', which have barely been alluded to in Hoch's paper and by Kolm in the discussion, or the inter-generational choices involved in the dynamic analysis of resource conservation and the use of social overhead capital.

The analysis of decision procedures also requires further research. From the presentation and discussion of Haefele's model, one can only conclude that our conceptual tools for analysing preferences and understanding social decisions are too dependent on our concepts of observed market behaviour. The many non-marketable goods and services involved in any environmental analysis call for more conceptual progress and scientific imagination before we can be sure that our conclusions are not founded on our own preferences rather than on those of society.

Finally, on the distributional issues, I should like to observe that it is not always clear that environmental policies should always be distributionally neutral, instead of being used as a tool of re-distributive policy. In practice, it seems that most governments take the latter attitude (as they have often done in the past with respect to other kinds of public goods). This should perhaps influence the orientation of our thinking in this field. I realise how difficult it is to do good distributional economics without introducing arbitrary value judgements; but I wonder whether the issues raised by environmental problems will ever be treated in a relevant manner without tackling that question more substantially.

Discussion of Paper by Henry Tulkens

Kolm agreed that it was time someone commented further on the time aspect of environmental problems. We knew that they differed very widely in the following respects: some environmental disruptions ceased when their causes stopped (noise); while others had long-lasting or even permanent effects (destruction of a unique site or of a species). But we had to be careful to consider all effects: noise could have durable consequences through physiological or psychological disruption. Of course, long-lasting effects raised especially difficult problems for their economic evaluation, since future generations were not here to tell us what importance they attached to a given aspect of their environment.

But, beyond economics, this raised very important and difficult philosophical problems. The tastes of future generations were largely determined by the values we gave them through culture, education, etc. If our only criterion was happiness (or *bonheur*, which was better), then we had to take this into account while shaping the tastes of future citizens. For instance, if we could teach them that Notre-Dame of Paris was not what we usually think of it but simply a funny heap of old stones, we could replace it with a nice parking lot and everybody would be happier. We make things difficult for future generations by giving them tastes which are costly to satisfy. Of course, some of these tastes may be genetically, rather than culturally, determined but, although the distinction is difficult to make, culture does have some effect. Do we therefore have the right to make future citizens unhappy by inculcating them with hard-to-satisfy tastes simply because they correspond to our own tastes?

Foster replied that we could all think of ways of measuring people's preferences under different conditions. The real problem consisted of a widely held belief that *the culture* was a tradition carried on by a minority who chose to interpret certain cultural artifacts as 'merit wants' of the people. Planning was, furthermore, a profession largely engaged in by people who had a strong sense of overriding notion that the culture must be carried on. There were thus areas, like the preservation of the countryside, where one could not say very much except that there were people who had sufficient power – and a feeling of justified 'merit wants' – to preserve these things.

On the question of manipulating preferences through education, *Foster* thought that there was little historical evidence to suggest that sustained changes in preferences could be achieved through education.

Kneese suggested a slightly different approach to the question of inter-generation preferences. He argued that one of the problems of utility maximisation models was that one could not accurately predict preferences and utilities over very long periods of time. It might therefore be better if we tried to develop other types of objective functions in these models. For example, we could perhaps keep the options of future generations open by not engaging in activities which produced irreversible effects.

Foster agreed with Kneese and commented that the problem of preferences, viewed in the context of traditional utility maximisation models, was insoluble. We could readily predict what the cost of rebuilding Notre Dame, or of demolishing an urban motorway, would be but we could not predict, even in probabilistic terms, what the preferences of future generations were likely to be and hence what the net social costs of keeping or of demolishing Notre Dame would be.

Mäler disagreed with this analysis. He felt that we already took account of the preferences of future generations by expressing them in our own preference functions. Utility maximisation models should, therefore, continue to maximise the utility of present generations since these models already implicitly subsumed the utility of future generations.

Uzawa did not think that the conference should be quite so complacent about the present generation of utility maximisation models. During the past 20 years a great deal of the natural and social environment in Japan, and in Tokyo in particular, had been systematically destroyed in the name of economic growth, efficiency and progress. He felt very strongly that our utility maximisation models had to be adapted, as he had tried to do in his paper, to allow explicitly for the consumption of natural resources and environment. It was only by doing this that one could realistically draw conclusions about the implications of alternative public policies.

Lave requested permission to speak on behalf of the Philistines. He felt that the conference was in danger of losing its objectivity. It was true that some people complained that too many historic buildings were torn down, but there were others – who had not yet been represented at the conference – who felt that too few were torn down. It was not merely a question of *conserving* resources, it was a question of conserving and investing them wisely. We should not, therefore, allow our misgivings about the preferences of future generations to result in the wasteful creation (or preservation) of public goods that were only desired by a few.

Hoch sympathised with this point of view. He felt that 'merit wants' often involved the expression of vested interests, e.g. that people wanted something for nothing. For example, national parks generally involved a subsidy to high-income groups. They were the principal users of such facilities, yet they paid very little for their use. He felt that this was true of many appeals to virtue (i.e. that they were upper-income subsidies).

Rothenberg felt that one could interpret the well-being of future generations as the satisfaction that the present generation derived from various provisions for the future. He agreed with Lave that he preferred to live in 1970 than in 1700, but why? How many of us, he asked, were responsible for the benefits we were able to enjoy in 1970? We were all born with absolutely nothing – having earned nothing and deserving absolutely nothing – yet we were given absolutely everything based upon thousands of years of human effort on our behalf. He felt that we therefore owed to the future a great deal more than we had been given ourselves.

This suggested that we should (a) recognise that we were only here at the sufferance of those who went before us and that those who come after us would be similarly beholden to us, and (b) he would not presume that his judgments today should commit future generations to the range of possibilities open to the present generation. So many of our preferences were superficial, casual and capable of being altered by the flimsiest changes of fashion. We should, therefore, allow for the possibility that others might be wiser and better informed than we were and should, therefore, preserve the options of any future generations.

Micro and Macro Models, Environmental Effects, and Growth versus the Environment: Rainer Thoss

I have been asked to say a few words about micro and macro models, environmental effects, and growth versus the environment. I should like to begin with what has been referred to as the antipathy between the micro and the macro model-builders. I do not personally believe that such antipathy exists. During the course of this conference it has become apparent to the macro builders that micro-economics is a very powerful device for solving theoretical welfare problems, but I think it has also become apparent that for certain practical problems aggregated models are often the most suitable. The term 'aggregated' does not of course imply that the models necessarily rely on national aggregates: it is rather that they utilise grouped data of one kind or another.

In spite of some apparent conflict, one of the main outcomes of this conference has been to enable the micro and macro modellers to gain a better appreciation of what each was attempting to do. As a macro modeller, I have certainly become more aware of the need to structure macro models along the same lines as the micro ones. This means of course that much of the prospective work in this field will have to follow the lead given by Tinbergen, Theil and Frisch whose models have been classified as flexible target policy models. This is probably essential if the macro models are to be applicable to general micro welfare analysis.

The most important link between the micro and the macro analysis seems to be the introduction of ambient standards. The role of the imputed price of these standards (for calculating optimal charges) has been clearly demonstrated at this conference. The imputed price of an environmental standard is thus one of the common features that links the micro and the macro models. The macro model, whose solution defines the best primal solution, should thus go hand in hand with a micro analysis of the optimal charges required to achieve this overall optimum. This represents an integration of both micro and macro approaches.

On the question of 'environmental effects' the discussion at this conference has shown how difficult it is to estimate empirical damage functions. The discussion of ambient standards was again relevant in this context since I believe that, until we can effectively estimate these damage functions, a crude ambient standard provides the best substitute to counter environmental degradation. Indeed, the resources required to estimate empirical damage functions need not be expended if an effective ambient standard can be devised.

Let me finally turn to the apparent conflict between growth and environmental quality. I think everyone agreed that there was no real conflict in the short run. It was likewise the composition of output, and not its volume, that was important when questions of environmental quality were being examined. Future research should thus seek to develop more disaggregated growth models (intermediate models half way between the present macro and micro models) unlike those used by Meadows and Associates which cannot show the opportunity cost of different environmental policies but can only diagnose an overall conflict of goals. Of course to compute the price of changing the composition of output (i.e. the price of environmental quality improvement) we need a model that explicitly allows for the resources required to bring this improvement about (clearly our measure of national output should also include some measure of the positive value of these quality improvements).

In the long run there is a further conflict of goals. Any diversion of current resources to improve environmental quality reduces the overall level of investment and hence of physical consumption in the future. This was one of the main conclusions of Uzawa's paper which suggested that further research was required on production functions to quantify what goods and services we have to forego in the future for the sake of environmental quality improvements now.

Discussion of Paper by Rainer Thoss

Førsund argued that the usefulness of macro models for planning and for formulating economic policy needed no elaboration. The question was whether it was any use studying environmental pollution at the aggregated level of a macro model. He thought that it was and advanced the following four reasons:

(a) It meant that the process of gathering all the relevant data on pollution was structured within a balanced, meaningful framework for the understanding of possible economic developments.
(b) The scale of the problem was emphasised facilitating comparisons with other economic magnitudes.
(c) The direct and indirect repercussions of each policy measure would be analysed, showing how the composition of goods and services was likely to be affected.
(d) In a dynamic setting (as in the M.S.G. model) the tradeoff between traditional growth and environmental quality (as a function of discharge of residuals) could be exposed.

Finally, as economists, he argued that it was also relevant to consider the resources required to obtain useful results in the various types of pollution research. Utilising an operational macro model, the returns from a modest input of human capital clearly justified the investment.

Kneese observed that economic concepts had a tremendous capacity to integrate the work of different disciplines. For example, an idealised production function which could be applied to resource problems, or a model for optimising the allocation of resources, could be put into a quantitative framework which would then provide a coherent conceptual framework within which many disciplines could work in a highly effective manner. The model discussed in the Russell paper [2] (e.g. the residuals model for the Delaware river) provided an opportunity for professions from various disciplines to work in a highly integrated fashion. The team included an ecologist, a systems engineer, a specialist in operations research, two economists and a political theorist. The need for this multi-disciplinary team had emerged from the modelling technique chosen and there seemed to be great scope for using economic models to integrate such disciplines.

Lave noted that one aspect of economic models that was rarely discussed, particularly in relation to distributional problems, was that undue accuracy often compromised against the acceptability of a policy. If we specified who gained and who lost (and by how much) too precisely then vested interests very often undid the purpose of the analysis. General conclusions, which did not specify things too precisely, often gained much wider public acceptance for unpopular policies.

Mrs. Lave argued that one of the difficulties about the growth versus environment debate was that, by and large, the people who were most likely to benefit from improvements in the environment were high-income groups, while the people most likely to benefit from growth were the lower-income groups. The debate about growth versus environment thus had important distributional effects. For example, the cost of the new U.S. auto legislation was going to fall most heavily on the poor – they would be paying higher prices for cars and they were generally least concerned about the environment.

Strøm commented that if a significant part of the pollution problem was connected with physical stocks of residuals in the environment, or if the damage done by pollution was dependent upon the discharge of residuals in the past, then it was necessary to study pollution within a dynamic model. He thought that these stock problems accorded with our actual experience. As far as the actual presence of physical stocks was concerned, it was sufficient to mention residuals like lead, mercury and cadmium, or pesticides like DDT. The other case, which amounted to more or less the same thing – depending on how the environment was defined – could be illustrated by the potential harmful effects of cigarette smoking. The disease was attributable to past rather than present behaviour. Even a heavy smoker would take a long time to recover after he had stopped smoking.

At a given point in time the potential externalities caused by the discharge of residuals belonging to an accumulating category cannot therefore be controlled by regulating the discharge of residuals now. The externalities occur at a later point in time. This complicates both the theoretical and the applied work in the field of environmental studies. If one neglects these stock problems, however, a significant part of the real-world pollution problem is overlooked.

Kolm disagreed with Mrs. Lave. He felt that people always assumed that environmental management would have distributional effects that would be regressive. However, many of these effects could be progressive. High-income groups were able to buy themselves a satisfactory private environment. When we spoke of public parks we meant that they were available to poor and rich alike. A public park could thus be thought of as a positive form of income distribution.

Uzawa pointed out that there were similarities between the management of the environment and labour management. At the end of the 19th century one of the most serious socio-economic problems was the exploitation of labour. However, when labour legislation, which sought to regulate working conditions, was first introduced it was violently opposed on the grounds that it would halt all economic growth. In fact, this did not happen and people eventually realised that good working conditions were complementary to long-term economic growth. The same was true of the debate on growth versus the environment. They were not contradictory but complementary.

Mäler agreed with Uzawa. He also suggested that there seemed to be a connection between the quality of the natural environment and the working (factory) environment.

This had not been discussed during the conference. In Sweden people had observed that damage to the natural environment could often be predicted from a knowledge of the factory environment. If workers were harmed by the use of some chemicals, then the natural environment would probably also be harmed by the same chemical although the effect of the damage took longer to appear.

Environmental Policies and the Role of the Economist in Influencing Public Policy: Karl-Göran Mäler

I have been asked to initiate a discussion on environmental policy measures. The following measures have already been discussed during this conference:

1. Markets   Polluters and pollutees negotiate over the amount of pollutants and over any compensation.

2. Property rights   The government (or any other authority) issues rights to use the capacity of the environment to assimilate waste. The price of these rights is determined by supply and demand.

3. Effluent charges   A tax on the amount of waste discharged into the environment.

4. Bribes   A subsidy based on the reduction of waste discharges.

5. Effluent standards   Limits on the amount of waste that can be discharged into the environment.
6. Regulations   Requirements that firms and other polluters use certain waste treatment or production processes.

I shall offer a very brief discussion of each of these measures.

1. As environmental quality is a public good, a market solution will incur extremely high transaction costs. The government must, therefore, intervene to establish the desired markets. Moreover, there are the familiar problems connected with public goods, e.g. non-convexities, etc. It does not, therefore, seem to be efficient to rely on a market solution.

2. Given the capacity of the environment to assimilate waste, one can conceive of negotiable rights to use this capacity. It is not clear, however, how the supply of these property rights can be determined when there are trade-offs between different pollutants, e.g. between biochemical oxygen demand (B.O.D.) and thermal pollution. It is likewise not clear what price should be charged for these rights when they are issued. If a non-equilibrium price is charged, one hopes that the markets are sufficiently stable for the price to converge. In such cases the scarcity rent of the assimilative capacity will be socialised, and no compensation will be paid.

3. We know that charges are efficient from the point of view of information requirements and that charges will result in a least-cost combination of waste discharges. Rents are socialised. There are two main problems with charges: (i) how can one find a good approximation for the optimal charge? and (ii) how can one devise a trial-and-error procedure to ensure that the actual charge converges on the optimal charge?

4. Bribes can be made equivalent to effluent charges, but they are generally more complicated. In some cases, as Kolm has pointed out, bribes are necessary. For example, when non-convexities are involved, a subsidy must be used. The same is true when there are very high monitoring costs. If it is feasible to let the polluter prove that he has reduced his pollution, on the other hand, bribes may be superior to charges.

5. Effluent standards generally require more information than effluent charges and are not as efficient. They generally also require the same monitoring of waste flows. Rents are not socialised.

6. Regulations may be used when it is impossible to meter the waste flow.

Let me now turn to the impact on technological development. Many people regard the incentives to change the present technology as the most important effect of different environmental policy measures. Since we know very little about the causes of technological development, except that economic incentives play some role, it is almost impossible to offer a satisfactory analysis of the effects of environmental policies on technological development. A very simple comparison between standards and charges is illustrated in Fig. 8.2. In this diagram *MC* represents the marginal cost of reducing waste discharges. The horizontal axis measures the amount of waste discharged. Assume now that it is possible to develop a new technology giving the marginal cost curve *MC'*. With a charge, *OH*, the firm

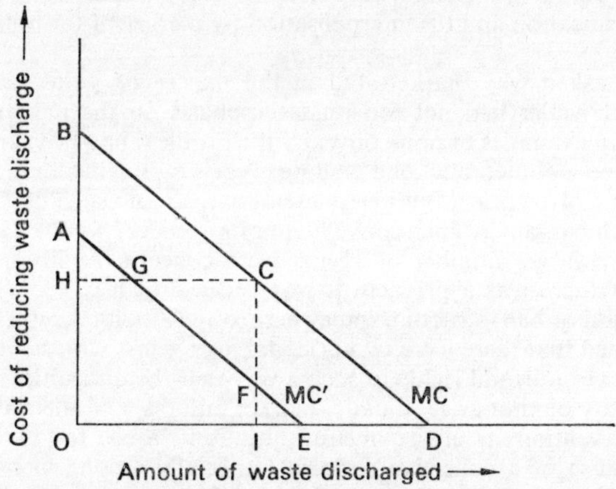

Fig. 8.2   Effect of environmental policy on development
of new technologies

will then save $(OHCDO - OHGEO) = CDEGC$ by adopting the new technology. With an emission standard *OI* on the other hand, the firm will save $ICD - IFE = CDEFC$ by adopting the new technology. This is less than the *CDEGC* saved by introducing the charge *OH*. Charges, therefore, seem to be more effective in encouraging technological improvements.

The conclusion of my discussion is, therefore, that charges seem to provide the best basis for any environmental policies. This leads to an important issue, not discussed in this conference so far. How do we convince our politicians and our authorities that economists can contribute to the improvement of environmental policy? There is no simple answer. However, I shall offer a few comments, based on my own experience. If you meet a politician individually it is relatively easy to convince him that, for example, a charge should be levied on the discharge of wastes. But when he is back in his natural habitat, he is subject to much stronger pressures. The problem is that the

economist can only argue on an intellectual level while other pressure
groups have much stronger incentives.

Moreover, in Sweden at least, the civil servants in the government
are mainly lawyers and I think that many lawyers are completely
impervious to economic argument.

Discussion of Paper by Karl-Göran Mäler

Kolm said that Mäler's six categories did not exhaust the methods of
dealing with pollution. There were also other ways of dealing with it. He
suggested two further categories: (i) voluntary restraint in conjunction
with civic education and (ii) internalisation by merger of the polluters and
pollutees.

He then asked why markets fail in the matter of pollution control.
He felt that Mäler had put too much emphasis on the matter of non-
convexity. In examples of non-convexity the problem had been specified in
terms of one polluter and one pollutee, each supplying or demanding
pollution at a given price. But in a bilateral bargain, no such unit price was
given to each bargainer. Then convexity, or its absence, becomes irrelevant.
And when a larger number of agents were concerned with a pollution
question, if there was a problem it was because of a collective concern
('public' good or bad) structure somewhere, often among the victims. Even
if we assumed that there were two price-taking agents, it was still possible
to define environmental rights in such a way that the difficulties caused by
non-convexity did not arise. Finally, market failures, and the rationale for
public intervention in environmental questions, arose for two reasons:
firstly, because of a collective concern (especially among victims); and,
secondly, because of the absence of any definition of rights. The latter
reason was attributable to the fact that environmental goods were usually
free and this incited the parties to fight politically for these rights before
resorting to free exchange between themselves.

Mäler replied that internalisation, by nationalising (socialising) a firm,
was not really an environmental policy measure because even a state-
owned firm could pollute. The government had, in one way or another,
to impose controls on the amounts of waste the socialised firm discharged
and it was these controls that took the form of regulations, standards, etc.

Kneese did not see why, in some cases, bribes should be superior to
charges because the burden of proof was put on the polluters. It was possible to put the burden of proof on the polluter with charges as well.

Førsund observed that Mäler had not mentioned the problems associated
with the high adjustment costs of a fixed technology. For example, charges
tended to penalise established firms (with a fixed technology) in relation to
new entrants.

Lave, returning to the role of the economist in the social decision process,
suggested that there was some confusion about the proper role for the
economist to play. We had no comparative advantage in attempting to
change society's values or to impose our own values upon society. Our
proper role was to determine the implications of programmes – implications

which would not otherwise be evident – and to advise society on the best way to achieve its objectives. It was the function of the economist to spell out the implications of child labour laws, occupational health and safety laws, and the implications of a shorter work week. The conventional wisdom of the late 19th century was that these were costly luxuries: time had shown that they led to higher productivity and higher incomes.

Benefit–cost analysis had similarly shown that pollution abatement constituted a net benefit to society. Our task was to publicise this fact: not to wander off and dream of imposing our tastes for many more public goods on a reluctant society. We should recognise that markets are able to reflect inter-temporal choices. For example, assume that there was a bottle of wine that would remain undrinkable for 200 years. Would it have a zero price because no person presently alive could enjoy it? Of course not. The bottle would be valuable in 199 years time; in 150 years time; and so on up to the present. The market allocation of durable goods (and resources) was an efficient one, given our state of knowledge. Any conservationist views can be expressed as our uncertainty, our inability to realise the consequences of our preferences, or as too high a rate of discount. We could argue about each of these, but he felt that growth theory had more or less eliminated all three as important possibilities.

We were, therefore, left with the possibility that we were climbing the wrong mountain, that the curves we were dealing with were the wrong shape, or that markets were too inefficient. He doubted that we could think of any gross changes, along the above lines, that would make us better off. He felt that we should instead return to the task of benefit–cost analysis to sketch out the implications of environmental policies and decisions related to the provision of public goods. His own prejudice was that society allocated too little effort to study. It should be part of our function to remedy this since the costs of study were so small in relation to the benefits of improved decisions.

Further Empirical Work and the Causes and Consequences of Changes in Urban Structure:
Edwin S. Mills

I have been asked to comment on two things: one is further empirical work and the other is the causes and consequences of changes in urban structure. Especially with the first I think it is a well-known theorem that any second-rate economist can ask more questions than the rest of the profession can answer. I shall, therefore, confine myself to two issues which I find particularly interesting.

When I look at the empirical studies which have been presented here, I am struck by the fact that the work on the cost side of alternative means of abatement is fairly far advanced. There are still problems of course but the whole field is fairly well developed. The cost side is very much further advanced than the benefit side which is usually dealt with by assuming an absolute standard or goal and then finding the cheapest way of meeting it. Relatively little has been said about how you decide what environmental quality improvements are sufficiently beneficial to be worth it. This is clearly the more difficult side of the balance sheet, but it is the side where economists have a real role to play in this field and to which they should devote some attention. We have one excellent study of water quality improvement on the Delaware and the profession has clearly got its money's worth out of that study. It is quoted again and again and is virtually the only economic study in the environmental area that has had any measurable influence in the public arena in the U.S. In spite of its influence, and the rigour of its analysis, it is nevertheless one in which the procedure for estimating the benefits cannot be easily defended. It was extraordinarily crude and I think we should have some more such estimates. Except for some work by Lester Lave in measuring the benefits from improving air quality there is little else that I know of that can be called 'careful' by the standards that economists tend to use. Both for air and water pollution, and also for congestion which has received a fair amount of attention at this conference, a good deal more research on the benefit side would be justifiable. I also think that, in a sense, doing such work provides one of the keys which unlocks the door to the political process. I think the politicians and the business people always ask what they are going to get from cleaning up the environment, i.e. what are the benefits of doing so.

Very closely related, and this is a very special prejudice of my own, is how externalities really work. We have had a good deal of discussion of fees versus subsidies and standards. I found this very educational and I hope I've learned something. There has been a lot of theoretical work done in this area but I think everyone would agree that externalities which remain after markets have done their work are very much a matter of transactions costs which are not in the interest of private parties to incur. The public sector is, therefore, called on to do something. But underlying all of this is some presumption that the transaction costs incurred in the public sector are small relative to those in the private sector. I think this is probably a reasonable presumption but we need some study of what kinds of transactions costs are involved with what kinds of public sector policies – fees, subsidies, standards, etc. We have not really studied the kinds of transactions costs that private parties incur and the kinds of transactions costs that arise in various kinds of public policies in a very systematic way.

Let me now conclude briefly with some comments on urban structure. It is a good question to what extent environmental problems, however defined, are intimately related with the process of urbanisation. Clearly in some broad sense they are but it is not clear how specific this relationship is. Furthermore, it is not very clear to what extent environmental problems cause changes in urban structure or to what extent the opposite causation is responsible; changes in the structure of urban areas cause, or possibly even cure, some environmental problems. The twin observations that a lot of people have made and which I think are the basic ones about urbanisation in the last century or so in the industrialised world are the rapid growth of urbanisation, i.e. massive migration of people to urban areas, accompanied by a rapid and very large decentralisation of urban areas. These two get confused in a lot of popular discussion. To some extent environmental conditions may affect both of these trends in the future. To some extent the trends will certainly affect environmental problems.

Now for a few personal observations. In the United States now, and in many West European countries, about three-quarters to four-fifths of the people live in good-sized metropolitan areas. This suggests that in the U.S. the process of very rapid urbanisation is about to come to a halt. It has come to a halt, or vastly slowed down, in some of the Northern European countries in much the way that I suspect it will do in the U.S. It is rather unlikely that the rural areas will literally be emptied out and if that is correct we cannot have a very much larger proportion of the population living in metropolitan areas. That has substantial implications for the likely future of environmental problems in urban areas. I also agree that growing environmental problems in urban areas are themselves an influence on the rate of growth of urban areas and this is what Hoch's paper was all about. I personally find that empirical work among the most interesting and provocative pieces of work I have seen in the field of urbanisation and environment in recent years.

To some extent environmental problems, especially congestion problems, have been the cause of decentralisation in urban areas. It is a very reasonable hypothesis and many people believe that downtown congestion and pollution have, in the U.S., led to suburbanisation of both housing and employment. Whether or not that is true, decentralisation has had an effect on environmental problems. To some extent it decentralises the discharge of residuals which is an improvement and it clearly reduces the amount of congestion that would otherwise have occurred. In fact, the conjecture has been made in the U.S. that the most serious urban congestion problems are likely to cure themselves in the next decade or two. Total employment in downtown areas in U.S. has been virtually stagnant since World War II and it is unlikely to increase whether we have congestion charges, effluent fees or subsidies. There is very little that the public sector can do to influence this. The great increase in congestion has come about mainly from the substitution of cars for other forms of locomotion like buses, walking or subways which require less land for rights-of-way per commuter. Given the difficulty and expense of increasing the right-of-way in built-up areas this substitution has led to congestion. This process is now more or less complete. About 90 per cent of commuting in the U.S. is by car and that can hardly increase much in coming years. In fact it will probably decrease slightly as public transit improves. This suggests that congestion will not get any worse in most U.S. cities in the foreseeable future. Now this process is clearly not complete in many European cities. It obviously depends on public policy towards the use of the automobile, the alternatives available and especially the quality and cost of public transportation. In general it suggests, however, that the relationship between the congestion of the environment and urban structure may be a somewhat complex one whose future is difficult to predict on the basis of forecast trends since the Second World War.

Discussion of Paper by Edwin S. Mills

Kneese observed that by the end of this century the world would probably have 7 billion people and that by the middle of the next century his most probable estimate was that it would have 15 billion people. Nearly all of this increase was going to take place in Asia, Latin America and Africa. These countries already had some of the world's largest cities and their growth rates were unbelievable. São Paulo, for example, was growing at a net rate of 300,000 people per year and, although there might be some cause for complacency in the developed world, he thought that we could be much less optimistic about the developing countries. We must expect to see monster cities developing during the next 30 years and this was an area where urban economists interested in the environment should now turn their attention.

Rothenberg agreed that this was a very serious problem. One needed to pay more attention to the urbanising process as leading to environmental problems because of the very heavily mortgaged use of space, and the competition for space, over time.

REFERENCES

[1] Rothenberg J. and Heggie I. G. (eds), *The Management of Water Quality and the Environment* (Macmillan, 1974), Chapter 3.
[2] *Ibid.*, Chapter 7.
[3] *Ibid.*, Chapter 4.
[4] *Ibid.*, Chapter 5.
[5] *Ibid.*, Chapter 1.

Index

Entries in **bold** type indicate papers presented by participants: entries in *italics* indicate contributions made by participants to the discussions

Ackley, G., 88

Adams, F. G., Seneca, J., and Davidson, P., 103 n.

Agences Financières de Bassins, 151

Ambient standards, xix

D'Arge, R. C., Kneese, A. V., and Ayres, R. A., 48 n., 271 n.; and Kogiku, K. C., 23, 48 n.

Arrow, K., and Scitovsky, T. (eds), 20 n.

Atleberg, S., 21 n.

Ayres, R. A., d'Arge, R. C., and Kneese, A. V., 48 n., 271 n.

Baumol, W., 170 n., and Oates, W., 189, 209 n.

Beckerman, W., *134–8, 142–3, 144–5*

Behrens, W., Meadows, Do. and De., and Randers, J., 48 n.

Biochemical oxygen demand (B.O.D.), 74–7

Bjerkholt, O., and Longva, S., 47, 48 n.

Black, R. J., *et al.*, 235 n., 270 n.

Bolwig, N. G., *275*

Bower, B., and Kneese, A. V. (eds), 48 n., 104, 132 n., 187 n., 188 n., 189, 209 n., 271 n.

Boyd, J. H., and Mohring, H., 170 n.

Bramhall, D. F., and Mills, E. S., 103 n.

Brody, A., and Carter, A. P. (eds), 48 n.

Brown, G. W., 270 n.

Buchanan, J., 170 n.

Bucksteeg, K., 133 n.

Burt, W. V., 270 n.

Carter, A. P., and Brody, A. (eds), 48 n.

Case studies in water mangement, influence of direct regulation, 101

Chemical polymers, 97

Chen, C. W., 270 n.

Clawson, M., 145

Clawson-Knetsch method-critique, 146–7

Cleaning technology, 166–7

Cobb-Douglas production function, 26

Common property resources, x, 4

Compensation, 284

Cootner, P., and Löf, G. O. G., 233, 270 n.

Cost-benefit analysis, xv

Dale, T., 21 n.

Dales, J. H., 144, 147 n., 270 n.

Davidson, P., Adams, F. G., and Seneca, J., 103 n.

Davis, R. K., 94, 95 n., 99 n., 103 n.

Delaware River Basin Commission, 225

Delaware River study, xv, 79–88

Delay, W. H., and Seadors, J., 270 n.

Delson, J. K., and Frankel, R., 233, 270 n.

Deutsches Institut für Wirtschaftsforschung, 111

Dissolved oxygen (D.O.), 74–7

Distribution, xix, 167–9, 282

Di Toro, D. M., O'Connor, D. J., and Thomann, R. V., 147 n., 270 n.

Dorfman, R., and Jacoby, H., 270 n.; and Thomas, B. (eds), 270 n., 271 n.

Dreze, J., and de la Vallée Poussin, D., 276, 277 n.

Duttweiler, D. W., *et al.*, 270 n.

Economists and public policy, 284–5

Efficiency, 6–7

Effluent charges, xvi, xvii, 7, 41–7, 193–203

Effluent standards, xvi, xvii, 193–203

Effluents, degradable and non-degradable, 74–6

Emission standards, xv

Empirical studies, 285

Environmental capital, 16

— deterioration, 3–4

— disruption, 5

Environmental effects, 283, 290–3
— improvements, cleaning and dilution, 158–62
— policy, 151–71, 172–88, 192, 208
— —, Mäler model of optimal, 190 ff.
— quality, ix–xi, xiv, xvii, 25, 224–72, 273–7
— rights, xvi
Evaluation and consolidation panel, xviii

Fair, G. M., Geyer, J. C., and Okun, D. A., 103 n.
Federal Water Pollution Control Act, 79–81
Federal Water Pollution Control Administration, 103 n.
Feller, W., 209 n.
Fiacco, F. A., and McCormick, G. P., 147 n., 271 n.
Fiering, M. B., 103 n.
Flexible-target fixed target policy, 108
Ford, D., and Leontief, W., 26, 48 n.
Førsund, F. R., xiii, *177–87, 222, 291, 296;* and Strøm, S., xiii, xiv, **21–69, 70–2**
Foster, C. D., *222, 275–6, 288*
Frankel, R., and Delson J. K., 233, 270 n.
Frisch, R., 290

Geyer, J. C., Okun, D. A., and Fair, G. M., 103 n.
Gibson, J. G., and Kavanagh, N. J., 146 n.
Grandmont, J.-M., 158
Granular carbon, 97
Grava, S., 133 n.
Growth, xix
Growth model, multi-sectoral, 26
— models and the environment, 283–4
— rates, disparity in sectoral, 26
— versus the environment, 290
Gunther, W., 116, 133 n.

Haefele, E. T., 271 n., 277 n.; Russell, C. S., and Spofforf, W. O., xiii, xviii, 199, 224–72, 273–7
Harwitz, M., and Mohring, H. D., 188 n.
Hayden Boyd, J., 179, 187 n.; and Mohring, H. D., 188 n.
Heggie, I. G., **281–5**: and Rothenberg, J. G. (eds), 277 n., 300 n.

Hestenes, M. R., 209 n.
Hicks, J., *19, 20, 70, 172–5*
Hoch, I., *71, 289*
Holzworth, G. C., 271 n.

Industrial allocation, optimal changes in, 121–30
Industrial structure, growth and residual flows, 21–69, 70–2
Information requirements for environmental policy, 189, 192
Input-output analysis, problems of, 70, 71, 72
— —, fixed proportions, 113
— —, model, xiii
Interdependencies of economic activities, 25
Investment in assimilative capacity, xi
Investment policy, control of, 193
Isard, W., *et al.*, 147 n.

Jacoby, H., and Dorfman, R., 270 n.; and Thomas, B. (eds), 270 n., 271 n.
Jansen, E., 21 n.
Johansen, L., 26, 45, 48 n.
Johnson, E. L., 88, 103 n., 271 n.

Kavanagh, N. J., *70, 145–6*; and Gibson, J. G., 146 n.
Keeler, E., Spence, M., and Zeckhauser, R., 23, 48 n.
Kehr, D., and Teichmann, H., 133 n.
Kelly, R. A., 271 n.
Kneese, A. V., xiii, xvi, 17 n., *19, 20,* 21, *71, 72,* **73–103**, 132 n., **134–47**, *213–17,* 228, 286, *288, 292, 296, 300*
—, Ayres, R. A., and D'Arge, R. C., 48 n., 271 n.
— and Bower, B. (eds), 48 n., 104, 132 n., 187 n., 188 n., 189, 209, 271 n.
— and Löf, G. O. G., 233, 271 n.
— and Smith, S. C. (eds), 103 n.
Kogiku, K. C., and D'Arge, R. C., 23, 48 n.
Kohlschütter, H., Meinck, F., and Stooff, H., 133 n.
Kohn, R., 271 n.
Kolm, S.-Ch., x, xii, xvii, *72, 142, 143, 145,* **151–71**, **172–88**, **217–22**, 223 n., *276, 287–8, 293, 296*
Kraus, M., and Mohring, H., 170 n.

Lave, J., *292*
Lave, L., *143, 289, 292, 296–7*
Leontief, W., 48 n., 70; and Ford, D., 26, 48 n.
Lerner, A. P., 142, 147 n.
Lévy-Lambert, H., 170 n.
Linear programming, xiv, xv, 82
Löf, G. O. G., and Cootner, P., 233, 270 n.; and Kneese, A. V., 233, 271 n.
Longva, S., and Bjerkholt, O., 47, 48 n.
Low-flow augmentation, 96

Machlup, F., 18–19
Madison, J., 271 n.
Mäler, K. G., xvi, xix, 171 n., 185 n., **189–212, 213–23,** *288–9,* **293–7**
Malinvaud, E., *276,* 277 n.
Marginal social costs, 7
Margolis, J. (ed.), 169 n., 188 n.
Martin, D. O., and Tikvart, J. A., 147 n., 271 n.
'Materials balance' approach, 21
McCormack, G. P., and Fiacco, F. A., 147 n., 271 n.
Meadows, Do. and De., Randers, J., and Behrens, W., 26, 48 n.
Meinck, F., Stooff, H., and Kohlschütter, H., 133 n.
Micro-straining, 97
Micro versus macro models, 283, 290, 291–3
Mills, E. S., xix, *18, 70–1, 141–2,* **297–300;** and Bramhall, D. F., 103 n.
Mishan, E. J., 188 n.
Models of water pollution, xv
Mohring, H. D., 178; and Harwitz, M., 188 n.; and Hayden Boyd, J., 170 n., 188 n.; and Kraus, M., 170 n.
Müller, A., *223*
Multi-sectoral growth model, 31
— — —, equations and notations, 55–7
— — —, industrial sectors, 50–5

Natural capital, 16
Nora, S., *et al.*, 170 n.
Norwegian Ministry of Finance, 21 n., 26

Oates, W., and Baumol, W., 189, 209 n.

Objective functions, 282–3
O'Connor, D. J., Thomann, R. V., and Di Toro, D. M., 147 n., 270 n.
Okun, D. A., Fair, G. M., and Geyer, J. C., 103 n.
Ozolins, G., and Smith, R., 233, 271 n.

Pareto optimum, 20
Pasquill, F., 271 n.
Penrose function, 12, 16, 17
Peskin, H. M., 31, 48 n.
Phelps, E. B., and Streeter, H. W., 133 n., 147 n., 272 n.
Pollution, xii, xvi, xviii, 17; controls, 284
Potomac river study, xv, 93–101; conclusions from, 100–1
Powdered carbon, 97
Preferences, xix, 282
Private and social capital, optimum investment in, 11–14
Prud'homme, R., *71, 143–4, 273–5*
Public goods, x
Public policy objectives, 108
Public treatment, xv

Qualitative returns to scale, xii, 151–71, 172–88; diagrammatic demonstration, 176
Quality of the environment, management of, 224–72, 273–7

Ramsey, F. R., 3, 9, 12, 17 n., 20 n.
Randers, J., Behrens, W., and Meadows, Do. and De., 48 n.
Recreation benefits, 87–8
Recycling, 24
Reproductive social capital, 16
Residuals, xii, xviii
—, accumulation of stocks of persistent, 36
— charges experiment, 37–47
— discharge coefficient, 27
—, disparities over time, 34
— disposal service of the environment, 22
—, export and import, 28
—, interdependencies, 104–8
—, list of, 49–50
—, measurement of, 30
—, quantities by industrial sector, 158–69
—, spatial distribution of, 47

Residuals generation, multiregional model of, 118; Ruhr model of, notation, 130–2

Rothenberg, J. G., **ix–xx**, 6, 17 n., *19, 20, 71, 72, 141*, 188 n., *275*, 289, *300*; and Heggie, I. G. (eds), 277 n., 300 n.

Ruhr area, maps, 106–7

Ruhr Basin, model for water quality management, 104–33

Ruhr model of residual generation, notation, 130–2

Russell, C., *70, 138–41, 144, 146–7*, 189 n.; and Spofford, W. O., Jr., 21, 23, 24, 48 n., 104, 132 n., 271 n., 286; Spofford, W. O., Jr., and Haefele, E. T., xiii, xviii, 199, **224–72, 273–7**

Samuelson, P. A., 5, 11, 17 n.

Schaumberg, G. W., Jr., 88, 103 n., 272 n.

Schmidt, U., 117, 133 n.

Schreiner, P., 26, 48 n.

Scitovsky, T., and Arrow, K. (eds), 20 n.

Seadors, J., and Delay, W. H., 270 n.

Self-purification capacity of the river, 119

Seneca, J., Davidson, P., and Adams, F. G., 103 n.

Sewage treatment cost, 116–18

Slyngstad, K., 21 n.

Smith, R., 233, 272 n.; and Ozolins, G., 233, 271 n.

Smith, S. C., and Kneese, A. V. (eds), 103 n.

Smith, V. L., 23, 48 n.

Sobel, M. J., 81, 103 n.

Social overhead capital, xii; optimum allocation of, 7–9; optimum management of, 3–20

Social utility, 12

Spence, M., Zeckhauser, R., and Keeler, E., 48 n.

Spofford, W. O., Jr., 92 n.; and Russell, C. S., 21, 23, 24, 48 n., 104, 132 n., 271 n.; Russell, C. S., and Haefele, E. J., xiii, xviii, 199, **224–72, 273–7**

Spurkeland, S., 27, 48 n.

Stäglin, R., and Wessels, H., 133 n.

Step aeration, 97

Stochastic hydrology, 77–8

Stochastic systems, optimisation in, 99 ff.

Stooff, H., Kohlschütter, H., and Meinck, F., 133 n.

Streeter, H. W., and Phelps, E. B., 74–6, 115, 133 n., 145, 147 n., 230, 272 n.

Strøm, S., 48 n., 292; and Førsund, F. R., xiii, xiv, **21–69, 70–2**

Strotz, R. H., 169 n., 178, 188 n.

Subsidies versus charges, xvii

Swedish Environmental Protection Board, 193, 195 n.

Teichmann, H., and Kehr, D., 133 n.

Theil, H., 133 n., 290

Thomann, R. V., Di Toro, D. M., and O'Connor, D. J., 147 n., 270 n.

Thomas, B., Dorfman, R., and Jacoby, H. (eds), 270 n., 271 n.

Thoss, R., xix, *72, 223, 276*, **290–3**; and Wiik, K., xiii, xiv, **104–33, 134–7**, 287

Tikvart, J. A., and Martin, D. O., 147 n., 271 n.

Tinbergen, J., 133 n., 290

Treatment, xii

Treatment policy, optimal, 205–8

Trent River Study, 145

Tulkens, H., xix, *19–20, 175*, **285–289**

Upton, C., 188 n.

Urban structure, 285

Uzawa, H., x, xii, **3–17, 18–20**, *276*, 287, 289, 293

Vallée Poussin, D. de la, and Dreze, J., 276, 277 n.

Victor, P. A., 26, 48 n.

Waste-assimilation capacity, 96

Waste loads, random, 203–8

Water management, economic data for, 105

Water quality, application of economic analysis to the management of, 93–103

Water quality, improvements, costs of alternative programmes, 90–3

Water quality improvement, efficiency comparisons of alternative programmes, 90–3

Water quality indicators, 114–16
Water quality management, an empirical linear model, 104–33
Water quality objectives, 82–6
Wessels, H., amd Stäglin, R., 133 n.

Wiik, K., and Thoss, R., xiii, xiv, **104–33**, **134–7**, 287
Wright, W. A., 272 n.

Zeckhauser, R., Keeler, E., and Spence, M., 48 n.